**LIEUTENANTS
AND LIGHT**

Robert E. C. Davis

Mapping the US Army Heliograph Networks in Late Nineteenth-Century Arizona and New Mexico

UNIVERSITY OF NEW MEXICO PRESS | ALBUQUERQUE

LIEUTENANTS AND LIGHT

© 2025 by the University of New Mexico Press
All rights reserved. Published 2025
Printed in the United States of America

Library of Congress Cataloging-in-Publication Data
Names: Davis, Robert E. C., 1961– author
http://id.loc.gov/authorities/names/n2025024292
http://id.loc.gov/rwo/agents/n2025024292
Title: Lieutenants and light : mapping the US Army heliograph networks in late nineteenth-century Arizona and New Mexico / Robert E. C. Davis. Other titles: Mapping the US Army heliograph networks in late nineteenth-century Arizona and New Mexico
Description: Albuquerque : University of New Mexico Press, 2025. | Includes bibliographical references.
Identifiers: LCCN 2025006436 (print) | LCCN 2025006437 (ebook) | ISBN 9780826368409 cloth | ISBN 9780826368416 paperback | ISBN 9780826368232 epub Subjects: LCSH: United States. Army—Communication systems—History—19th century | Heliograph—History | Military bases—Arizona—History—19th century | Military bases—New Mexico—History—19th century | Indians of North America—Wars—Arizona http://id.loc.gov/authorities/subjects/sh86005985
Classification: LCC UG582.H4 D38 2025 (print) | LCC UG582.H4 (ebook) | DDC 358/.248—dc23/eng/20250604
LC record available at https://lccn.loc.gov/2025006436
LC ebook record available at https://lccn.loc.gov/2025006437

Founded in 1889, the University of New Mexico sits on the traditional homelands of the Pueblo of Sandia. The original peoples of New Mexico—Pueblo, Navajo, and Apache—since time immemorial have deep connections to the land and have made significant contributions to the broader community statewide. We honor the land itself and those who remain stewards of this land throughout the generations and also acknowledge our committed relationship to Indigenous peoples. We gratefully recognize our history.

Cover illustration adapted from map courtesy of the author
Cover image courtesy of the State Archives of North Carolina
Designed by Isaac Morris
Composed in Adobe Jenson, Trade Gothic Next, and Vendetta

For Andrea

Contents

LIST OF ILLUSTRATIONS —— ix

PREFACE —— xiii

ACKNOWLEDGMENTS —— xv

CHAPTER 1. Heliographs, Command and Control, and Mapping Science —— 1

CHAPTER 2. The 1882 System —— 30

CHAPTER 3. Fort Bowie —— 37

CHAPTER 4. Fort Huachuca —— 59

CHAPTER 5. Fort Cummings —— 77

CHAPTER 6. Fort Bayard —— 88

CHAPTER 7. The Northern and Southern Heliograph Lines to Fort Stanton —— 102

CHAPTER 8. The Sierra Ancha —— 123

CHAPTER 9. San Carlos and the Gila Valley —— 136

CHAPTER 10. Fort Verde and Whipple Barracks —— 154

CHAPTER 11. The Pinaleño Mountains —— 165

CHAPTER 12. Fort Lowell —— 182

CHAPTER 13. Epilogue —— 191

APPENDIX 1. Profiles —— 197

APPENDIX 2. Fort Selden —— 215

APPENDIX 3. Table of Connections —— 220

NOTES —— 229

REFERENCES —— 274

INDEX —— 289

Illustrations

Figures

FIGURE 1. A heliograph on display at Fort Bowie, Arizona —— 4
FIGURE 2. Number of heliograph network stations established per month in 1886 —— 8
FIGURE 3. Messages sent within the 1886 heliograph network —— 8
FIGURE 4. Blueprint map of the heliograph system of 1890 —— 17
FIGURE 5. Annotated photograph of Fort Bowie, facing south —— 42
FIGURE 6. Part of the Bureau of Land Management's General Land Office 1882 map of Bowie, Arizona —— 42
FIGURE 7. Colorado Peak as seen from the west —— 44
FIGURE 8. Heliographic station No. 3 at White's Ranch —— 51
FIGURE 9. Decision card showing present-day Mount Wrightson's various names —— 63
FIGURE 10. View from Calabasas to today's Mount Wrightson and Josephine Peak —— 65
FIGURE 11. View from the heliograph station atop the ridgeline south of Fort Stanton —— 111
FIGURE 12. View from the 1889 Fort Stanton heliograph station to the Sierra Blanca station —— 111
FIGURE 13. View to the north from Mount Reno —— 130
FIGURE 14. Photographs of Mount Turnbull —— 143
FIGURE 15. US Army Signal Corps expenditures on "military signaling" between 1886 and 1894 —— 193
FIGURE 16. Members of the First North Carolina Infantry and Michigan Signal Corps at a heliograph station —— 193
FIGURE 17. Number of times the words *heliograph*, *buzzer*, and *wireless* or *radio* are mentioned in the various chief signal officer annual reports —— 194

ILLUSTRATIONS

Maps

MAP 1. Heliograph systems, stations, and connections of 1882–1893 —— 2
MAP 2. Extent of the 1886 operational heliograph coverage —— 11
MAP 3. Visibility diagrams —— 23
MAP 4. The 1882 heliograph system —— 31
MAP 5. Fort Grant heliograph station —— 33
MAP 6. Ash Butte heliograph station —— 34
MAP 7. Dos Cabezas heliograph station —— 35
MAP 8. Chapter 3 heliograph stations and connections —— 38
MAP 9. Heliograph stations at and near Fort Bowie —— 39
MAP 10. Stein's Pass heliograph station —— 41
MAP 11. Colorado Peak heliograph station —— 45
MAP 12. Old Camp Rucker heliograph station —— 46
MAP 13. Heliograph station at Emma Monk —— 47
MAP 14. Camp Henely heliograph station —— 48
MAP 15. White's Ranch heliograph station —— 52
MAP 16. Range ring analysis of White's Ranch —— 53
MAP 17. Henry Forrest heliograph station —— 55
MAP 18. Limestone Mountain heliograph station —— 57
MAP 19. Chapter 4 heliograph stations and connections —— 61
MAP 20. Fourr's Ranch South heliograph station —— 66
MAP 21. Fort Huachuca heliograph stations —— 67
MAP 22. Fourr's Ranch North heliograph station —— 73
MAP 23. Antelope Springs heliograph station —— 75
MAP 24. Chapter 5 heliographs and connections —— 78
MAP 25. Fort Cummings heliograph stations —— 79
MAP 26. Heatley's Well heliograph station —— 82
MAP 27. Lake Valley heliograph station —— 83
MAP 28. Hillsboro heliograph station —— 85
MAP 29. Deming heliograph station —— 87
MAP 30. Chapter 6 heliographs and connections —— 89
MAP 31. Fort Bayard heliograph station —— 92
MAP 32. Pinos Altos, New Mexico, heliograph station —— 93

ILLUSTRATIONS

MAP 33. White House, New Mexico, heliograph station —— 94
MAP 34. Alma, New Mexico, heliograph station —— 96
MAP 35. Siggins's Ranch, New Mexico, heliograph station —— 97
MAP 36. Lyda Spring (Mule Spring) heliograph station —— 99
MAP 37. Hachita Mining Camp heliograph station —— 101
MAP 38. Chapter 7 heliograph stations —— 103
MAP 39. Fort McRae heliograph station —— 105
MAP 40. Dripping Spring heliograph station —— 107
MAP 41. Nogal heliograph station —— 108
MAP 42. Fort Stanton heliograph stations —— 109
MAP 43. Sierra Blanca heliograph station —— 113
MAP 44. San Nicholas Peak heliograph station —— 114
MAP 45. Little Florida heliograph station —— 115
MAP 46. Rincon, New Mexico, heliograph station —— 118
MAP 47. San Andreas heliograph station —— 119
MAP 48. Tularosa heliograph station —— 121
MAP 49. Chapter 8 heliograph stations and connections —— 125
MAP 50. Mount Reno heliograph station —— 127
MAP 51. Lookout Peak heliograph station —— 129
MAP 52. Mazatzal and Spur heliograph stations —— 133
MAP 53. Fort McDowell heliograph station —— 135
MAP 54. Chapter 9 heliograph stations and connections —— 137
MAP 55. Triplets heliograph station —— 138
MAP 56. San Carlos heliograph station —— 139
MAP 57. Fort Thomas heliograph station —— 141
MAP 58. Mount Turnbull heliograph stations —— 144
MAP 59. Saddle Mountain heliograph station —— 145
MAP 60. Table Mountain heliograph station —— 147
MAP 61. Areas matching Lieutenant Murray's common visibility on April 5, 1890 —— 151
MAP 62. Heliograph stations atop the Pinal Mountains —— 153
MAP 63. Chapter 10 heliograph stations and connections —— 155
MAP 64. Whipple Barracks heliograph station —— 157
MAP 65. Bald Mountain (Glassford Hill) heliograph station —— 158

ILLUSTRATIONS

MAP 66. Baker's Butte heliograph station —— 159
MAP 67. Fort Verde heliograph station —— 161
MAP 68. Squaw Peaks heliograph stations —— 163
MAP 69. Chapter 11 heliograph stations and connections —— 167
MAP 70. Alpina and Heliograph Peak heliograph stations —— 171
MAP 71. Eggleston's RM station at Merrill Peak —— 175
MAP 72. Murray's RM station at Grand View Peak —— 181
MAP 73. Chapter 12 heliograph stations and connections —— 183
MAP 74. Fort Lowell heliograph station —— 184
MAP 75. Mountain Spring heliograph station —— 186
MAP 76. Rincon Peak heliograph stations —— 189
MAP 77. Fort Selden heliograph station —— 219

Preface

This book began as what I thought would be a simple write-up of a couple dozen heliograph station locations scattered around the southwestern United States in the late nineteenth century. (To date, I have reasonably determined the locations for eighty-six heliograph stations and am certain there are more I have missed.) The US Army used these heliographs, or mirrored communication devices, to keep in contact with forts, posts, camps, and outposts.

Small-scale maps (maps with little detail) of some of the heliograph stations and their connections are available. My goal was to create a more detailed map of the networks formed by the linked heliographs—a map where the heliograph locations are based on the recorded observations and notes of the officers, mostly lieutenants, who emplaced the heliographs, coupled with modern geographic modeling and analysis.

In the nineteenth century, as today, young officers exhibited boundless determination and energy that made the whole system work. In this case, the "whole system" refers to the greatest heliograph network ever built, which spanned eleven years and, at its largest geographic extent, covered an area the size of Kansas. Hundreds of soldiers and dozens of officers made the systems function, sending and receiving thousands of messages. Some of the officers involved later reached the highest ranks in the US Army, some left the military for business careers or public office, and one attempted to abscond with the Presidio of San Francisco's commissary funds (and indeed he did, but he later returned them). After the Indian Wars, most of the lieutenants participated in operations throughout the world. Lieutenant Richard Paddock, who met his wife, Grace (John J. Pershing's sister), at Fort Stanton, New Mexico, died in China. Grace died a few years later, leaving their two young children orphaned; the extended Pershing family looked after them.

Among the almost forty young officers who helped build these heliograph systems, sixteen would go on to become general officers. Two became chief of staff of the US Army, three earned the Medal of Honor, and one became a US senator. One officer was killed on San Juan Hill, another in action in the Philippines. Many served in Cuba, most served in the Philippines, and eight served in World War I.

While these heliograph systems were closely tied to the army's conflict with the Apaches, this is not a tale of that tragic conflict but rather an exploration of the heliograph systems and the people who built them. Although the heliograph played a key role as a reliable and efficient communication tool during the Geronimo campaign, the army's use of heliographs in the Southwest began in 1882 and ended around 1916 (although use was not continuous).

These systems demonstrated that a heliograph network could be deployed rapidly, provided a reliable and integrated communication system, saved resources, and extended a commander's authority and control over military movements and forces in the field. Moreover, the success of the heliograph in the Southwest solidified the heliograph as a trusted part of the army's communication suite. Given the army's use of heliographs from the Indian Wars through World War I, a closer examination of the people, places, and events surrounding the development of heliograph networks in the 1880s and 1890s is warranted.

Robert E. C. Davis
October 30, 2024
Tucson, Arizona

Acknowledgments

I would like to extend my heartfelt thanks to my wife, Evangeline, for her proofreading, patience, and understanding during my frequent trips into the hills while writing this and for her unwavering support in general. I am immensely grateful to the Arizona Historical Society library in Tucson for its invaluable assistance. My friend Colonel Larry Holcomb, USMC (retired), deserves special recognition for dedicating much time to helping me brainstorm an appropriate title for this work, as well as for his insights on history and the military.

Special thanks to my neighbor Jim Terlep, who in his mid-seventies joined me on an ascent of a rather steep, high hill that held the location of a long-forgotten heliograph—Jim was prepared for the ascent; I was not. Michael Dennis of the US National Oceanic and Atmospheric Administration's National Geodetic Survey offered helpful insights into line-of-sight geometry across and above the surface of the earth. Many thanks to Rudy Stricklan, a land surveyor, for advice on datums, coordinate reference systems, and such. Robert A. Wooster's edits and suggestions were invaluable in improving the final manuscript.

I extend my thanks to Garrett Leitermann of the Bureau of Land Management for patiently fielding my many questions about Fort Selden. Special recognition goes to Stephen Gregory of the Fort Huachuca Museum, who not only offered to guide me to the 1886 heliograph network location but also provided valuable insights into the 1890 heliograph stations. Lastly, to the editors at the University of New Mexico Press, especially Michael Millman, Anna Pohlod, and Peg Goldstein, I offer my sincere appreciation and thanks for their tireless patience and professionalism.

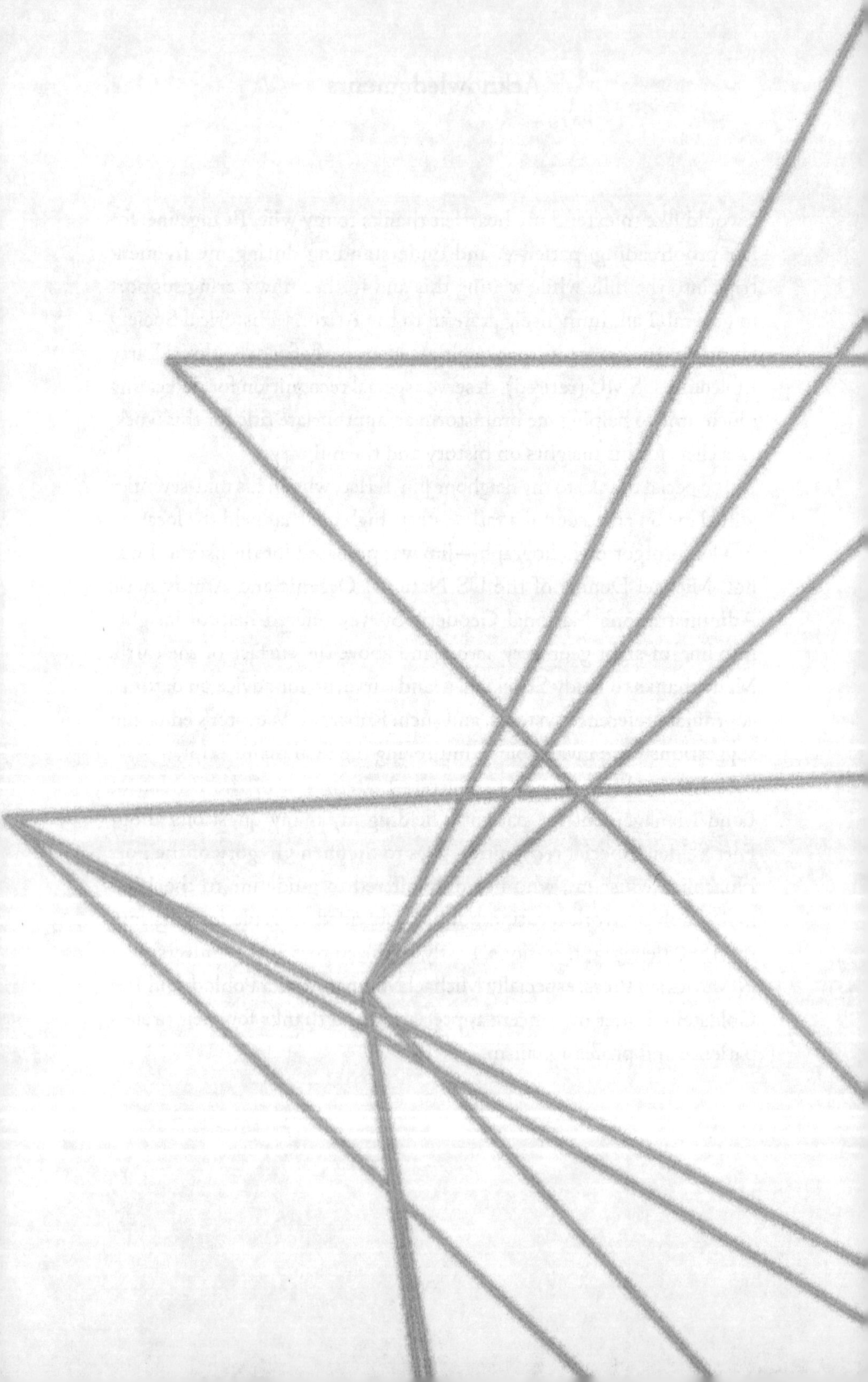

Chapter 1

Heliographs, Command and Control, and Mapping Science

IN THE LATE NINETEENTH-CENTURY SOUTHWEST, THE US ARMY EMPLOYED a device called a heliograph, a signal mirror, to communicate among its detachments, outposts, and forts. Contemporary small-scale maps reveal locations of the heliographs and their intended connections. Some of these heliograph locations are well-known; others appear as large dots on a small map, where the scale makes their locations doubtful. Adding uncertainty, many geographic names have since changed or been forgotten. Present-day Stanley Butte was once Saddle Mountain, and Mount Ord was Reno Peak. Some of the names belie certainty: "Heliograph Peak" in the Pinaleño Mountains suggests a clear location for heliograph operations. On closer examination, however, the area near today's Grand View Peak, far from Heliograph Peak, was the center of army heliography atop the Pinaleños. Presented here is an effort to better identify where these heliographs were, who placed them there, how they were connected, when they were employed, and, in the case of the system established by General Nelson A. Miles in 1886, how they affected operations.

A heliograph is a signaling device using a mirror (see Figure 1). A heliograph station is a place staffed by soldiers and holds the equipment necessary to connect with other heliograph stations. Mirrors at one station reflect the sun's light toward distant stations. Soldiers use Morse code to send and receive messages between stations. On some heliographs, a key attached to the mirror allowed operators to make a slight shift in the mirror's orientation. More often, however, screens

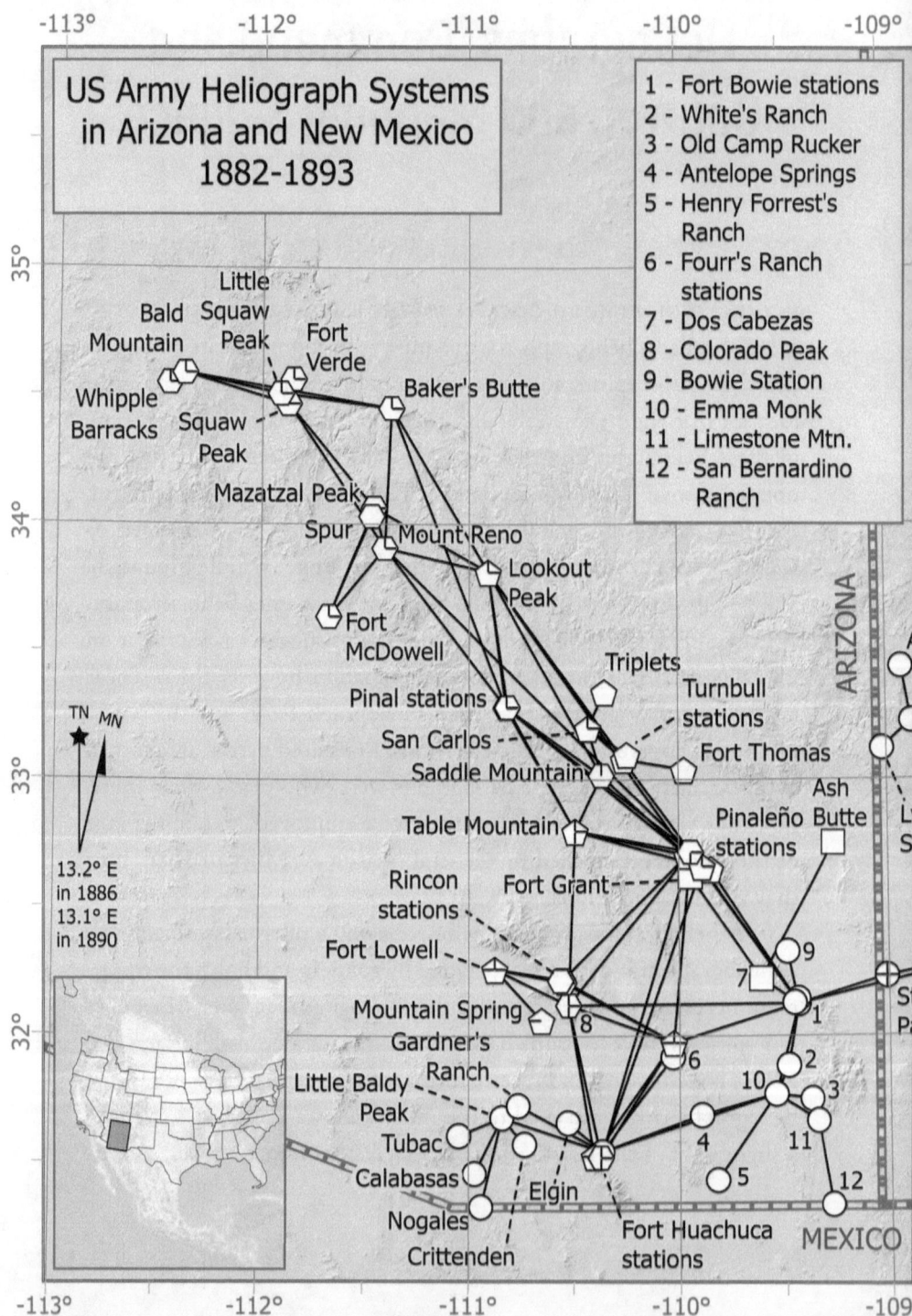

Map 1. The heliograph systems, stations, and connections of 1882–1893.

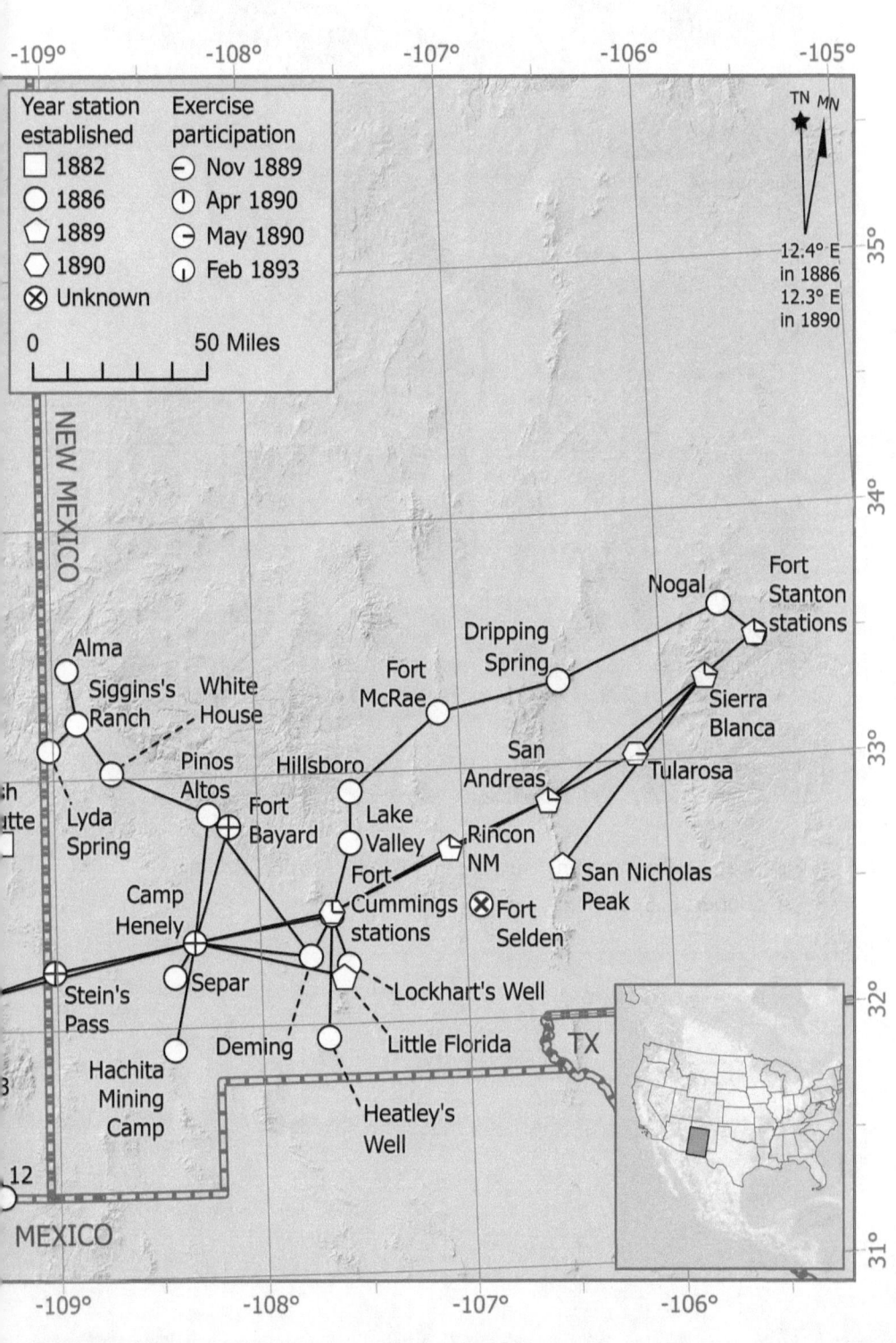

Figure 1. A heliograph on display at Fort Bowie, Arizona. Photo by Robert E. C. Davis.

acting as a shutter created coded signals. Heliograph stations linked together to form a network or a system of communication.

In the mid-nineteenth century, the British began working in earnest on developing a reliable apparatus for use as a heliograph. Sir Henry C. Mance invented a heliograph that was employed by British forces in Afghanistan, Africa, and India.[1] The simple, lightweight Mance heliograph was tested for four years in India and was found to be effective—under the right conditions—at ranges of up to 50 miles. In the late 1870s, the Mance heliograph was used in India during the Jowaki campaign and in Afghanistan, connecting locations such as Jalalabad and Peshawar, Korst and Bannu, and Grishk and the Khojak Pass.[2]

It is not surprising, then, that heliographs also became part of the US Army's communication tools around the same time, in the mid-1870s. Training for officers on the use of heliographs started in 1877 at the Signal Corps school located in Fort Whipple, Virginia (later renamed Fort Myer).[3] Lieutenant Frank C. Grugan, an instructor at the Signal Corps school, who patented his own heliograph in 1881, showcased a heliograph at the International Exposition of 1876 in Philadelphia.[4] General Nelson A. Miles, noted for his involvement in the Geronimo campaign of 1886, mentioned in his *Personal Recollections and Observations* that he employed heliographs borrowed from the Signal Corps to establish communication between Fort Keogh and Fort Custer in the late 1870s and between Vancouver Barracks and Mount Hood in the early 1880s.[5] The 1879 US Army *A Manual for Signals* devotes a chapter to heliograph use.[6] By 1881 the Signal Corps had started experimenting with various types of heliographs with the aim of improving their utility for field operations.[7]

Lieutenant Marian P. Maus, an enterprising infantry officer and recipient of the Medal of Honor for his 1886 actions in Mexico, constructed a network of heliographs in southeastern Arizona in 1882. He successfully connected Fort Bowie, Fort Grant, and scouting camps along the upper Gila River near Clifton, Arizona. Maus addressed the challenging problem that vexed multiple rotations of future heliograph practitioners—the formidable barrier posed by the almost 11,000-foot

Pinaleño Mountains, which separated much of the Gila Valley from Forts Bowie and Grant.[8] Maus's work is often overlooked, overshadowed by the more substantial effort led by General Miles, the department commander, in 1886.[9]

Miles created a heliograph network covering large swaths of Arizona and New Mexico as part of a larger system to coordinate efforts against Geronimo. Miles appointed two officers, Second Lieutenant Alvarado M. Fuller of the Second Cavalry in Arizona and First Lieutenant Edward E. Dravo of the Sixth Cavalry in New Mexico, to create the interconnected network of heliograph stations.[10] Their 1886 heliograph network extended from Nogales, Arizona (south of Tucson), to Fort Stanton in New Mexico (northeast of Alamogordo). In addition to the stations set up by Fuller and Dravo, other heliographs were dispatched with cavalry troops as well. In a correspondence dated 1926 between then retired Colonel Dravo and Charles Gatewood (the son of the Charles B. Gatewood from Geronimo's surrender), Dravo mentioned that all the observation posts listed on the field operations map had a heliograph.[11] These heliographs allowed troops and outposts to connect with one of Fuller's or Dravo's more permanent heliograph stations and therefore their chain of command.

Fuller, a prior enlisted lieutenant with years of experience (and a futurist writer whose *A.D. 2000* is still available today), started setting up heliograph stations at and near Fort Bowie in late April and then shifted his efforts toward Fort Huachuca. By early June the area around those two forts included ten stations covering most of southern Arizona east of the Santa Cruz River. The New Mexico stations started operating on May 9 with Camp Henely and Separ (a rail station with a link to the telegraph). Dravo, another experienced officer, proceeded to connect stations and camps nearer the international border. Fuller, on June 27, set up the station at Stein's Pass, thereby connecting the Arizona and New Mexico heliograph networks into one. By the end of June, the majority of the 1886 stations were set (see Figure 2). The 1886 system continued to grow in New Mexico after Geronimo's September 1886 surrender with a line to the east to Fort Stanton.[12]

The 1886 heliograph network accomplished four things. First, it facilitated the efficient movement of information. During daylight hours, it excelled at transmitting information across a dispersed force, enabling the sending and receiving of thousands of militarily relevant messages in a large operational area. Second, the heliograph's extended range—double that of signal flags—provided commanders with greater flexibility in positioning, commanding, and controlling forces.[13] Third, the increased range led to resource savings, as fewer stations were required to carry out the same tasks, reducing demands on personnel, logistics, and time. Finally, the heliograph system was integral to the commander's operational concept, becoming an essential part of planning and executing operations against the enemy.

During its operational span, Miles's heliograph network successfully transmitted just under three thousand messages (see Figure 3).[14] Integrated into a broader communication infrastructure that included rails, roads, telegraphs, mail, and couriers, the heliograph network played an important role in disseminating information: intelligence about Apache movements, status updates, logistic requests, and weather reports. It also kept commanders connected with their chains of command.[15]

Because of the heliograph, in late May 1886, Captain Adna R. Chaffee, a future chief of staff of the US Army, could position a detachment in the choke point at Heatley's Well, New Mexico, while still maintaining communication using a single visual link to Fort Cummings, 34 miles to his north—over twice the distance allowed by signal flags. He then moved the detachment northeast to Lockhart's Well, where he quickly reestablished communications via heliograph. In late June, Captain Gustavus C. Doane's troop on the western slope of Cochise Stronghold was 32 miles distant from Fort Huachuca but still connected via a single heliograph link. The troop at Bisbee connected to Fort Bowie 50 miles to the north with two heliographic links relayed together. Several signal flag links would have been needed to cover the same distance; indeed, the soldiers with heliographs at Hachita Mining Camp, Hillsboro, Alma, Camp Henely, Old Camp Rucker, Emma Monk, and Antelope Springs were all much farther from a fort than flag signaling allowed.

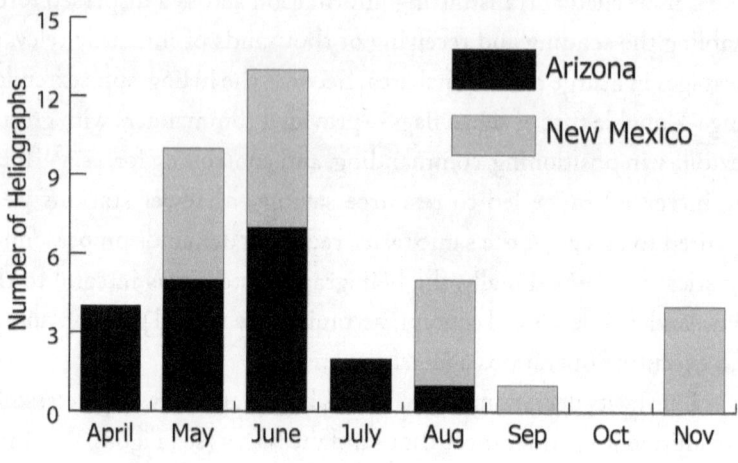

Figure 2. Number of heliograph network stations established per month in 1886.

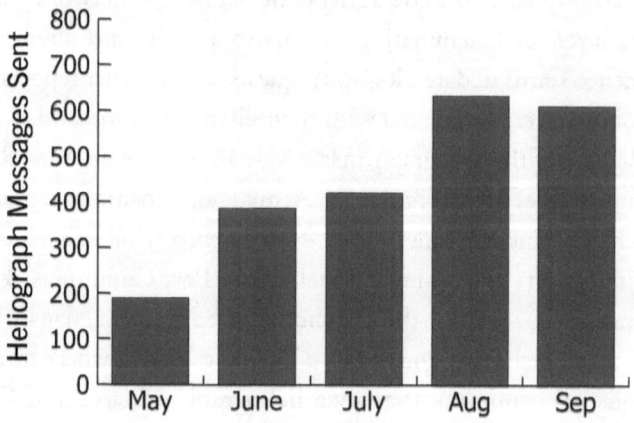

Figure 3. Messages sent within the Arizona part of the heliograph network. Fuller recorded two sets of message counts: one in his textual report and another in his tabular report. The numbers from the tabular report are shown here. Dravo provided two tables: one covering the period before July 31, 1886, and another for August 1886, totaling 754 messages sent. Data from these tables shows that altogether, 2,994 messages were sent between May and August 1886.

The extended range of the heliograph gave troop commanders more options for positioning—over twice as many.[16]

If the visibility area for all forts or observation stations used during Miles's campaign against Geronimo extending out 31.8 miles (the median line of sight across all heliograph networks) is mapped, the area in which a cavalry troop commander, equipped with a heliograph, could maintain communication with his chain of command becomes clear. This system, in operation until Geronimo's surrender, covered approximately 8,900 square miles. This afforded troop commanders significant freedom of movement and positioning.[17]

In contrast, the area covered by flags and their shorter range would have been considerably smaller (using the same number and position of stations): 4,200 square miles—less than half. If just the heliographs' point-to-point connections were supplanted by a signal flag network, approximately twice the number of stations would have been needed— sixty-three signal flag stations instead of thirty heliograph stations.[18] Such a flag-based system would have required twice the signal equipment and trained operators. While that would have been burdensome enough, the underlying cost of placing any sort of remote communication station (flag or heliograph) was logistics and security—both of which would necessarily increase as well. All of these would have drained resources from cavalry and infantry units.[19] Furthermore, the time needed to pass information through a flag network would have doubled. One might guess that such a resource-intensive flag-based system would not have been attempted in the first place, requiring Miles to modify his vision of how operations against the Apaches would unfold.

Miles received instructions from Lieutenant General Philip H. Sheridan to prevent hostilities from spreading among friendly Apaches and to conduct relentless operations to capture or eliminate hostile Apaches.[20] Miles transformed Sheridan's guidance into his own concept, which he transmitted in his one-page General Field Order No. 7 of April 20, 1886.[21] In this order, he emphasized that the primary goal of the troops was to capture or eliminate any hostile Apache bands found in the region. He also instructed that information should be relayed via signal detachments stationed on high peaks to monitor Apache movements and transmit messages between camps.

Miles further stressed that once field commanders had knowledge of hostile Apache movements, they were to pursue the enemy relentlessly until they were either relieved by another unit or the hostile Apaches had been captured or killed. Commanders were also instructed to transmit information on hostile Apaches across the chain of command. Miles provided guidance on force positioning, logistics, reporting requirements, and relations with local citizens. Miles's order laid out a framework of communications, information management, procedures, and decision making—he described what is now referred to as command and control.[22] Fuller and Dravo built a communication network; Miles built a system, an integral part of which was the heliograph network, designed to locate the enemy, communicate, distribute information, and direct the movement of forces to pursue the enemy until its individuals were captured or killed.

Miles's plan was pushed into action only seven days after it was signed when on April 27 (prior to the heliograph deployment and only fifteen days after Miles took command) the Apaches came north from Mexico into the Santa Cruz Valley.[23] Under pressure from the army, the Apaches were pursued back into Mexico.[24] Again in late May, the Apaches crossed the border into Arizona and split their force: Geronimo headed back to Mexico, and a group led by Naiche (the second son of Cochise and the chief of the Chiricahuas),[25] sprinted north toward Fort Apache.[26] Naiche, after avoiding an ambush near Fort Apache, reversed course and returned to Mexico through the Patagonia Mountains.[27] However, before crossing into Mexico, at camp in the Patagonia on June 6, Naiche was surprised and attacked by Lieutenant Robert D. Walsh (who later served in World War I as a brigadier general).[28] Walsh was sent to intercept Naiche by Captain Henry W. Lawton, leader of a battalion at Calabasas.[29] Naiche and his band escaped Walsh's attack and fled into Mexico.[30] During Naiche's gambit up to Fort Apache and back, he was pursued by no less then fourteen detachments of US

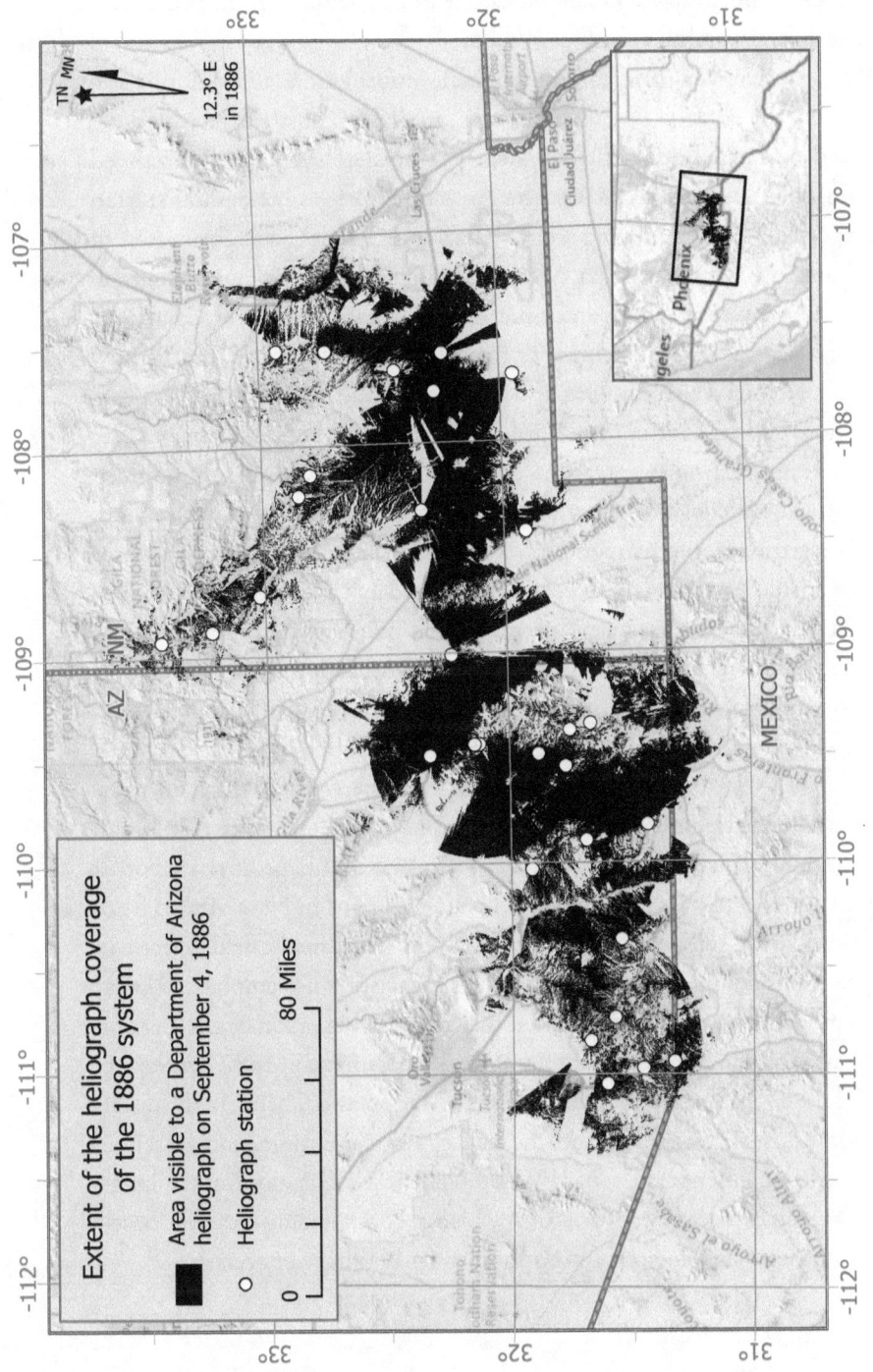

Map 2. Extent of the 1886 network's operational heliograph coverage.

forces, who, following General Order No. 7, "[took] up the pursuit ... until [relieved by] a fresh command."[31]

In late May and early June, "helio information" allowed forces to strike the Apaches "in their second camp[,] ... on their retreat[,] ... and near the border."[32] The "near the border" event likely occurred on June 5, when soldiers at the Antelope Springs heliograph station spotted Apaches moving south. The soldiers sent that information to Fort Bowie and Fort Huachuca. Fort Huachuca relayed, via heliograph and finally courier, this information to Captain Lawton at Calabasas, who sent several detachments to intercept the Apaches; this is how Lieutenant Walsh came to attack Naiche's camp on June 6.[33] Naiche, after slipping south of the border, rejoined Geronimo on June 9.[34] By this time, Lawton's force was in pursuit. It followed Geronimo deep into Mexico, leaving the umbrella of the heliograph network behind. Nevertheless, the heliograph network continued to support a barrier to hostile Apache activity north of the border,[35] even expanding in July, August, September, and November.[36] While Lawton did not use the heliograph as he moved in Mexico, he was still supported by the other communication means until the goal of General Field Order No. 7 was achieved with the Apache surrender in September 1886.[37]

The use and development of heliograph networks in the Southwest continued after Geronimo's surrender. In November, Lieutenant John J. Pershing, who became commander of the American Expeditionary Forces during World War I, established a heliograph line in New Mexico from Fort Bayard to Fort Stanton. The next year, Miles held a field exercise in September and October. During this exercise, Miles emphasized that "commanders would occupy their districts of observation by the location outposts, signal and heliograph stations."[38] Beginning in 1887, lieutenants explored throughout Arizona searching for potential heliograph station locations, aiming to expand the heliograph communication system if needed. Lieutenants Leonard Wood, William A. Glassford, and Clough Overton explored large areas of Arizona, recommending locations based on observation, logistic supportability, and military concerns.

Glassford, a signal officer who joined the army as a private and later became a leader in army aviation, surveyed possible heliograph station sites from Whipple Barracks (near Prescott) to Willcox, Arizona. His focus was connecting "posts in this Territory and especially from the Mogollon to the Sierra Madre Mountains."[39] Wood, a Harvard-educated medical doctor who was a trusted lieutenant of Lawton's and eventually became chief of staff of the US Army, reported on potential connections from Mount Thomas (now known as Mount Baldy) in the White Mountains through the Sierra Ancha to the Gila Basin.[40] Overton, a cavalryman who later served in Cuba and was subsequently killed in the Philippines, concentrated his efforts in the Tonto Basin. They each gave detailed reports describing the terrain and proposed station locations and connections for an expanded heliograph system.

In 1888 the army published the manual *Instructions for Using the Heliograph of the United States Signal Service*.[41] Also in 1888, General Adolphus W. Greely, the army's chief signal officer, convened a board with the task of selecting or designing a "standard heliograph" for army use.[42] Captain Grugan, Lieutenant Maus, and Lieutenant Frank Greene, a member of the Signal Corps, were assigned to this board by Greely. The board determined that none of the heliographs currently available met the army's requirements, leading them to develop the "service heliograph." The service heliograph was light, compact, and sturdy enough to withstand the punishment caused by "almost any circumstances."[43] It featured a two-mirror configuration, allowing operators to use it effectively regardless of the sun's position relative to the target. In other words, operators could still signal to the front even with the sun behind them. In addition to the board's recommendation, the Signal Corps saw a significant increase in funding for army signaling, with expenditures more than doubling, from about $2,000 to $5,000, between 1888 and 1889.[44] The new service heliographs were manufactured and provided to the Department of Arizona "as rapidly as appropriations would permit" and in time to be used in a series of exercises starting in November 1889.[45]

In the fall of 1889, Major William J. Volkmar, serving as the adjutant general of Brigadier General Benjamin H. Grierson's Department of Arizona staff,[46] proposed and took the lead in organizing a large-scale exercise scheduled for May 1890.[47] The aim of this exercise was to expand the existing heliograph system created in 1886. Volkmar, a seasoned cavalryman and veteran of the Civil War,[48] drawing upon his experiences and those documented by Wood, Glassford, Overton, and others, planned to extend the 1886 heliograph system northward to Whipple Barracks and establish an additional line eastward to Fort Stanton. To prepare for the May event, exercises were held in November 1889 and again in April 1890.[49]

In anticipation of these exercises, Volkmar issued specific directives to the acting signal officers (ASOs) of the various heliograph divisions.[50] These directives encompassed a range of important considerations, including the selection of heliograph types for specific stations, the necessity of having commissioned officers present at certain stations, designated times for establishing connections with other stations, the potential reduction of stations based on successful long-range connections, protocols for handling a "through" message (a message relayed by intermediate stations between the sending station and the intended recipient), and reporting requirements.[51] Of particular significance to this work, Volkmar required that "immediately upon effecting communication between adjacent stations ... actual magnetic bearings will be noted."[52] These measurements would be used to help identify many station locations.

Volkmar's initial intentions for the November exercise were established in General Order No. 25 of November 2, 1889.[53] This order outlined a system of thirty-six heliograph stations spread among fourteen divisions.[54] However, by November 6, Volkmar had reduced to eight the required number of stations, with the possibility of four more. Due to the murder of guards and the escape of imprisoned Native convicts from San Carlos, it was thought that heliograph signals north and northwest of the Pinaleño Mountains (sometimes referred to as the Graham Mountains or Mount Graham) would increase already high tensions.⁵⁵ What remained for the November exercise were Camp Henely, Fort Bayard, Stein's Pass,⁵⁶ Fort Bowie, Fort Huachuca, Mount Graham, Cochise Stronghold (near Fourr's Ranch), Fort Lowell, and an intermediate station between Lowell and Cochise (present-day Pistol Hill). The officers establishing these stations described their efforts and those locations within Volkmar's report of the exercise.⁵⁷ The exercise started on November 18 and continued for five days.

Between the November and April exercises, lieutenants were sent to the field to further refine terrain assessments to determine optimal locations for heliograph stations. Lieutenant Mallard F. Eggleston explored peaks near San Carlos, Lieutenant Charles W. Fenton surveyed the area between Whipple Barracks and Fort Verde, while Lieutenant Thomas H. Slavens assessed the terrain between Fort Lowell and San Carlos. Lieutenant George H. G. Gale examined the landscapes around Fort Lowell, Fort Huachuca, Fort Bowie, and Fort Grant. In December, Eggleston and Lieutenant John M. Neall spent some cold weeks high in the Pinaleño Mountains, working to find suitable heliograph station sites. During the same period, Lieutenant Alexander L. Dade explored the Table Mountain area on the range to the west. In December and January, Lieutenants Henry W. Hovey and Richard B. Paddock mapped out a new line of heliograph stations extending to Fort Stanton, New Mexico. In January, Gale ascended today's Rincon Peak to evaluate its suitability for a station. The collective efforts of these officers laid the groundwork for a much more extensive exercise in April, which involved the operation of twenty-five heliograph stations.⁵⁸ For the purposes here, the April exercise, along with the work that preceded it, established the majority of the heliograph station locations used in future exercises and cemented the structure of the heliograph system.

The April 1890 concerted practice, or exercise, focused on that part of the system that was not used and tested in November. The goal of the May exercise was to integrate "sections of the line [that] had been previously examined in detail and . . . prove the system's efficiency as a whole."⁵⁹ The May exercise was a success. It was also important enough that General Greely was present. Greely reported to Secretary of War

Redfield Proctor in 1890 that "the most important event in connection with the Signal Corps of the army has been the unprecedentedly successful establishment and maintenance of an elaborate system of heliograph signaling in the Department of Arizona."[60]

During February 1893, the army held another heliograph exercise in southern Arizona and New Mexico. This exercise tested the army's ability to rapidly deploy heliograph teams to establish a communication network.[61] First Lieutenant Greene led the exercise. He was the department's signal officer under General Alexander M. McCook, the department commander, and one of the few regular signal officers involved with the southwestern heliograph systems.[62]

Greene directed that stations be established simultaneously at nine of the 1890 heliograph station locations. These were Mount Graham, Table Mountain, Bowie Peak, Colorado Peak, Fourr's Ranch North, Fort Huachuca, Stein's Pass, Camp Henely, and Fort Bayard.[63] One of Greene's goals was to test the department's ability to quickly establish a network, without reconnaissance and with only a couple days' notice. The detachments sent to these stations fell in on the same locations using the 1890 data and were provided "blueprint maps of the heliograph system of this department on which the direction and bearing of connecting stations are clearly shown; these, with the carefully erected stakes left by the parties last occupying the points."[64] Based on this, it is doubtful that any new station locations were established for the 1893 exercise.

In 1887, the year after the heliograph's showing in Miles's campaign against the Apaches, high demand for the heliograph's capabilities prompted Greely to lament that he was unable to provide the supply of heliographs demanded by the western commanders.[65] The increased distance between communication stations is what made the heliograph popular.[66] According to Greely, the heliograph's ability to signal farther than flags made it a necessary tool for campaigning in the trans-Mississippi west.[67] Extended communication range leads to an extended span of control over forces in the field. The ability to extend the reach of command and control was (and is) highly prized by the military. Using heliographs, a cavalry troop commander was able to exploit the

Figure 4. Blueprint map of the heliograph system of 1890 (Source: Grierson, Annual Report of Brigadier General B. H. Grierson).

inherent mobility of his organization over a much larger area while still keeping contact with a higher headquarters. The heliograph's increased range also allowed the signal officer at the higher headquarters, using fewer resources, to create a larger network for the operations of field commanders.

It is not surprising, then, that it was infantry and cavalry officers who designed the heliograph networks of the Southwest to satisfy the needs of commanders. Maus, an infantryman, created his 1882 heliograph system to connect two forts and the telegraph system with Colonel Albert G. Brackett's forces "in active campaigns against the Indians"[68] along the upper Gila River. Miles's 1886 network, constructed by Lieutenants Fuller and Dravo, both cavalrymen, nested prominently in Miles's plans. Major Volkmar, also a cavalryman, building upon the 1886 system and using the field experience and ingenuity of several junior officers, brought the heliograph network to its largest extent during the May 1890 exercise. A heliograph-equipped field commander operating within Volkmar's heliograph network would have been able to connect with twenty-five stations spanning an area the size of Kansas. The field commander's connection to the heliograph network also connected him with the telegraph. General Greely himself demonstrated this from Bowie Peak during the May exercise, when he sent a "heliogram" through the network to Fort Whipple, Arizona. At Fort Whipple, Greely's message was transposed and sent via telegraph to Washington, DC.[69]

The southwestern heliograph networks, of which the 1890 exercise was the highwater mark in terms of size and complexity, showed the army and the world that a large heliograph system could be deployed rapidly, provided a reliable and integrated communications system, saved resources compared to other visual signals, and extended a commander's authority and direction of military movements and forces in the field.[70] Moreover, the progression of operational use and exercises led the army to accept the heliograph as an important communication platform. In 1891, the year after the large May exercise, the officer in charge of the Signal Corps Division of Military Signaling, Lieutenant Thompson, declared that "signaling with the *heliograph* is so attractive a method of communication, the range is so great, and the results, when conditions are favorable, so clear and decided that general interest has been developed in the use of the instrument."[71] In 1894, buoyed by additional funding as the Signal Corps shed its Weather Service responsibilities, the army inventory of service and station heliographs was three hundred—enough that the older models were recalled.[72] Heliographs were part of the curriculum at West Point, and each of twenty-two US colleges with professors of military science had a set of heliographs.[73] Heliographs were funded and were in the field, in classrooms, and in army manuals.[74] Field commanders and their lieutenants, through operations and exercises, had pulled the heliograph from being a curious technology to being a pillar of army communications.

Six years after the 1893 exercise, Greely summarized that the operations of the modern signal service rested on three pillars: visual, electric, and aviation (balloons). First among these three was visual, consisting of flags and heliographs.[75] While heliograph use tapered off after 1900, as late as 1914 a US Army heliograph station was providing communication from and to the headquarters of the US Expeditionary Forces from the top of the Terminal Hotel in Vera Cruz, Mexico.[76] Later still, in 1916, heliographs were used in Texas as part of the Mexican Expedition.[77] Correspondence dated June 21, 1918, suggests that during World War I, heliographs were still being used by the US Army to transmit personnel information in France.[78] The heliograph's use in France marked forty-one years (1877–1918) of heliograph adaptation, training, and field service—well over a quarter of the army's history to that point.

The heliograph stations used during the operations and exercises discussed here were only a fraction of the total number of stations. Many other stations were established, tested, and used but were not part of the Geronimo campaign or the larger exercises; Fort Thomas, Triplets, and Mount Turnbull are examples. The stations mapped here are those for which there is documented evidence suggesting that a heliograph was set up at the mentioned location and connected to another heliograph station within a larger network. In a few cases, while a direct flash-to-flash connection is not explicitly documented, contemporary reports record azimuths to other stations and historical maps confirm these stations' network placement, providing strong supporting evidence for their inclusion. An exception to this rule is Fort Selden in New Mexico, where there is little to indicate a connection to the larger network; however, there is sufficient circumstantial evidence, including a presidential proclamation, to warrant the inclusion of Fort Selden's heliograph station and analysis as an appendix.

Not mapped are several transitory stations used by detachments of soldiers in the field—for instance, those scouting camps near Clifton that Lieutenant Maus connected to in 1882. Similarly, the White's Ranch station, south of Fort Bowie, was positioned to connect with the "several troops of cavalry" in the canyons of the western slope of the Chiricahua Mountains.[79] Too little information is available to include those field detachment stations here.

Fixing a heliograph station's location on a map is at odds with the heliograph's attributes of being light, mobile, and easily moved to carry out the same work from different places. There may have been several stations set up at different times and places with the same mission. Most stations were simply a place on the top of a hill or a rock outcrop. While some stations did have improvements, such as platforms, a telegraph line, or wooden posts, the only essential requirement was visibility.[80] The station locations presented here result from a combination of descriptions, logic, and geospatial analysis, making use of the best available data and techniques. It is important to note that the locations given here are not archaeological sites in the traditional sense, where evidence of past human presence, such as spent shell casings, posts, or rock structures, are found. The locations given are where the lieutenants' descriptions and available data suggest that a heliograph was set up and used.

A map authored by Major Volkmar and produced after the May 1890 exercise gives general locations of many heliograph stations used between 1886 and 1890 (and, by extension, 1893). This map is a general reference showing major rivers, forts and camps, towns and cities, roads, railroads, telegraph lines, heliograph stations, and latitudes and longitudes.[81] The map also shows sightlines from station to station (see Figure 4). Unfortunately, there is no distinction between a sightline used in 1886 and one used in 1889 or later. For example, from Fort Stanton the map shows two sightlines emanating from a single point, one sightline to the Nogal station to the north and another to the Sierra Blanca station to the southwest. In fact, those sightlines did not extend from a single point in either space or time. There were two station locations set up at Fort Stanton, one in 1886 connecting to Nogal and another, at a different location, connecting to Sierra Blanca in 1889.

The azimuths of these sightlines are documented in a table located in the upper right corner of Volkmar's map. These azimuths were obtained using prismatic compasses capable of measuring to half-degree precision. During the May exercise, Volkmar instructed officers to record as many readings as possible and report the mean average of their various observations.[82] These carefully measured azimuths served the purpose of map correction in 1890 and remain integral to current geographic modeling efforts. However, within this table, some azimuths are marked with an asterisk, indicating that they were estimated rather than directly measured in the field. These estimated azimuths typically relate to 1886 stations that were not used in the 1890 exercise, preventing the ASOs from conducting "observations of a particular point ... repeatedly, on different days, and a *mean* of such readings reported."[83]

Given the scale of Volkmar's map and the presence of estimated azimuths in some cases, there are uncertainties regarding the relationships between certain stations, particularly those absent from the Spencer map[84] (the excellent map titled *Field of Operations against Hostile Chiricahua Indians* produced by cavalryman-turned-engineer Eugene J. Spencer).[85] The Spencer map includes most of the 1886 heliograph stations.[86] However, this map does not show heliograph stations in the 1886 line extending to Fort Stanton or any of the later stations not otherwise used in 1886.

Along with these maps and Volkmar's reports of the 1889 and 1890 exercises are a volume of reports describing the reconnaissance and occupation of individual heliograph stations. These reports were often rich enough with geographic information to show a precise location. Sometimes the geographic information of one station was reported in the description of another. For example, Lieutenant Carl Reichmann reported from Saddle Mountain that the Table Mountain station was located on the "southern edge of the left one of the two [table-shaped] mountains."[87]

It makes sense that the soldiers conducting the reconnaissance and selecting of the station locations used the highest point available, as that point, presumably, offered the best chance of initial success. Indeed, in his 1887 report, Lieutenant Glassford concluded that "stations should be made at summits of peaks, as far as possible, and circle of observation not sacrificed to convenience by getting in comfortable sheltered spots beneath mountain tops," also mentioning that high ranges reduce the "evils of refraction."[88] However, other concerns affected choosing a location: accessibility, forage for animals, water availability, timber clearing, and the military evaluation of the location with respect to concealment and defensibility. These other factors became clear in the reading of the various lieutenants' reports.

Despite the heliograph's mobility advantage, it was mobile only in as much as it could still carry out its station's assigned purpose. The location, therefore, required visibility to all other stations assigned. For instance, a station on the Pinal Mountains that needed to communicate with both Lookout Peak and the soldiers at the San Carlos Agency could only be in a handful of geographic areas where visibility to those other stations was simultaneously possible (shown in hatched black on Map 3). The requirement for observation was the major consideration for mapping the various station locations.

Supplementing the historical records penned by the officers setting up the heliograph systems, computer-generated visibility models, or diagrams, were constructed using elevation models with a resolution of 5 meters and 10 meters.[89] Visibility diagrams are maps showing areas that are visible from a particular point. These diagrams account for the curvature of the earth, atmospheric refraction, and the height of the instrument, 4 feet.[90] However, visibility information provided by

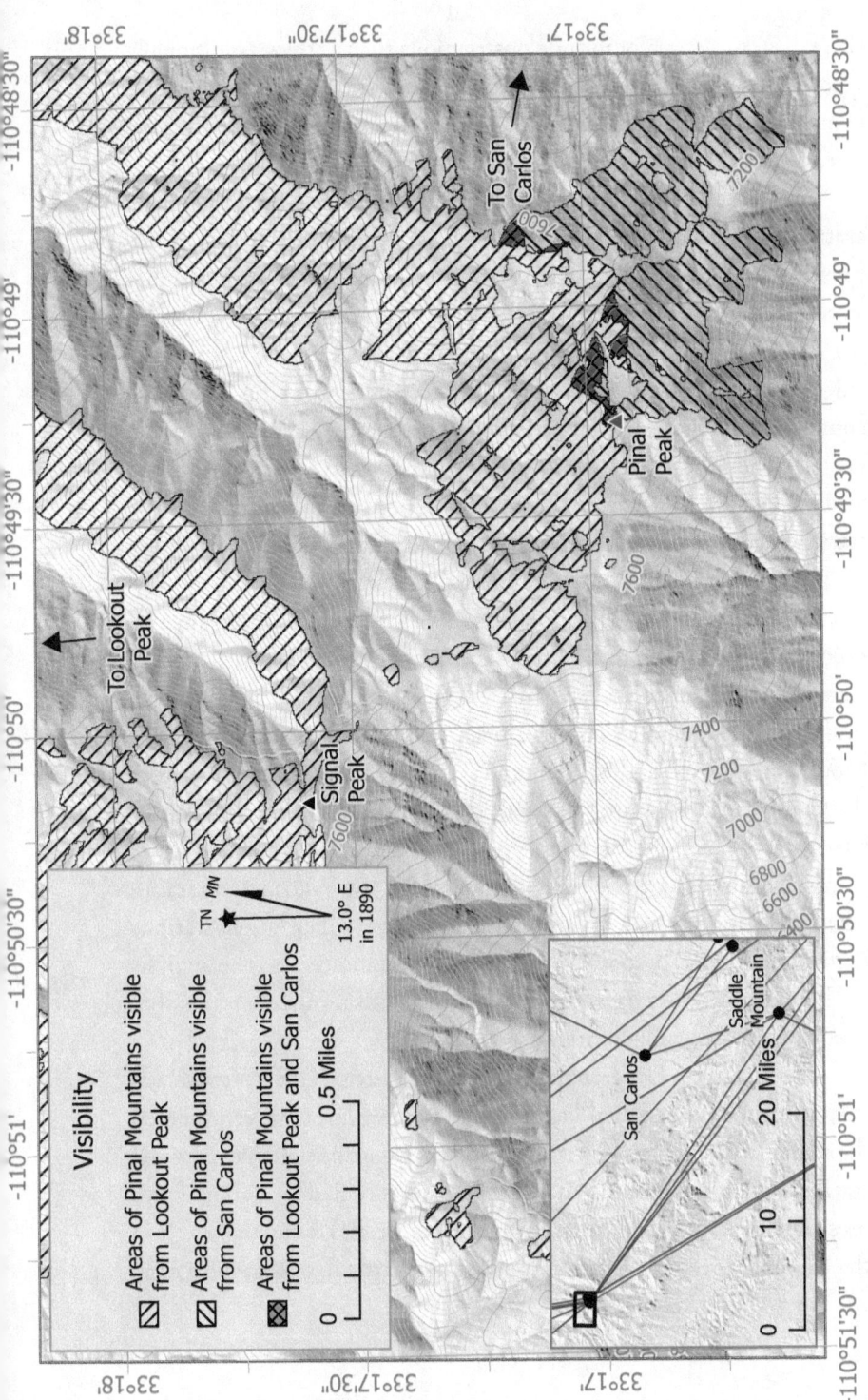

Map 3. Example of visibility diagrams.

these diagrams does not include obstructions such as trees. Additionally, where the 10-meter resolution models were the only ones available, some micro-terrain was missed. The underlying data for these models were produced by the US Geological Survey.[91]

As most of the stations were visible from more than one other station, the areas where multiple visibility diagrams overlap reveal potential sites for stations.[92] In cases where a station linked two other stations situated on opposite sides of a mountain range (which was often the case), the area of common visibility quickly narrows down to the ridge's summit in between. Indeed, in most instances where multiple connections were required for a heliograph station, the area of common visibility, the intersection, becomes smaller with each overlapping visibility diagram. Accordingly, the number of possible locations for a station becomes smaller as well. These small areas of common visibility are powerful indicators of where the stations were.

While a visibility diagram can be modeled from just about any two points, care was taken to select stations for the model whose locations were already reasonably established. A simple example is locating which peak from the complex of small peaks along the ridgeline of Bowie Mountain was used in the heliograph system. In this case, visibility from the Camp Henely station was used, as its location is easily determined based on the well-worded descriptions of several lieutenants. More complex examples are the three stations atop the Rincon Mountains. Their descriptions are, in large part, based on which other stations the lieutenant and corporal setting them up could see. So the analysis of these stations was based on the locations of six other stations that needed to be determined first.

For this study, Geographic Information System (GIS) software was used to create the visibility diagrams and to draw lines that represented the station-to-station azimuths and distances along the sightlines between stations.[93] These azimuths and distances, as specified by Volkmar and presented in tabular form on his map, and sometimes found separately in the lieutenants' reports, often converged in a particular geographic area, and in some cases the convergence of those lines alone was enough to help establish a location. In other cases, a more comprehensive analysis of the lines and distances was necessary.

In cases where a more detailed analysis was needed, a terminal point was plotted at the end of each azimuth-distance line from one station to another. It is worth noting that as the distance between connected stations increased, even a small error in compass reading could result in a substantial deviation when plotting such a point.[94] To account for this potential deviation, weights were assigned to these points based on the length of the line. For example, at the station on Fourr's Ranch North, some points marked the ends of relatively short lines, such as Colorado Peak (present-day Rincon Peak) at 31 miles away, while others marked the ends of longer lines, like the station on the Pinaleño Mountains, situated 53 miles away. To determine the weight for each point, the inverse of the distance was used, resulting in a weight of 1/31 for Colorado Peak and 1/53 for the station on the Pinaleños.

In total, there were six azimuth-distance lines with associated terminal points converging on the Fourr's Ranch North station.[95] Calculating the weighted mean center of these points established a single averaged point, which helped determine the most likely location—particularly in cases with multiple potential station locations derived from visibility analysis, such as in the Fourr's Ranch North example.[96] To avoid repetition throughout the text, the formula for calculating the mean center is here:

$$\overline{X}_w = \frac{\sum_{i=1}^n w_i x_i}{\sum_{i=1}^n w_i} \quad , \quad \overline{Y}_w = \frac{\sum_{i=1}^n w_i y_i}{\sum_{i=1}^n w_i} \quad ,$$

where $\overline{X}_w$ is the weighted mean X coordinate, $\overline{Y}_w$ is the weighted mean Y coordinate, w_I is the weight for each endpoint, x_I is the X coordinate for each endpoint, and y_I is the Y coordinate for each endpoint.[97]

Georeferencing is the process of aligning scanned images of historical maps with current maps or aerial photographs. By identifying reference points common to both, the scanned map is adjusted to match the scale and orientation of a modern map or photograph. The National Archives has digitally published detailed maps from the late 1800s for some of the forts mentioned in this manuscript. Where relevant, these maps were georeferenced using present-day features that existed when the older maps were created. For example, an 1884 map of Fort Bayard, New Mexico, shows a hospital and several officers' quarters that still stand today. Using these features, the 1884 map can be superimposed on current maps, revealing the locations of other 1884 features that no longer exist. In the case of Fort Bayard, the location of the "Engineer House," a landmark used to describe the location of the heliograph station there, was determined through georeferencing.

Regardless of the analysis tool employed, whether it was based on visibility, mean center calculations, or any other method, care was taken to ensure that the selection of heliograph stations for analysis matched the time frame and circumstances faced by the heliograph operators. For example, according to the Volkmar map, Fort Huachuca was depicted as connected to six locations: Mount Graham, Rincon Mountain, Colorado Peak, Fourr's Ranch North, Little Baldy Peak, and Antelope Springs. However, only three of these connections were used in 1890: Mount Graham, Colorado Peak, and Fourr's Ranch North. Little Baldy Peak, Antelope Springs, and another station near the Dragoon Mountains, Fourr's Ranch South, were exclusively used in 1886. These two sets of analytical data result in multiple different heliograph stations for the same "Fort Huachuca."[98]

Other credible geographic information was also used to aid the process of identifying heliograph stations. This information includes photographs, reports, field visits, period US Geological Survey and US Army topographic maps, and General Land Office (GLO) maps.

Two books hold the great majority of primary sources used in this manuscript: *Instructions for Signal Officers Heliograph Divisions & Stations, Department of Arizona 1890* and *Report of General Practice of the Heliograph System, Department of Arizona, in May, 1890*.[99] Both are essentially bound volumes of the reports of the lieutenants who reconnoitered and established the various heliograph stations. The army produced both over Major Volkmar's signature.

The reports in *Instructions for Signal Officers Heliograph Divisions & Stations* often recorded officers' thoughts on why they established, or didn't establish, the stations where they did. If this book had a list of chapters, it would include twenty: a circular written by Volkmar; General Order No. 2; "Reports of Preliminary Reconnaissances and Concerted Practice between Arizona Divisions of the Department Heliograph System During April 1890"; "Report of Concerted Practice Department Heliograph System, in November, 1889"; fifteen additional chapters, each a report (submitted between 1886 and 1890) by an officer involved with establishing the networks; and a chapter of tabular reports.

The two chapters following General Order No. 2 include several reports by individual officers in the field. In my citations of these, the officer writing the report is included in the citation. In all, there are thirty-two reports written by junior officers who led teams that planned, searched for, tested, emplaced, and improved heliograph stations throughout Arizona and New Mexico.

The *Report of General Practice of the Heliograph System, Department of Arizona, in May, 1890* is Major Volkmar's comprehensive report of that exercise and includes reports of the officers occupying individual stations. By and large, the stations used for the May exercise were already established and used in 1886 or for the November and April exercises, so there were few changes in station locations. Where there were changes, a fresh analysis of the location is given if needed.

Other notable sources include various reports to the secretaries of war as well as reports of the chief signal officer of the US Army; Rebecca Rains's *Getting the Message Through*;[100] Bruno Rolak's article "General Miles' Mirrors";[101] Roger Kelly's "Talking Mirrors at Fort Bowie";[102] a US Army publication studying military heliography written by Colonel John V. Mills;[103] George W. Cullum's *Biographical Register of the Officers and Graduates of the United States Military Academy*; General Miles's *Recollections*, published in 1896;[104] Brigadier General Grierson's 1890 report to the War Department;[105] and dozens of maps and sketches provided by the US Geological Survey, the Library of Congress, the National Archives, and the US Bureau of Land Management (BLM).

US National Oceanic and Atmospheric Administration (NOAA) data was used to find magnetic variation at the time and place for each of the stations. Magnetic variation ranged from 11.8° at what was then San Nicholas Peak in New Mexico to 13.8° in the region around Forts Whipple and Verde in Arizona. Knowledge of magnetic variation in the late 1800s enables conversion of magnetic azimuths given throughout officers' reports into today's true azimuths. Knowing the true azimuths facilitates representation of azimuth lines and distances on current GIS maps and facilitates analysis.[106] For a map depicting a single heliograph station, the magnetic declinations associated with the north symbols are calculated at the station. For maps with multiple stations, the magnetic declination is calculated for the center of the map.

Each chapter begins with a map that depicts the heliograph stations relevant to that chapter, along with their connecting links. The symbols representing the heliographs include the years they were established, with small hash marks superimposed on these symbols, akin to clock hands, indicating the exercises in which the station participated. Most maps showing specific heliograph station locations are at a 1:24,000 scale, though other scales are employed when they provide greater insight. The base layer of the maps varies depending on the context; in most cases, a hillshade (a type of shaded relief) is used because it effectively displays the terrain. Contour lines indexed with elevation are overlaid on the hillshade. In some instances, period GLO plat maps are used to show details that align more closely with the late 1800s. The majority of the layers on the maps are available either through the BLM or the US Geological Survey, with additional sources including the National Archives and the Library of Congress. Often, the name of the heliograph station matches that of a nearby fort or camp. In these cases, the heliograph label's text is slightly larger and uses a different symbol.

With respect to the development of the heliograph networks, this narrative follows a general chronology, beginning with the 1882 heliograph system, then the 1886 network, and finally the systems established in the late 1880s and early 1890s. Strict organization by historical military district or time was complicated by stations, such as Bowie Peak, that had heliograph connections spanning both military and state boundaries and remained in use across multiple systems over the eleven years these networks operated. Rather than adhering to military divisions or a strict timeline, I was guided in my approach by regional considerations and network connectivity.

Chapter 2

The 1882 System

IN 1882, LIEUTENANT MAUS CREATED A HELIOGRAPH SYSTEM CONNECTING Fort Grant; Fort Bowie; the upper Gila River near Clifton, Arizona; and possibly a southern line to Camp Price, east of Old Camp Rucker.[1] Maus, an infantry officer, had already distinguished himself by that time, earning a silver star in action at Bear Paw Mountains during the Nez Perce War.[2] In July 1882 he was posted at Fort Grant and "in charge of field signaling [in] S.E. Arizona."[3]

Maus needed two intermediate stations to connect Fort Bowie and Fort Grant with heliographs. The longest distance between any two stations, as reported by Maus, was 25 miles.[4] Various methods of visibility analysis have produced either too many or too few potential locations for these stations. However, a station at Government Peak and another in the foothills south of today's Stockton Pass would likely meet this requirement, as the distance between these two points is about 25 miles. These locations remain speculative, though, as several other points, either lower in the Dos Cabezas Mountains or higher in the foothills, could also have been suitable.

What is not speculative is that during Maus's return to Fort Grant (July 20–21), he climbed the highest point of the Dos Cabezas Range. There he set up a heliograph and exchanged several messages between Fort Grant and Fort Bowie. The highest point of that range is the Dos Cabezas Peaks, with the southern of the two tallest at about 8,350 feet. From this peak, all the ridge on which Bowie Peak sits is visible, along with Bowie Station, Fort Grant, most of the peaks of the Pinaleño Range, and almost the entirety of Sulfur Spring Valley to the south.

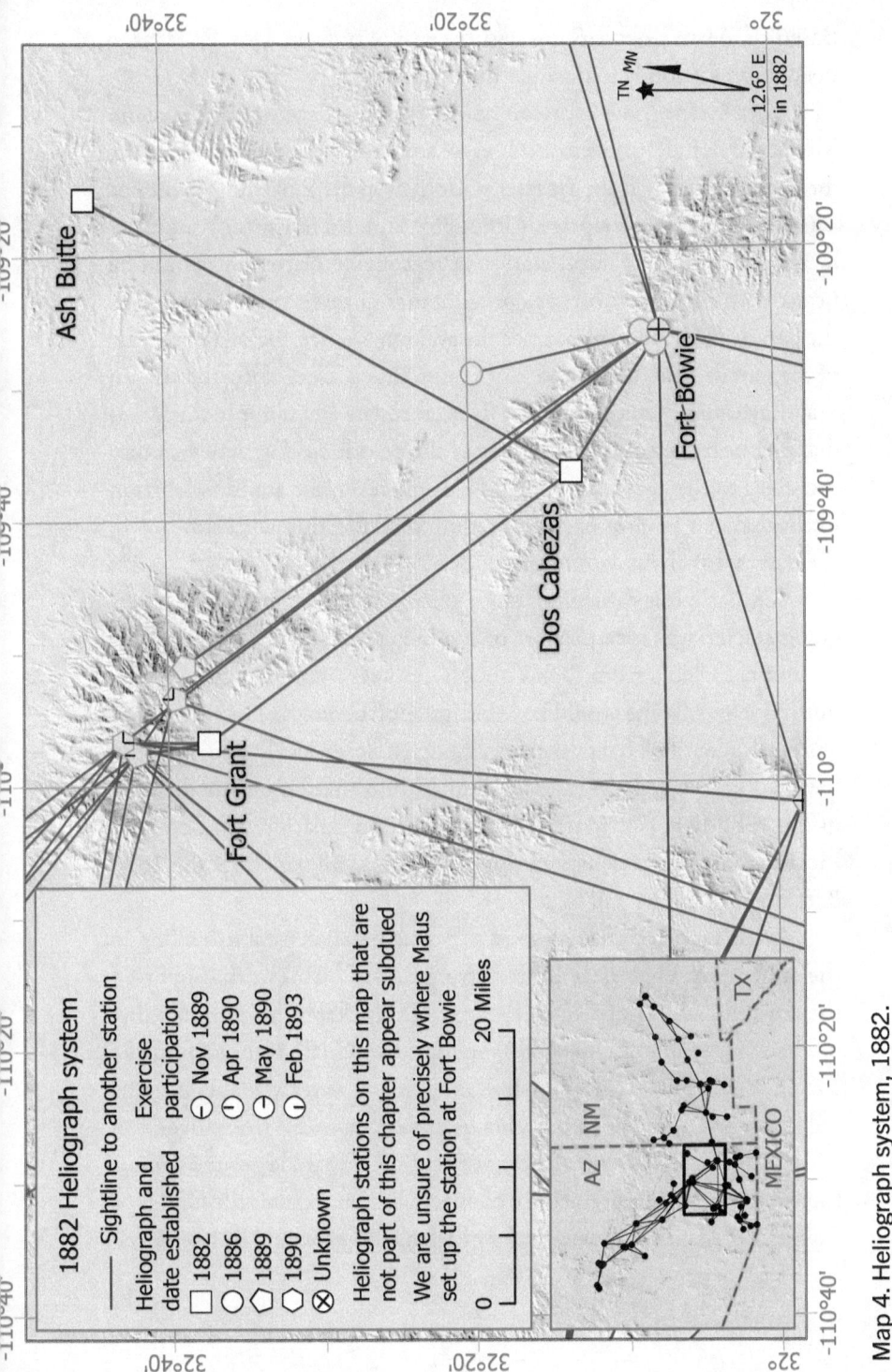

Map 4. Heliograph system, 1882.

Based on Maus's description and the visibility from Dos Cabezas, a heliograph station is mapped at this location.[5]

Fort Grant, now a prison at the western base of the Pinaleño Mountains, is fully visible to an observer atop the Dos Cabezas Peaks. The fort was also visible from the later stations atop the Pinaleño Mountains (discussed in a later chapter). Although Maus did not specify the exact location of the heliograph station, a reasonable placement would be near the commanding officer's or adjutant's quarters on or adjacent to the parade field.[6] However, since the adjutant's office, located northeast of the parade field, is close to Fort Grant Creek, trees along the stream may have obstructed visibility. Therefore the heliograph station is mapped near the southeast corner of the parade field, where visibility was clear to the area near Fort Bowie. However, the station's location likely changed as later exercises required connecting with stations in the nearby Pinaleño Mountains.

Similarly, it is difficult to state where the 1882 Fort Bowie heliograph station was set up. Most of the hills to the south of the parade ground are visible from Dos Cabezas, and very little is visible to the north. Helen's Dome would have been a good choice, as it is prominent and easily identified from its shape, though a closer, less lofty peak could have worked as well. With such limited information on the location of the 1882 Fort Bowie station, no further speculation will be made. However, in the next chapter, using additional information, the 1886 Fort Bowie stations will be mapped.

Maus reported that "as soon as possible I shall establish a line to the camp to the Gila (near Clifton) from Bowie."[7] This work supported the scouting efforts of Colonel Albert G. Brackett (the commanding officer of Fort Grant in July 1882) on the upper Gila.[8] Maus identified a station at Ash Butte (today's Ash Peak) that had water available and was only 10 miles from the upper Gila. Ash Peak is visible from Overlook Ridge, a short distance northwest of Fort Bowie's parade grounds, but if there remained a requirement to connect with Fort Grant through Dos Cabezas, then either Bowie Peak or Helen's Dome would be a required

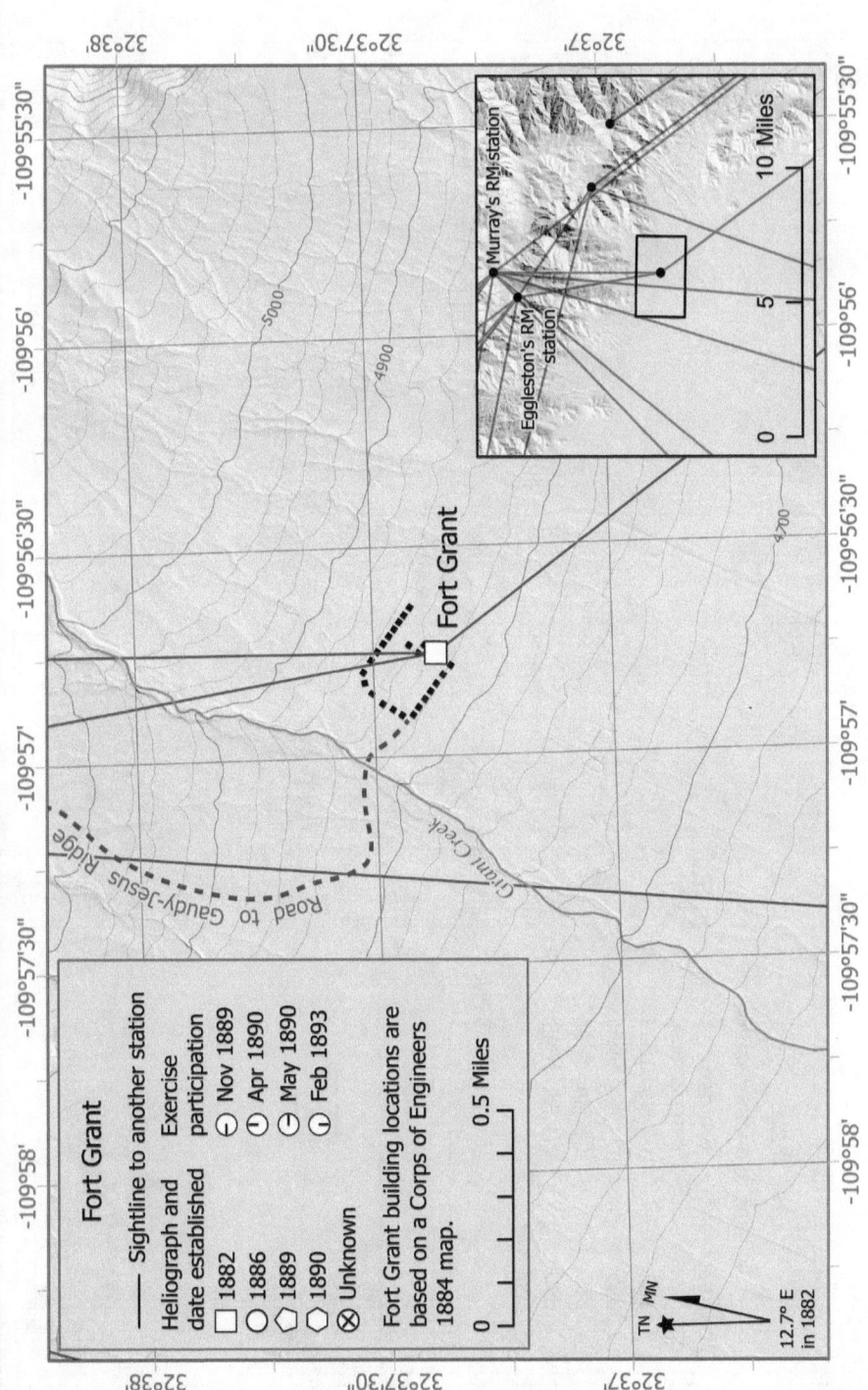

Map 5. Fort Grant heliograph station.

Map 6. Ash Butte heliograph station.

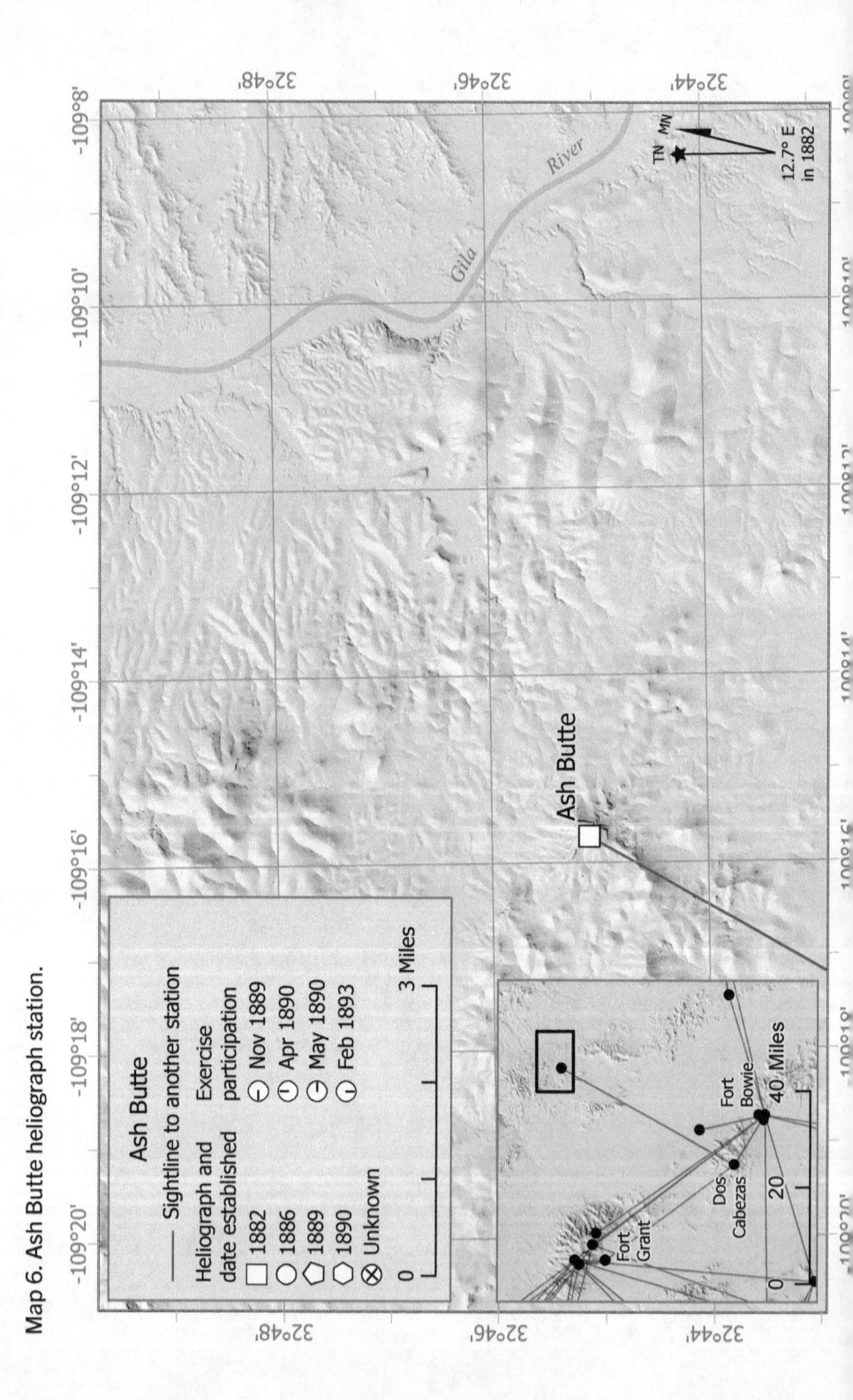

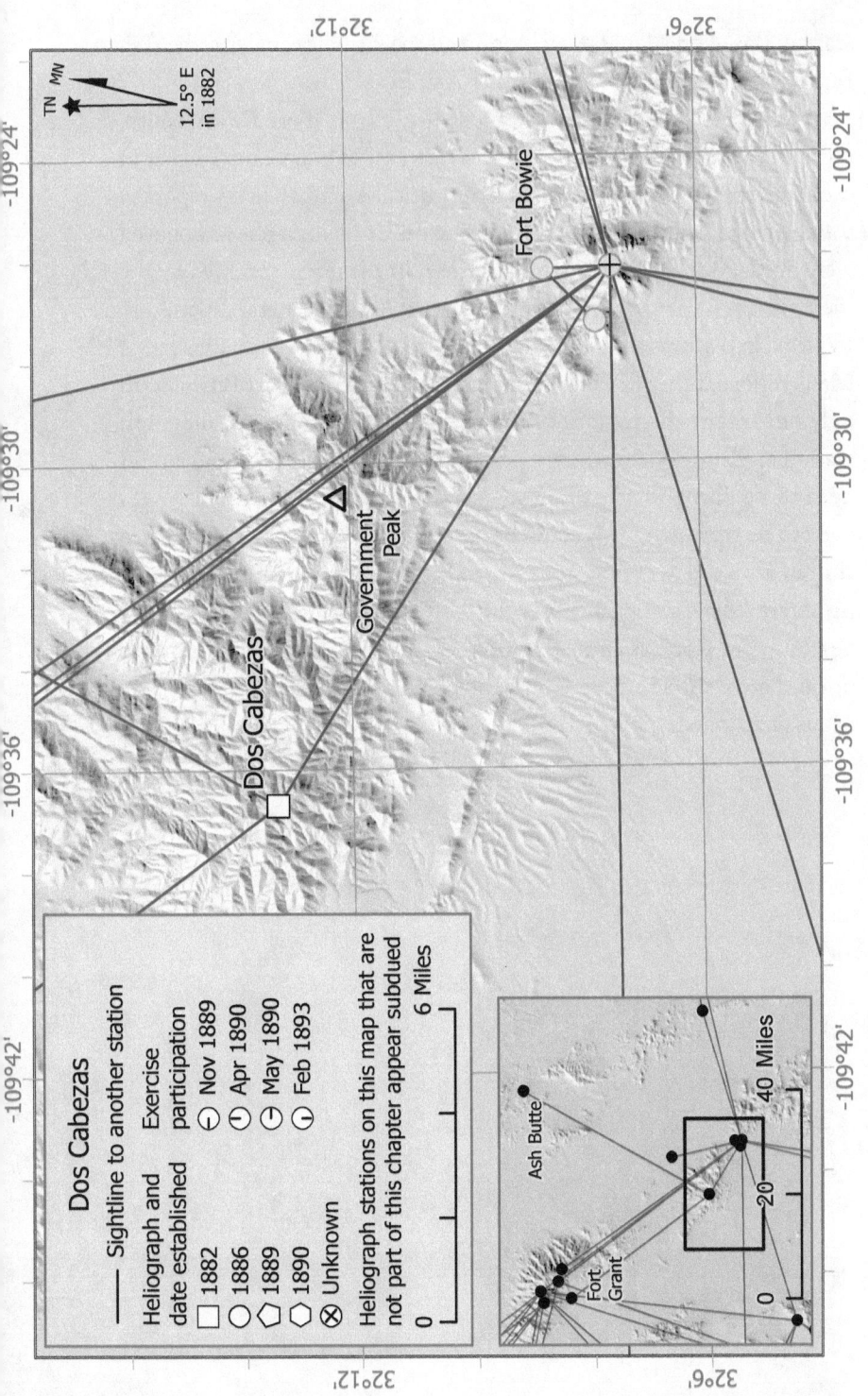

Map 7. Dos Cabezas heliograph station.

station. Based on Maus's description and the visibility analysis, the Ash Butte station is mapped atop today's Ash Peak.[9]

Maus also mentioned a proposed line from Fort Bowie south, terminating at Price (meaning Camp Price, east of Old Camp Rucker in the southern Chiricahua Mountains, not Price, Arizona). In his report to the chief signal officer, he requested an extra six heliographs he believed were needed for this line.[10] This southern line is not mentioned by the Department of Arizona commander, Brevet Major General Orlando B. Willcox, in his annual report, suggesting that it was never established.[11] Maus reported that he could reach Camp Price from Fort Bowie with only one intermediate station. If this were to be done, the heliograph station at Camp Price (which sits in today's Price Creek Valley) would need to be on an adjacent hilltop and the intermediate station would need to be very high. This could be accomplished with the Camp Price station atop today's 6,300-foot May Day Peak near Camp Price, with the intermediate station at either the 9,800-foot Chiricahua Peak or the nearby Paint Rock at almost 9,400 feet. As will be seen, a heliograph line did cross the southern Chiricahua Range in 1886 but not the way Maus envisioned.

Chapter 3

Fort Bowie

FORT BOWIE IS LOCATED IN THE NORTHERNMOST HILLS OF THE CHIRICAHUA Mountains, about 13 miles south of Bowie, Arizona. Bowie Pass, renowned for its role in the Bascom Affair,[1] is 2.5 miles to the west and separates the Chiricahuas from the Dos Cabezas Mountains. The fort was named after Colonel George Washington Bowie of the Fifth California Volunteer Infantry, who commanded the soldiers that built it in 1862.[2] Fort Bowie was established to protect the region's residents and to safeguard transportation networks.[3] Four heliograph stations were located at or very near this fort, with the heliograph at Fort Bowie playing a central role in General Miles's 1886 communications system, connecting New Mexico, Fort Huachuca, and smaller cavalry camps to the south. Overlapping with the telegraph system, the four stations were quickly reduced to the one on Bowie Peak.

Lieutenant Fuller began his account of establishing the Arizona heliograph stations with the following: "On the 29th day of April, I had the home station [Fort Bowie] working with No. 2 ... placed on Helen's Dome."[4] The 6,400-foot Helen's Dome was an excellent choice. From this peak, located southwest of Fort Bowie, all five connecting stations were visible: Fort Bowie, White's Ranch to the south, Emma Monk (the extreme northern point of the Swisshelm Mountains, named after a nearby ranch),[5] Stein's Pass in New Mexico, and Bowie Station. Helen's Dome was also easily recognizable due to its distinct shape. However, its location southwest of Fort Bowie and its elevation posed a disadvantage: the Dos Cabezas Mountains obscured the view from Helen's Dome to the Pinaleño Mountains. While this wasn't a concern in 1886, the Pinaleño Mountains became an important connection as the heliograph system expanded.

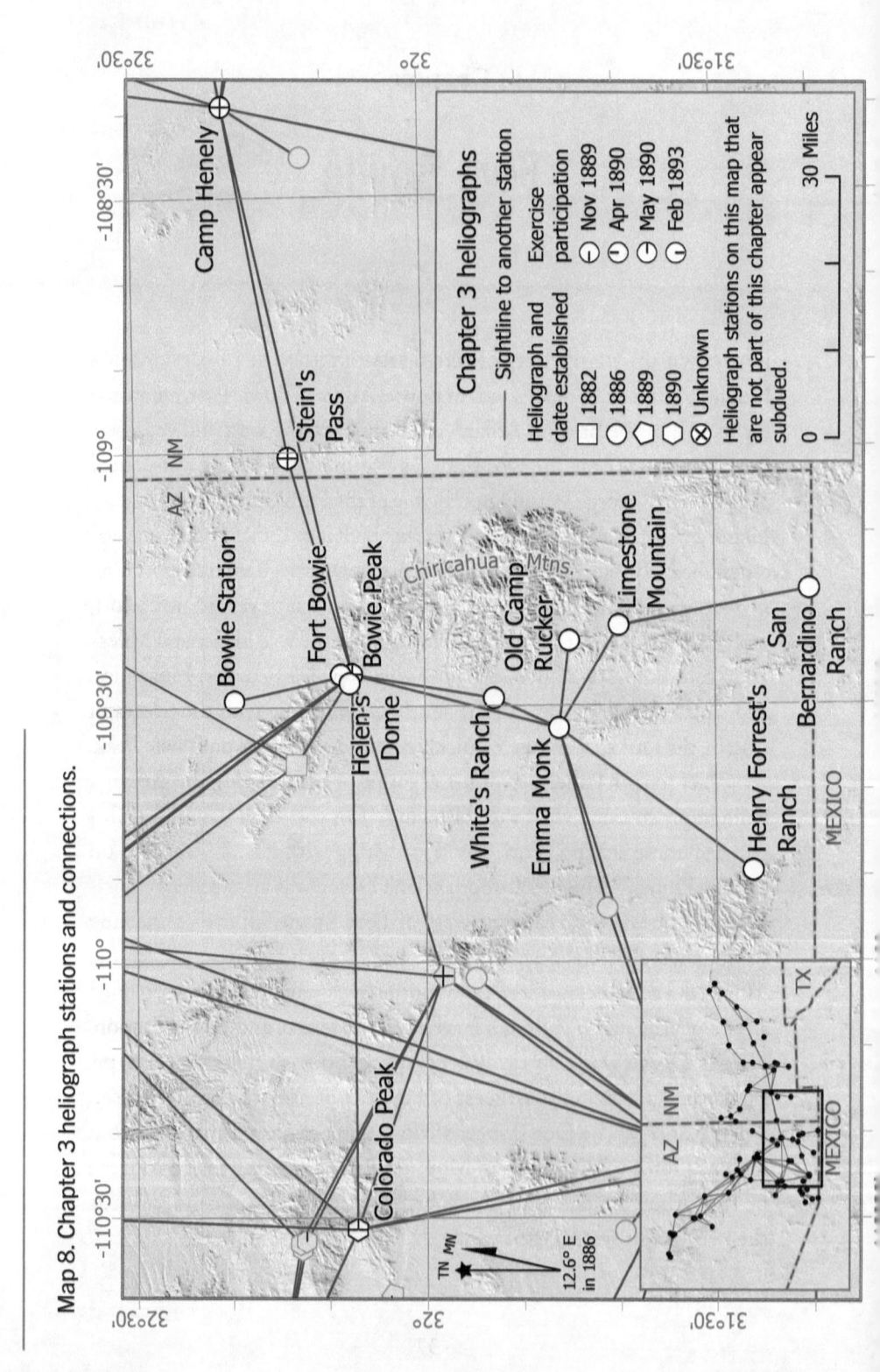

Map 8. Chapter 3 heliograph stations and connections.

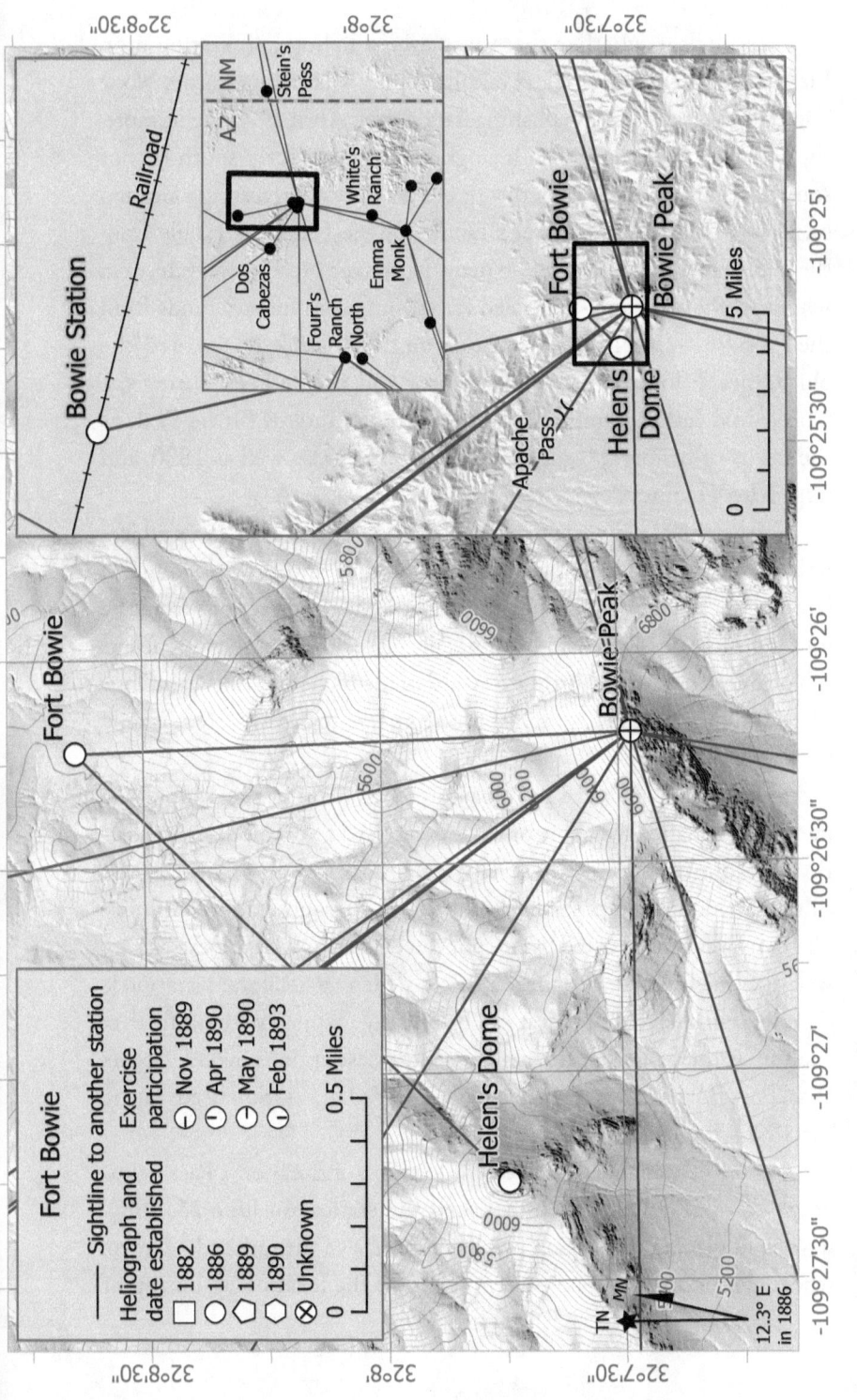

Map 9. Heliograph stations at and near Fort Bowie.

It's unclear whether there was ever a heliograph station atop Helen's Dome. By July, Fuller's tabular report was calling station No. 2 "Bowie Peak," with an establishing date also of April 29.[6] Furthermore, the Spencer map shows the heliograph station clearly south (if not south–southeast) and not southwest of Fort Bowie, suggesting a location on Bowie Peak. Unlike Helen's Dome, Bowie Peak was visible from the Pinaleño Range. In 1889, when Lieutenant Neall—a cavalryman who in 1899 was court-martialed for stealing commissary funds from the Presidio—established the station on Bowie Peak, he repaired "the old telegraph line (in use between post and Bowie Peak during the Chiricahua Indian campaign)."[7] All of this points to Bowie Peak as the nexus of heliography at Fort Bowie in 1886, as well as 1890, and not Helen's Dome.

Nevertheless, Fuller reported Helen's Dome as a station, and it is unlikely he confused the two, as Helen's Dome and Bowie Mountain have distinctly different shapes and are in different directions from the fort. It is quite possible that Fuller initially placed the station atop Helen's Dome only to move it shortly afterward to the higher Bowie Peak. Thus Helen's Dome as a heliograph station appears to have existed, if only for a very short time.

The station at Bowie Peak was situated on a small peak along the inverted U-shaped ridgeline of what is referred to on modern maps as Bowie Mountain. There are several small peaks along that ridge, and to narrow down the possible site of the heliograph station, visibility diagrams were constructed using connected stations whose locations are well documented. For instance, the Camp Henely heliograph station is described in such detail that it can be reliably mapped without further analysis. Other stations with similarly clear descriptions include Stein's Pass, Fort Bowie, Bowie Station, Colorado Peak (a station used during the 1890 exercise), and Emma Monk.

Three lieutenants described the location of the Stein's Pass heliograph station. Fuller, who established the station on June 26, 1886, described placing the station on the bluffs above the railroad.[8] Dravo, whose district Stein's Pass was in, reported the location as "on top of

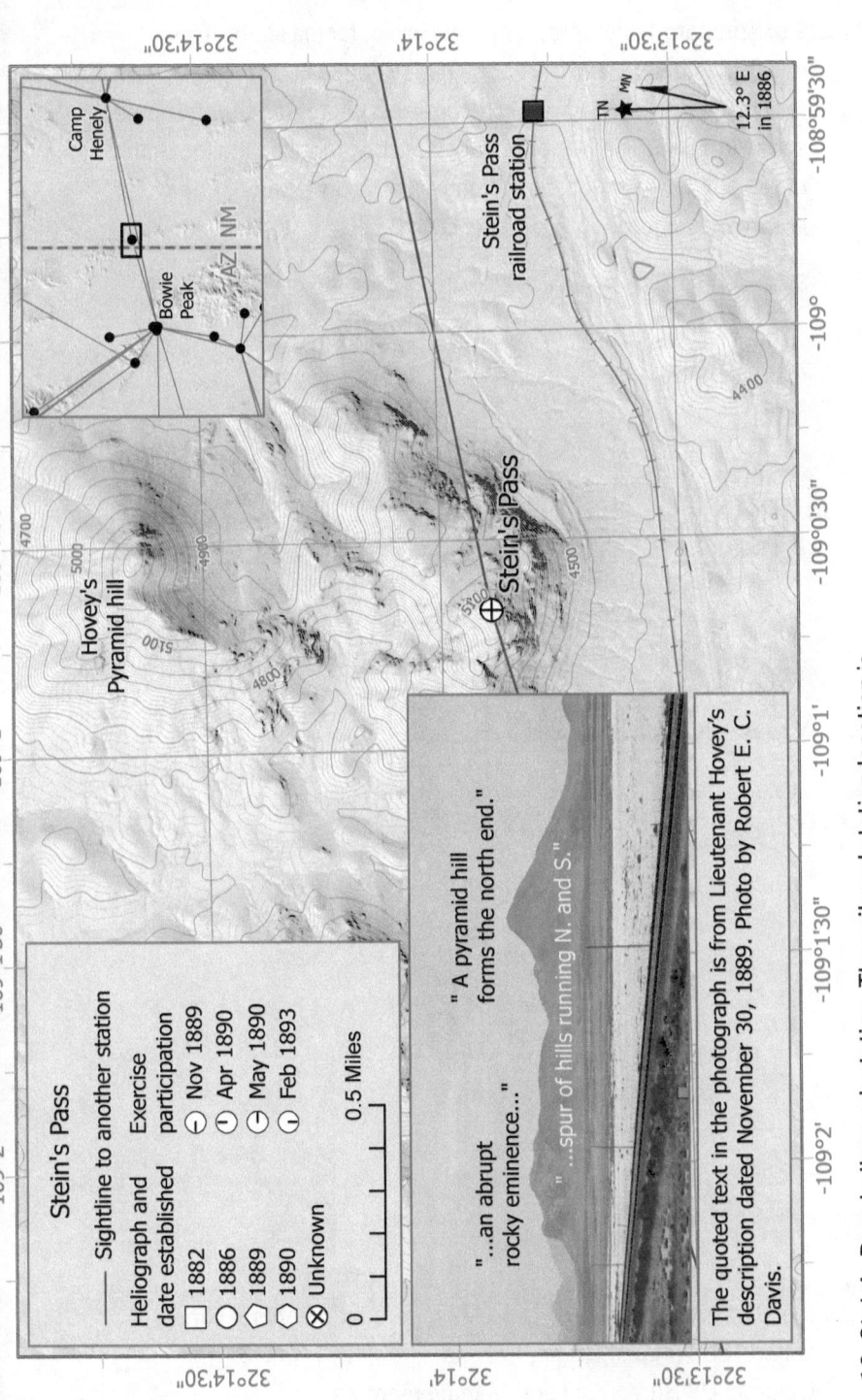

Map 10. Stein's Pass heliograph station. The railroad station location is based on the 1876 General Land Office plat.

Figure 5. Annotated photograph of Fort Bowie, facing south. This image was taken near the spot where Charles B. Gatewood reportedly captured a widely recognized photograph of the same scene in 1886. The National Park Service identifies the location of the adjutant's office, as indicated in the annotation. Photograph and annotation by Robert E. C. Davis.

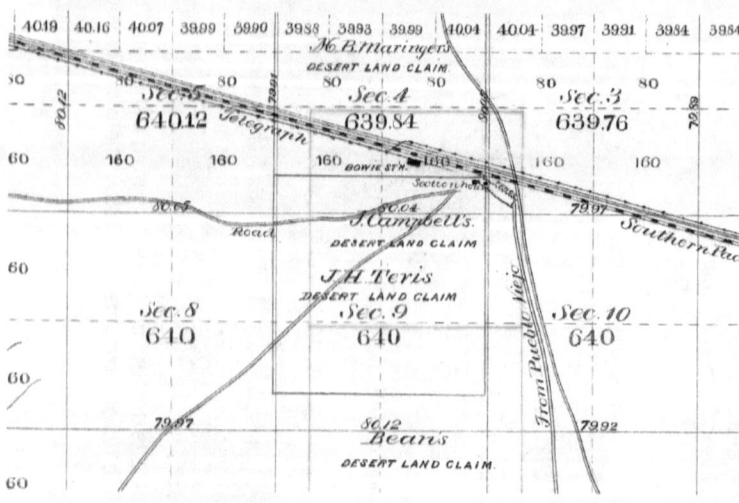

Figure 6. Part of the Bureau of Land Management's General Land Office 1882 map of Bowie, Arizona, showing "Bowie St'n." (Source: US Bureau of Land Management.)

high rocky ridge 3/4 mile west of R. R. station, and just north of R. R."[9] Lieutenant Hovey, the future commandant of Norwich College,[10] who reoccupied the station for the November 1889 exercise, provided the most complete description of the station's location: "on the east side [of Stein's Pass] a spur of hills runs nearly N. and S. A pyramid hill forms the N. end, then southerly a succession of lower hills terminating at the S. end in an abrupt rocky eminence. . . . On the highest point of this eminence the station was established."[11] This high point, at the southern end of the spur and atop the eminence, is where the Stein's Pass heliograph station is mapped.

The heliograph station on the grounds of Fort Bowie was located in front of the adjutant's office.[12] Fuller established the "Home Station" there on April 29, 1886.[13] A photograph, reportedly taken by Lieutenant Gatewood and included in Robert M. Utley's historical report for the National Park Service, marks the adjutant's office near the southwest corner of the parade grounds.[14] The National Park Service's 1980 *Historic Structure Report* similarly maps the office in this location.[15] GIS software and aerial imagery were used to map the location of the adjutant's office and the heliograph station, a short distance away. This heliograph station may not have been used much, as the telegraph line no doubt reduced its need. This station was not included in the 1890 or 1893 exercise, presumably for the same telegraphic reason.

The Bowie, Arizona, heliograph station was located at the town's railroad depot. Fuller established the station there on July 14, 1886.[16] The Bureau of Land Management's GLO map of 1882 gives us a fix on the railroad station (Figure 6).[17] The heliograph station is mapped based on the GLO map's "Bowie St'n" label.

Colorado Peak is today's Rincon Peak.[18] In March 1890, Lieutenant Gale, a cavalry officer and the ASO at Fort Lowell, was tasked with finding a heliograph station in the Sierra Colorado east of Tucson (the southern part of today's Rincon Mountains).[19] The high point in that part of the Sierra Colorado is as prominent today as it was then—a clearly defined and notable peak (Figure 7). At that time, the mountains just north of the Sierra Colorado were referred to as the Rincon Mountains.[20]

Figure 7. Colorado Peak as seen from the west. Photograph and annotations by Robert E. C. Davis.

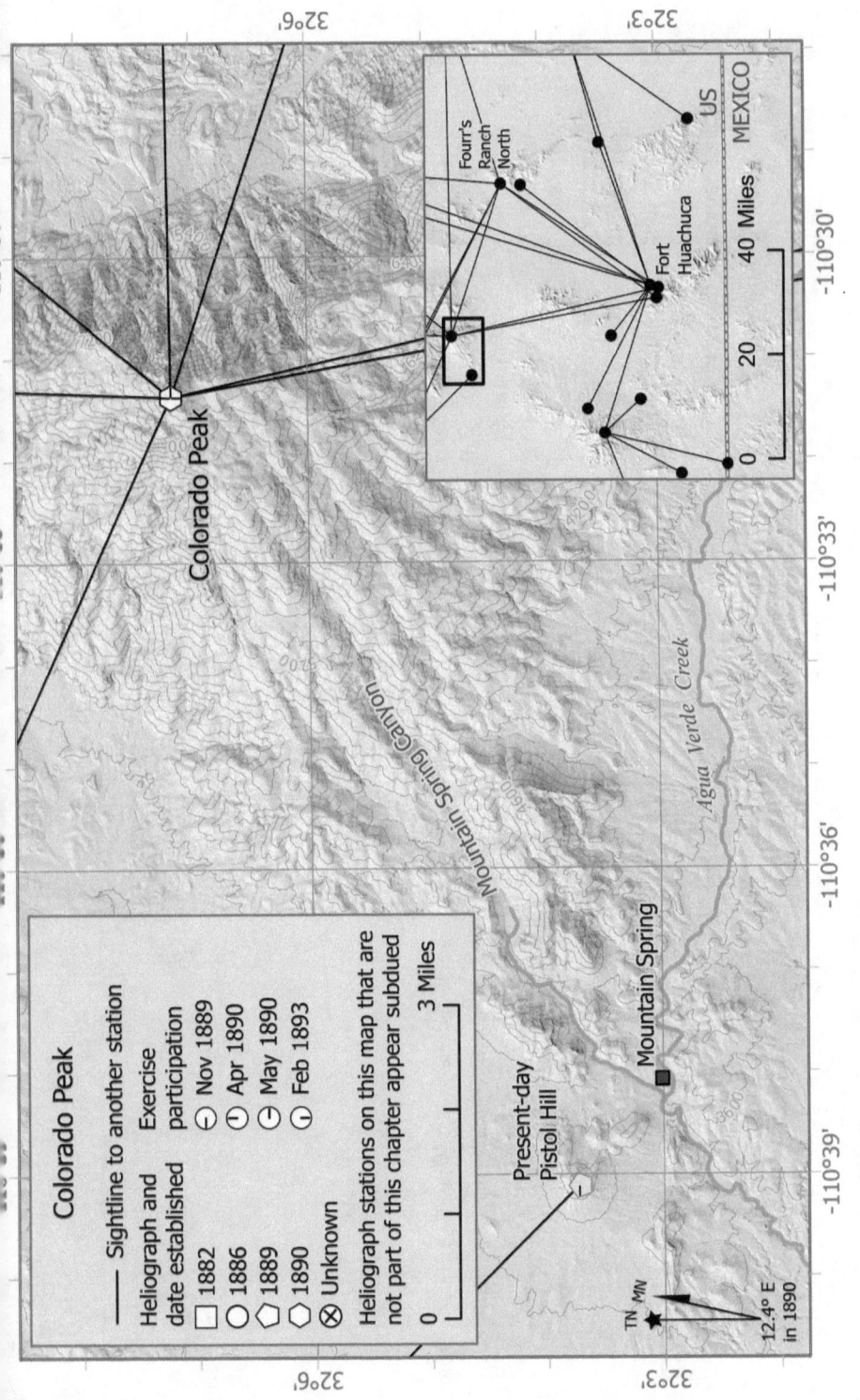

Map 11. Colorado Peak heliograph station.

Map 12. Old Camp Rucker heliograph station.

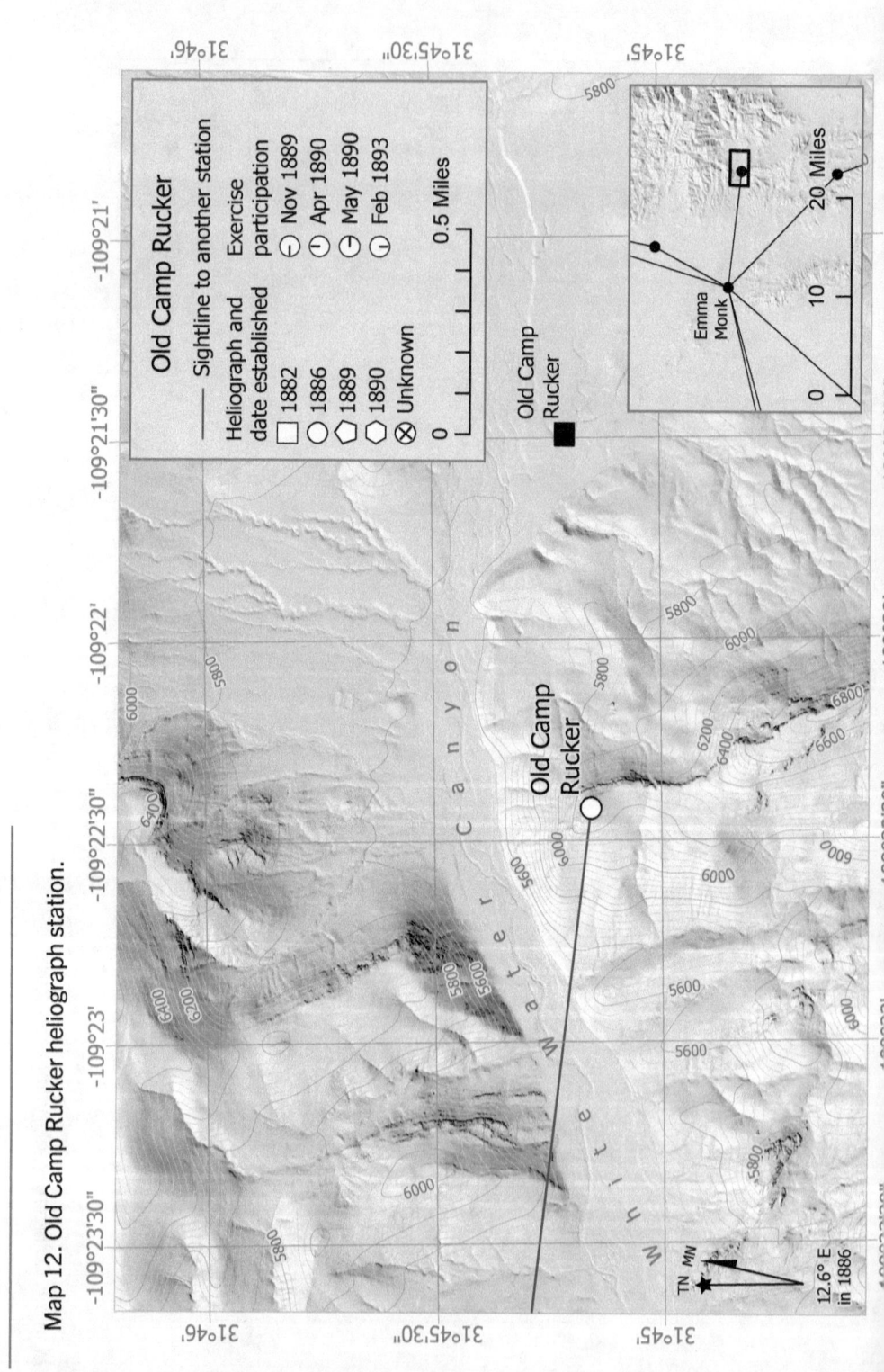

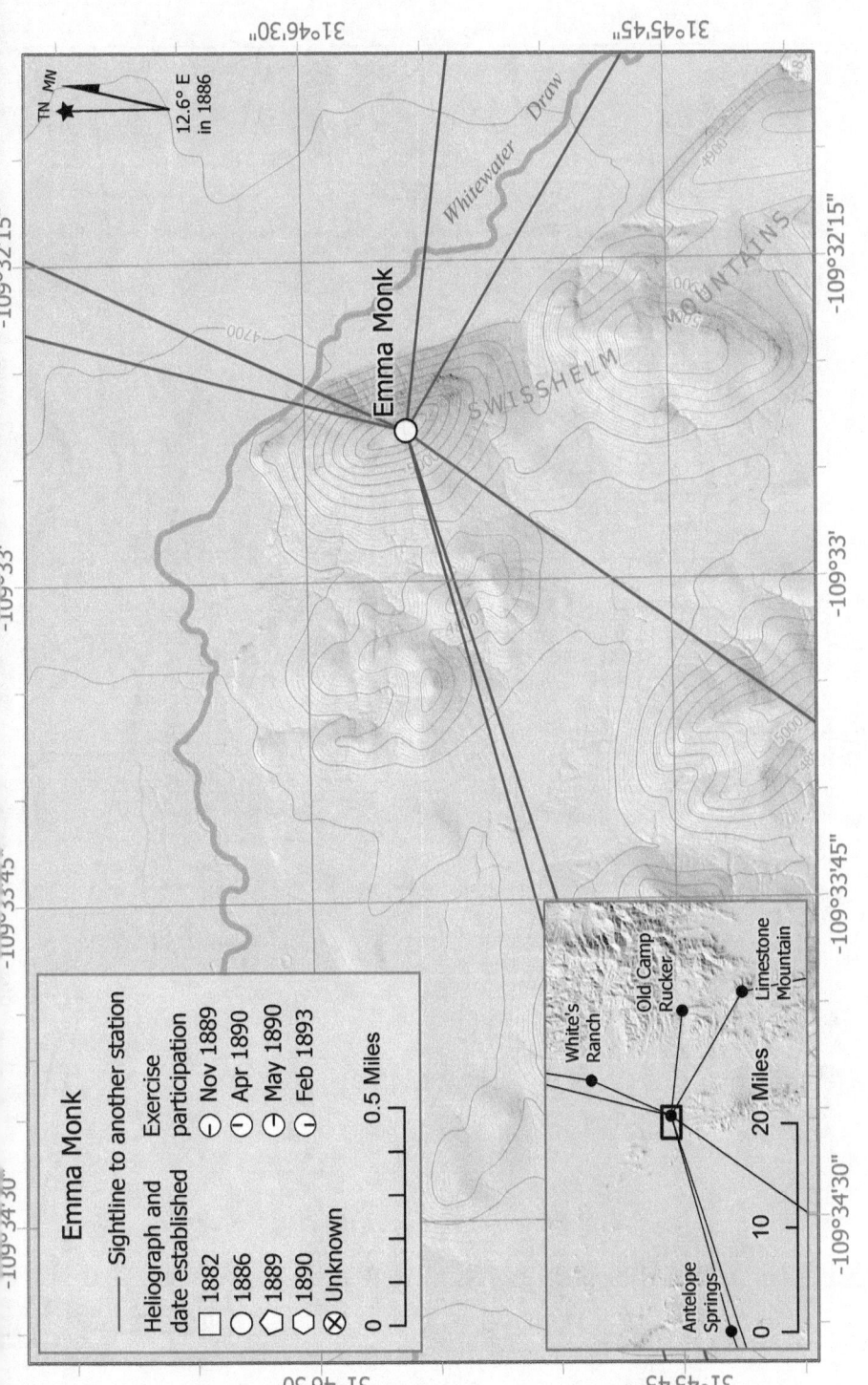

Map 13. Heliograph station at Emma Monk.

Map 14. Camp Henely heliograph station.

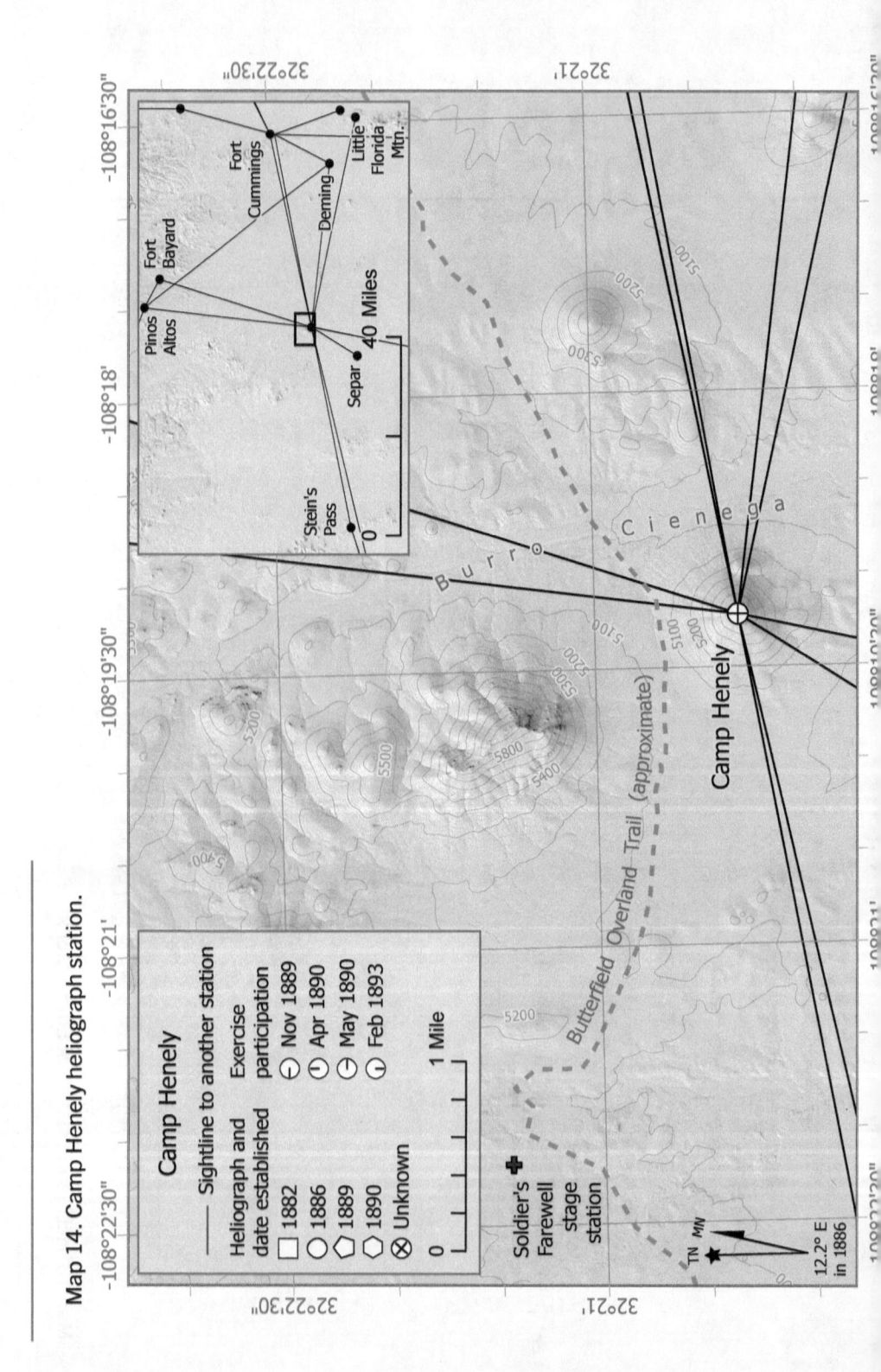

For his base of operations, Gale used the Mountain Spring stage at what is now Poste Quemada Ranch, just west of what was then called Mountain Spring Canyon (now Poste Quemada Canyon).[21]

Gale described the peak as being "about seven miles to the north of east of Mountain Spring" and as "flat, irregularly shaped and well above timber," and about 7,000 to 8,000 feet above Mountain Spring.[22] Gale reported that from this peak, the views were excellent for 65 miles in all directions, except to the north and northwest, where the views were restricted by the Rincon and Santa Catalina Mountains.[23] With the exception of the altitude, which is too high (Colorado Peak is only about 5,050 feet higher than Mountain Spring), this fits the description of today's Rincon Peak, 8.1 miles to the northeast of Poste Quemada Ranch. Gale established a heliograph on this peak on March 18, 1890.[24]

Fuller established the heliograph station at Old Camp Rucker on May 13, 1886. Located in the southern Chiricahua Mountains, Old Camp Rucker was well supplied with grass and water and housed several troops.[25] The camp was situated about 4 miles upstream from the mouth of White River Canyon (today's Rucker Canyon). Since the camp, primarily consisting of tents, had its view of the valley to the west obstructed by nearby hills, Fuller positioned the heliograph station on a hill "6,325 feet high, one mile from the camp below."[26] Just under a mile west (downstream) of the camp and adjacent (south) to the stream and road is a hill standing at 6,370 feet elevation with excellent views to the west. The station is mapped at the topographical crest of this hill.

The sole reason Fuller established a station at the northern point of the Swisshelm Mountains on May 9, 1886, was to connect to Old Camp Rucker.[27] Fuller reported, "I was compelled to put in a station on the north point on the Swisshelm mountains, as Rucker canyon station could only be seen from nearly a due west point."[28] Fuller put a sharper point on this description in his tabular report, where he stated that the location was at the "extreme northern point of Swisshelm Mts. A.T."[29] The northernmost point of the Swisshelm Mountains is a 550-foot hill (above the plain) next to today's Whitewater Draw and about 10 miles almost due west of the heliograph station at Old Camp Rucker. Fuller

CHAPTER 3

described this station's camp at the bottom of the hill, 500 feet below.[30] Based on this information, the heliograph station has been mapped on this hill adjacent to Whitewater Draw.

When Lieutenant Dravo received orders to establish the New Mexico heliograph system, he was serving as a witness to a court-martial, so he was delayed in taking to the field. Embroiled with his court-martial duties, Dravo sent Robert Sherwood, a member of the Signal Corps, to start setting up heliograph stations.[31]

On May 9, 1886, Sherwood established the Camp Henely heliograph station atop "a butte 2 1/2 miles S. E. from Soldier's Farewell station on old Overland Route."[32] About 2.5 miles southeast of the station was a cluster of three hills straddling the Burro Cienega (which maintains its name today). The station was atop a "pyramid butte W. side of valley coming from Burro Cienega."[33] To be extra clear, he mentioned that there was "a smooth pyramid butte . . . to the E. on the opposite side of the valley, and a very rough, rocky hill to the N. of the signal butte."[34] Based on these apt descriptions, the Camp Henely heliograph station is mapped atop the western pyramid-shaped butte, present-day Bessie Rhoads Mountain.[35]

With the locations of these heliograph stations established (Camp Henely, Stein's Pass, Fort Bowie, Bowie Station, Colorado Peak, and Emma Monk), a more precise understanding of the Bowie Peak location can be determined by overlapping visibility diagrams from these stations.[36] The combined visibility analysis identifies two hilltops of nearly equal elevation, separated by only 350 yards. Using a high-resolution (1-meter) elevation model from the US Geological Survey, it was determined that the western hilltop is the highest of the two visible peaks, though only by a few inches. Based on visibility and higher elevation, this is where the Bowie Peak heliograph station has been mapped.

The White's Ranch station (Fuller's station No. 3) was located on the plain near Turkey and Rock Creeks, which flow from the Chiricahua Mountains to the east.[37] This heliograph station has the distinction of being photographed in 1886, with numerous copies of the image in circulation (see Figure 8).[38] Another unusual feature of this station is

Figure 8. Heliographic station No. 3 at White's Ranch. The photograph is overexposed, but with adjustments using image processing software, the Chiricahua Mountains can be discerned in the background, indicating that the photographer was likely facing east. (Source: National Archives, Record Group 111, NAID: 530897).

Map 15. White's Ranch heliograph station.

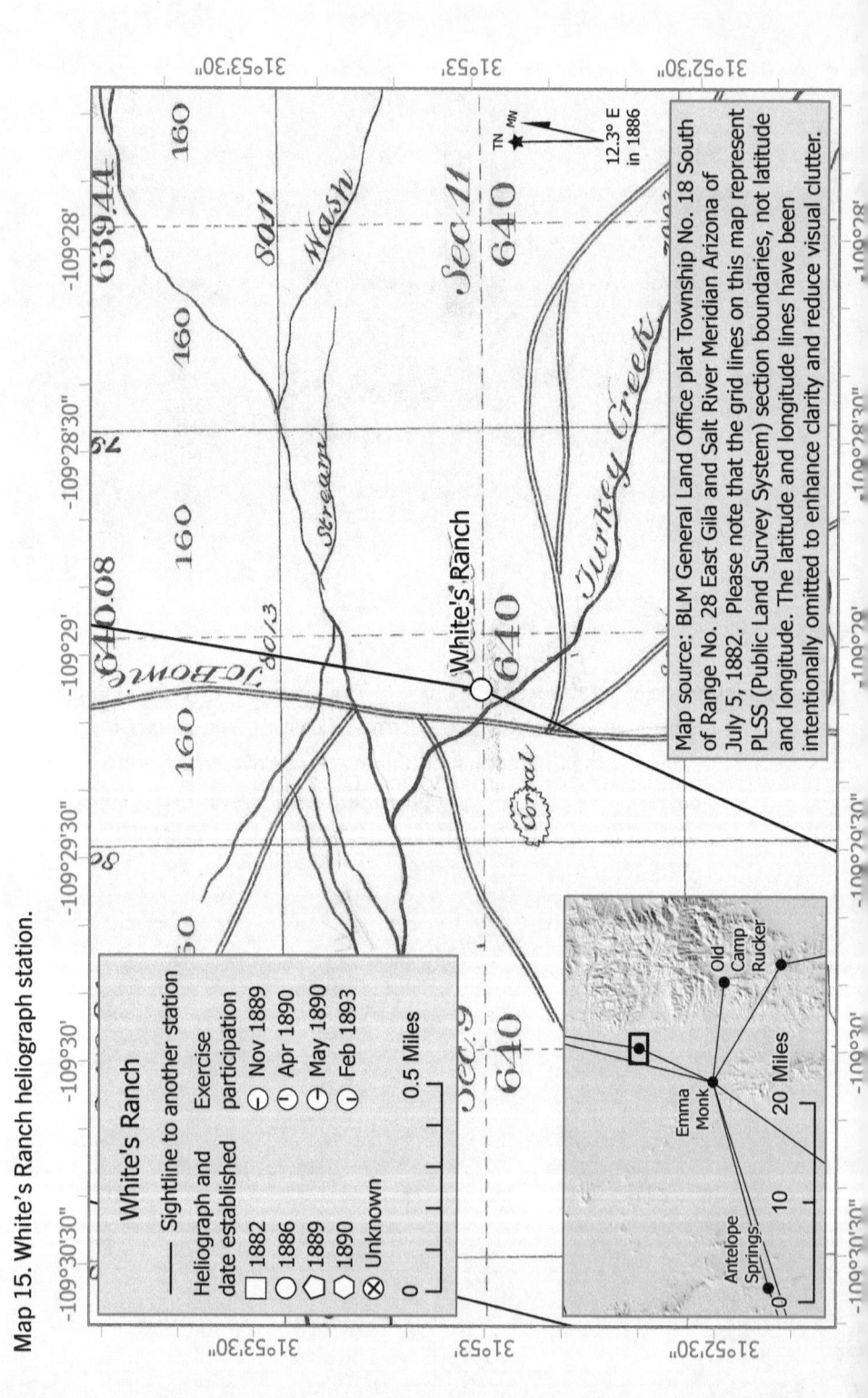

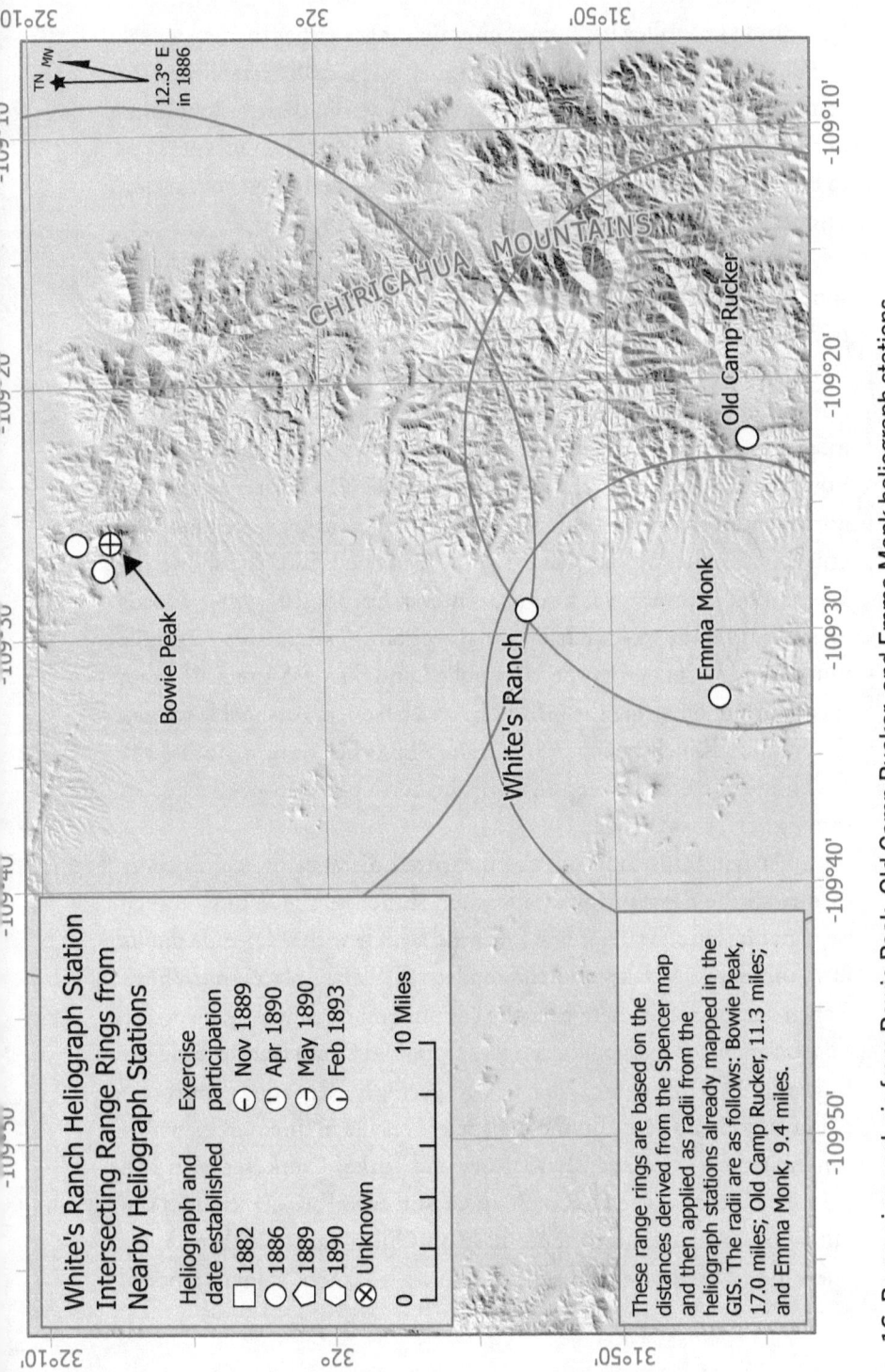

Map 16. Range ring analysis from Bowie Peak, Old Camp Rucker, and Emma Monk heliograph stations.

CHAPTER 3

its location in a valley between mountain ranges rather than on a peak or hilltop, where most heliograph stations were. Fuller established the station on April 20, 1886, "on the plain, and within three or four miles of the Chiricahua Mountains." The purpose of this plain location was to create a link between Fort Bowie and the "several troops of cavalry" camped in the canyons of the nearby Chiricahua Range.[39]

Except for the Bingham map of 1884, contemporary maps do not show a White's Ranch in this area. The Bingham map depicts "Whites Rch." along Turkey Creek, roughly midway between Fort Bowie and Old Camp Rucker on the eastern side of the Sulphur Spring Valley.[40] The Spencer map shows the heliograph station in roughly the same position; moreover, the Spencer map's greater detail shows a road leading to Fort Bowie just to the west of the station. The 1882 GLO plat map shows a north–south road marked "To Bowie" crossing Turkey Creek near the center of Section 10 (see Map 14). In fact, the GLO plat shows the intersection of three roads near the center of Section 10, with five roads intersecting in the western half of that section. (A section is a 1 x 1 mile unit of land as defined by the US Public Land Survey System.)[41] From this information, using the GLO plat as a base, it is reasonable to map the White's Ranch heliograph station along the western side of the road leading to Fort Bowie and along the north bank (to avoid repeated fording) of Turkey Creek.

Verifying this location, the measured distance on the Spencer map from the Emma Monk heliograph station to the White's Ranch heliograph station is 9.4 miles. Drawing a circle with a 9.4-mile radius from the Emma Monk station on modern GIS maps places the White's Ranch station somewhere near the circumference. Similarly, using the Spencer map, the measured distances from Old Camp Rucker and Fort Bowie are 11.3 miles and 17.0 miles, respectively. Repeating the process of drawing circles based on these distances results in three circles with a common intersection near Bowie Road and Turkey Creek (see Map 16).

The Henry Forrest's Ranch house and corral are depicted on the 1885 GLO map near the mouth of Mule Gulch, east of Bisbee. Fuller refers to this location as "Frost's Ranch, Bisbee Cañon."[42] In his reports,

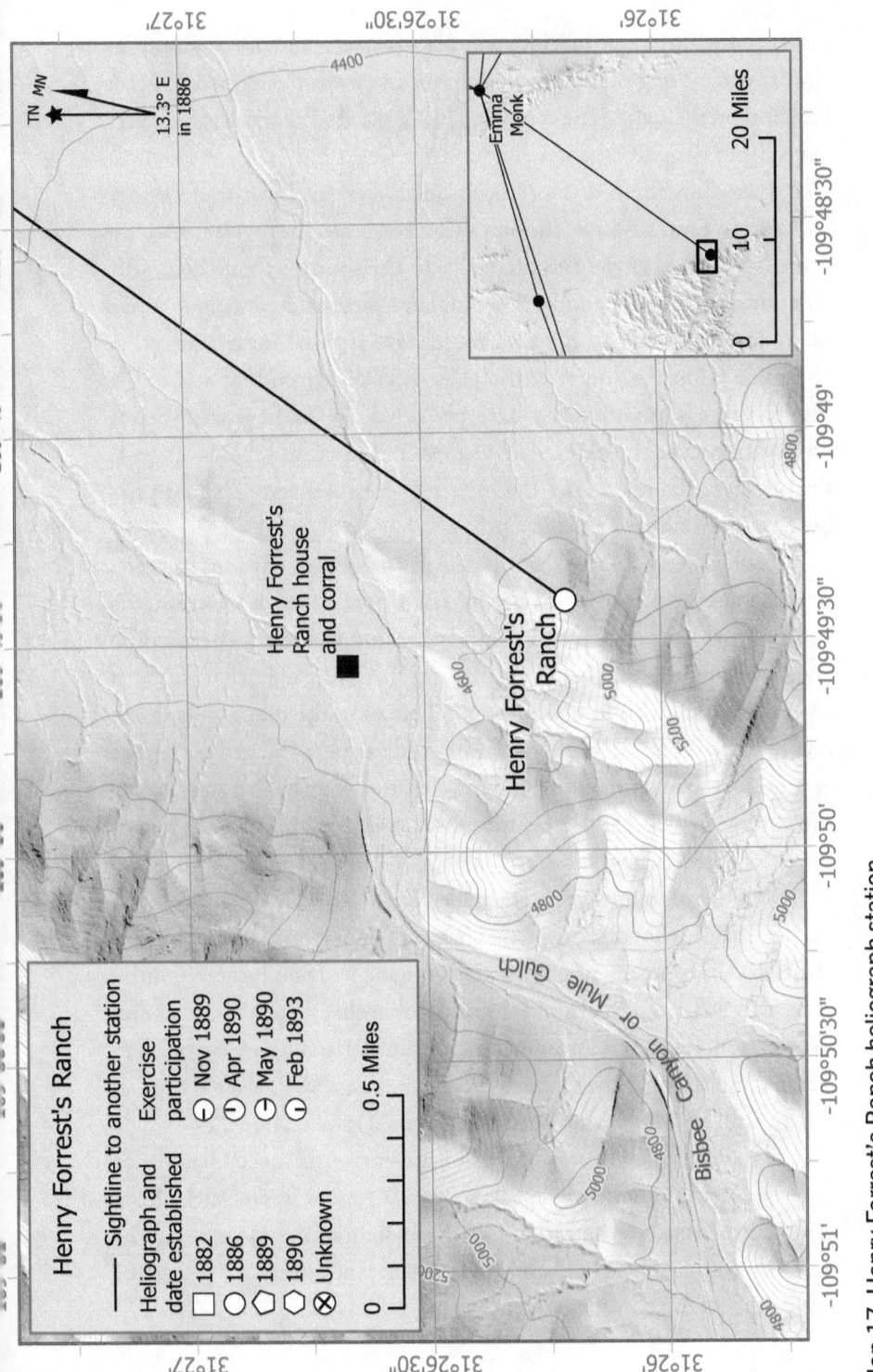

Map 17. Henry Forrest's Ranch heliograph station.

CHAPTER 3

Fuller marked the station's location at the eastern end of the mouth of the canyon, noting that the heliograph station was positioned 250 feet higher than the camp. The camp was likely situated near the same water source that supplied the ranch house.[43]

Based on the 250-foot elevation difference and the necessary visibility to Emma Monk (the only other connected heliograph station), a small level area at the end of a ridge to the south—about 500 yards from the ranch house and 243 feet higher—fits the description. While many low-lying hills in the area could have sufficed for a connection to Emma Monk (as most of the plain is visible from that station), a higher vantage point would have provided the soldiers with better observation over the plain. The nearest and most suitable location is the one described here, and the Henry Forrest heliograph station has been mapped on that small ridge.[44]

Leonard Wood noted that in late August 1886, as Lawton's battalion approached the US–Mexico border from the south with Geronimo's band in tow, Lawton traveled from their camp to the San Bernardino Ranch to connect with Miles via heliograph.[45] While much of the area around the ranch buildings offers good views to the mountains to the north, there is a small bluff a few hundred yards to the east on higher ground that would provide a bit more elevation for a heliograph station. However, even with added elevation, this area remains blind to any of the listed 1886 heliograph stations.

While it is not mentioned by Fuller, Dravo, or Miles in their reports on the 1886 heliograph system, Lieutenant Glassford, near the end of his 1887 report, mentioned a station used in 1886 located 9 miles south of Old Camp Rucker. He reported that this station had visibility to Skeleton Canyon, Cottonwood Canyon, Guadalupe Canyon, San Bernardino Creek, Silver Creek, Dos Cabezas, and Power's Ranch in the Swisshelm Mountains (near the Emma Monk station).[46]

Based on visibility, two possibilities arise to meet Glassford's criteria. First is an unnamed peak about 10.1 miles line-of-sight from Old Camp Rucker (the camp, not the heliograph site) or about 13 miles by trail.[47] This peak stands just over 6,500 feet and is west of

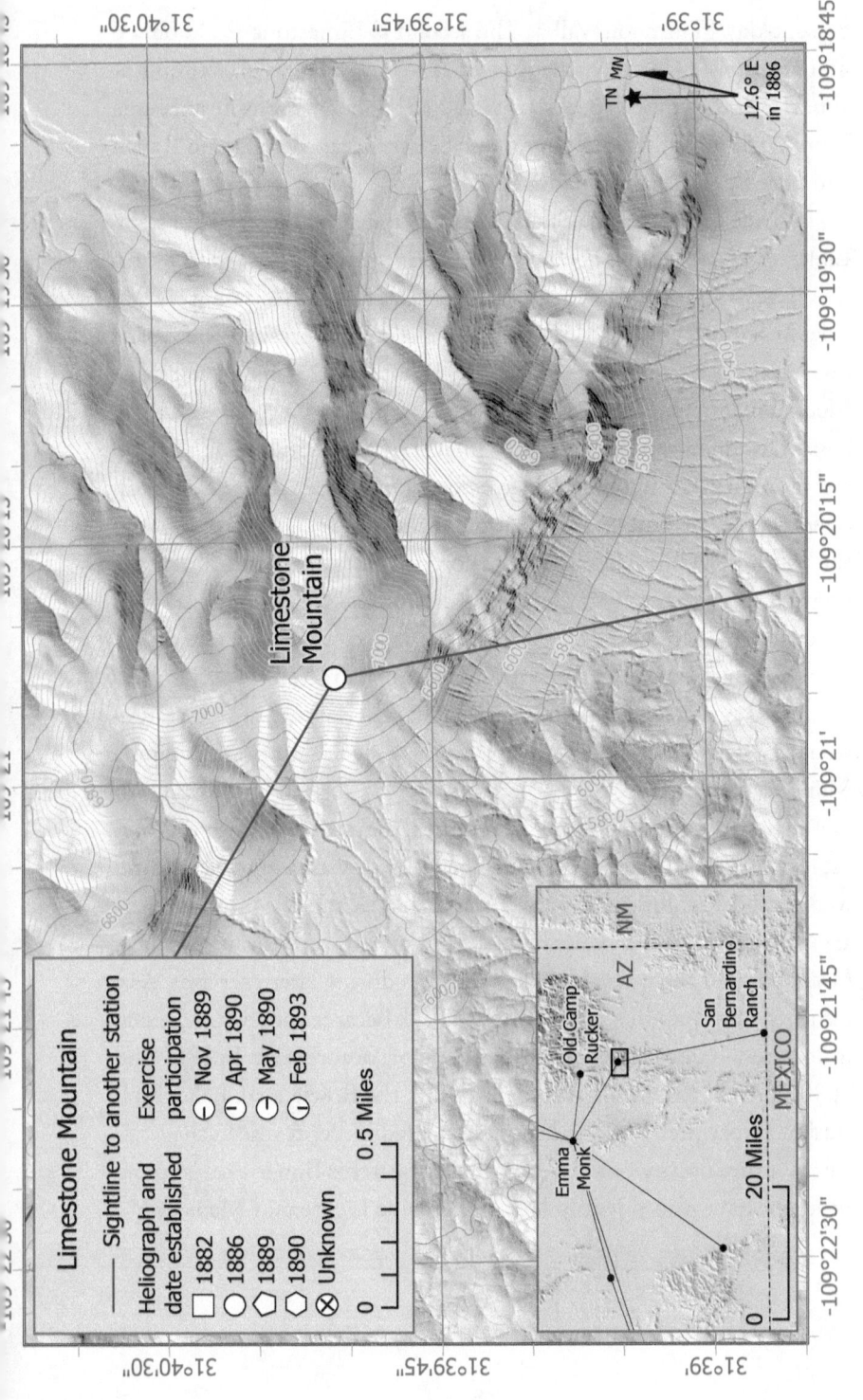

Map 18. Limestone Mountain heliograph station.

present-day Halfmoon Valley. The second is Limestone Peak, only 6 miles from Old Camp Rucker but close to 9 miles by trail, standing at almost 7,200 feet. Both peaks are visible to the heliograph site at Emma Monk, making either a potential link between San Bernardino Ranch and Fort Bowie (via Emma Monk).

While the unnamed peak is closer to 9 miles in a direct line of sight, Limestone Mountain is about 9 miles along the trail. In practical terms, the route to the unnamed peak passes by Limestone Mountain. Today, driving south from Old Camp Rucker along US Forest Service roads (essentially the same roads shown on period GLO maps), Limestone Mountain dominates the skyline after about 2 miles once you cross a small pass, continuing another 7 miles. With its higher elevation—almost 700 feet more than the unnamed peak farther south—Limestone Mountain offers a superior vantage point for observing the area, particularly into the San Bernardino Valley. As Glassford suggests, it is an ideal vantage point. The heliograph station is mapped at the topographic crest of Limestone Mountain.

It is likely this was the station Lawton contacted from San Bernardino Ranch at the end of August 1886. The soldiers at that station would then have relayed Lawton's messages through the station at Emma Monk to Fort Bowie.[48]

The heliograph stations at and near Fort Bowie were the first established in the 1886 system. Fort Bowie was a significant and well-established fort, and with the addition of the telegraph line the army extended to the heliograph station on Bowie Peak, it became a focal point, visited by high-ranking officers during later exercises. As a result, the heliograph station at Bowie Peak became somewhat fixed in its location, requiring other stations to adjust accordingly. In 1889 and 1890, considerable effort was made to link the Bowie Peak heliograph station across the Pinaleño Mountains. Had Fuller foreseen the challenges future users would face in connecting to his Bowie Peak station, he might have placed it atop Dos Cabezas, as Lieutenant Maus did.[49]

Chapter 4

Fort Huachuca

IN LATE MAY 1886, THE DEMONS OF THREE-DIMENSIONAL GEOMETRY were conspiring against Fuller's attempts to place a heliograph station where it could connect the town of Tubac, Arizona, with Fort Huachuca.[1] Departing Fort Huachuca on May 29, he headed to Elgin, several miles to the west, where he successfully signaled back to the fort.[2] Following Sonoita Creek, Fuller reached Old Camp Crittenden in late morning the following day.[3] In the afternoon of May 30, he moved into the eastern slopes of the Santa Rita Mountains, and although he found a location to signal back to Fort Huachuca, the site was otherwise unsatisfactory.[4]

The next day, May 31, Fuller went north to Gardner's Ranch and climbed a nearby hill but was unable to find a suitable location for a heliograph station. On June 1, accompanied by his orderly, Fuller pressed 6 miles west from Gardner's Ranch to the top of the Santa Rita Mountains. While this spot was ideal for heliograph communication, it was unusable due to a lack of water. About 6 miles west of Gardner's Ranch is Old Baldy, the highest point of the Santa Rita Mountains, though Fuller may have gone to a lower hilltop just to the north.[5]

At this point, Fuller's strategy shifted. Initially he had planned to place the heliograph station high in the Santa Rita Mountains, intending to signal across the crest from Huachuca to Tubac. However, it now seems his goal was to establish a station near Gardner's Ranch, connecting with Fort Huachuca and south to the Patagonia Mountains, then to Tubac; he would go around the mountain instead of over it. To that end, on June 2, Fuller established station No. 8 a mile west of Gardner's Ranch and headed south, via Crittenden Station (a railroad station distinct from and south of Old Camp Crittenden), where a cavalry troop was camped,[6] to the Patagonia Mountains.[7]

On June 3, Fuller reached the Patagonia Mountains and spent the day searching for a location where he could signal back to station No. 8 above Gardner's Ranch, but he was unsuccessful. The next two days proved equally unsuccessful, and on June 5 he left the mountains for Calabasas (north of present-day Nogales, Arizona), where there was a great deal of activity, as Lawton's battalion was there.[8] On June 6, Fuller's heliograph system alerted Lawton that Naiche was heading south for the Patagonia Mountains. After dispatching detachments to intercept, Lawton left Calabasas with his battalion the next day in pursuit of the Apache.[9] General Miles was also at Calabasas, along with Lieutenant Wood, who was arranging a logistics train to follow Lawton.[10]

On June 7 Fuller traveled to Tubac, where he set up a heliograph station "in the town."[11] He returned to Calabasas at noon on June 9. While there, he spoke with General Miles, who authorized him to hire a guide, animals, and civilian packers to make another attempt at siting a station in the Santa Rita Mountains.[12] Despite the activity in Calabasas and Lawton's pursuit south of the border, it is notable that Miles kept Fuller, an experienced soldier, focused on the task of building the heliograph network.

On June 10 Fuller went upstream to Crittenden Station, where he hired a guide, a packer, and four burros.[13] On June 11, Fuller, with this new assemblage of men and animals, returned to Gardner's Ranch, where he collected up station No. 8 and, with the help of the guide, moved everything to "Little Baldy Peak, 1–1/2 miles S. of Old Baldy Santa Rita Mts., A.T.," reaching the peak on June 12, 1886.[14]

Old Baldy Peak is known today as Mount Wrightson. According to the US Board on Geographic Names decision card, the name Old Baldy had been in use since at least 1880.[15] Supporting this, the National Geodetic Survey monuments atop Wrightson are labeled Baldy and Baldy 2.[16] Further solidifying the name, "Mt. Baldy" appears on the Spencer map just north of the heliograph station.[17] This heliograph station is located a little over a mile south of Old Baldy, now known

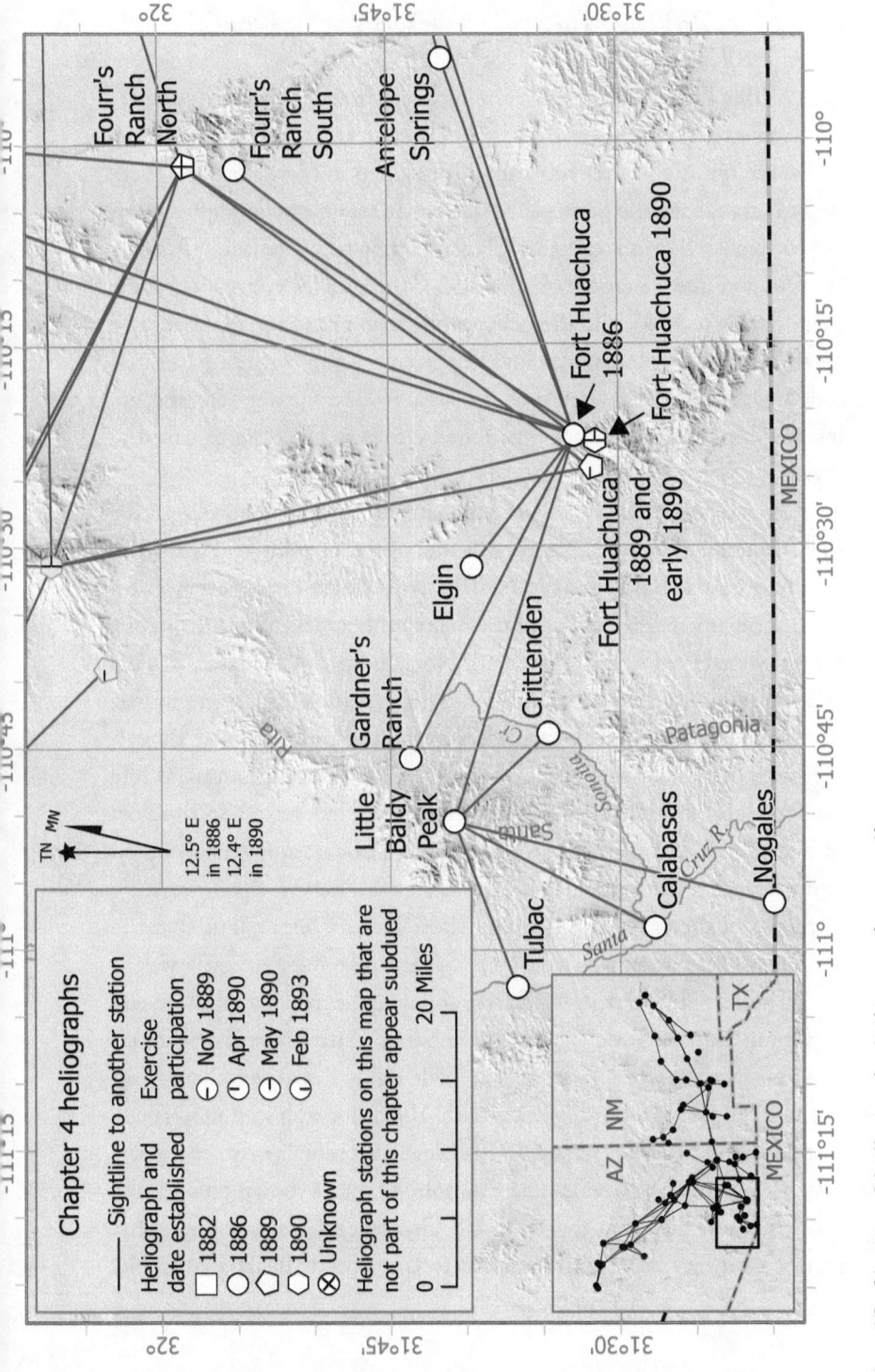

Map 19. Chapter 4 heliograph stations and connections.

as Josephine Peak, and is mapped on the topographical crest of today's Josephine Peak.

While Fuller was in the Sonoita Creek area in May and June, he made no attempt to establish a heliograph station at Crittenden Station. However, on July 9, after returning from a trip to New Mexico, where he had established the Stein's Pass station, he set up a heliograph station at Crittenden Station, tasked with connecting to Little Baldy Peak. A "Crittenden" site is marked on older US Geological Survey topographic maps dating to 1905, and the heliograph station has been mapped on a small hill near that location. However, given the good visibility to Little Baldy Mountain from much of the area around Crittenden Station, it's possible the station was positioned closer to the road to ease the logistics of supporting it.

Private William W. Neifert, who later became a meteorologist for the Department of Agriculture, was one of the operators of the 1886 station on Little Baldy Peak.[18] Neifert wrote about his experiences on the frontier, including his use of the heliograph. He recalled, "From our station we worked occasionally with Nogales [and] Calabasas."[19] This account identifies two additional heliograph stations that were in use. The individuals responsible for establishing them are not named, though they were likely members of the troops detailed to those locations. While Neifert does not tell us when they were contacted, based on the flow of his writing, it was after the Little Baldy Peak station was occupied and before the Crittenden Station was created. So these stations were likely established around mid-June 1886. Neifert does tell us that the station at Crittenden was led by an "infantryman named Lovejoy."[20]

The 1905 US Geological Survey topographic map places Calabasas at the junction of Sonoita Creek, the Santa Cruz River, the railroad, and a network of roads. The view to Little Baldy Peak is excellent from much of the area surrounding Calabasas. The heliograph station has been mapped just east of this junction. The Nogales station is mapped a short distance north of the border, near the railroad tracks shown on both the 1880 Eckhoff map and the 1905 US Geological Survey topographic map.[21] Visibility from both locations to Little Baldy Peak is favorable.

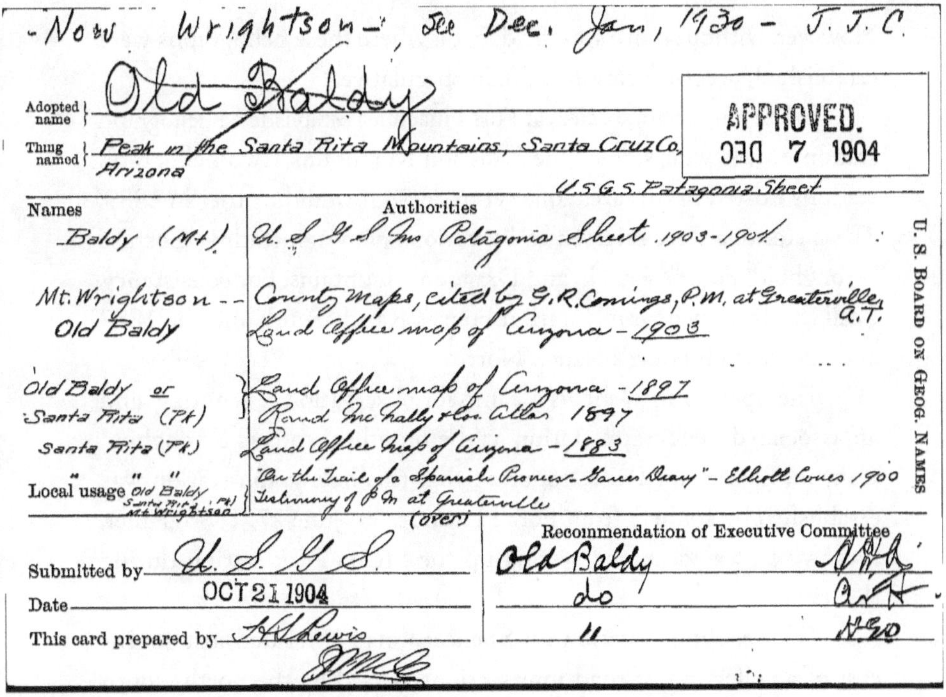

Figure 9. US Board on Geographic Names decision card showing present-day Mount Wrightson's various names (Source: US Geological Survey).

However, without more specific data on where these heliographs were established, precise locations remain speculative.

In late June 1886, soldiers at Fort Huachuca established a heliograph station on the west side of the Dragoon Mountains. Two heliograph stations existed in this area: one set up in 1886 and the other in 1889. These stations were referred to by various names, including Cochise Stronghold, Fourr's Ranch, and Dragoon Mountains. For consistency, I call the 1886 heliograph station Fourr's Ranch South and the 1889 and later station Fourr's Ranch North.

The Spencer map shows a "camp for observation & scouting" and an associated heliograph station at the mouth of today's Stronghold Canyon. Lieutenant Fuller reported that this heliograph station was established by soldiers from Fort Huachuca on June 27, 1886. Fuller, meanwhile, was occupied setting up the Stein's Peak station during that time.[22]

The heliograph station on Spencer's map is located just southeast of a collection of road intersections caused by the north–south Dragoon–Tombstone Road and the trail leading up Stronghold Canyon. These roads are identified on period maps.[23] The area is rugged, so the heliograph is mapped on the first reasonable hilltop that fits the Spencer map's layout of features.[24] Using the Spencer map, the scaled distance from Fort Huachuca to the heliograph was determined to be about 32 miles, consistent with the present-day distance to the heliograph's location.

Fuller reported the location of the 1886 heliograph station at Fort Huachuca as on "a small hill above parade ground" and established it on May 23, 1886. He also reported that the camp, Fort Huachuca, was at an elevation of 4,812 feet and the station at 4,912 feet for a difference of 100 feet.[25] If the common visibility is calculated from the two connected stations—Little Baldy Mountain (present-day Josephine Peak) and Fourr's Ranch South, large swaths of the fort are visible to both. By limiting those swaths to areas where the elevation is between 50 and 150 feet above the elevation of the parade grounds, only a few areas remain: the hill just to the east and adjacent to the parade grounds as

Figure 10. View from Calabasas to today's Mount Wrightson and Josephine Peak. Photo and annotation by Robert E. C. Davis.

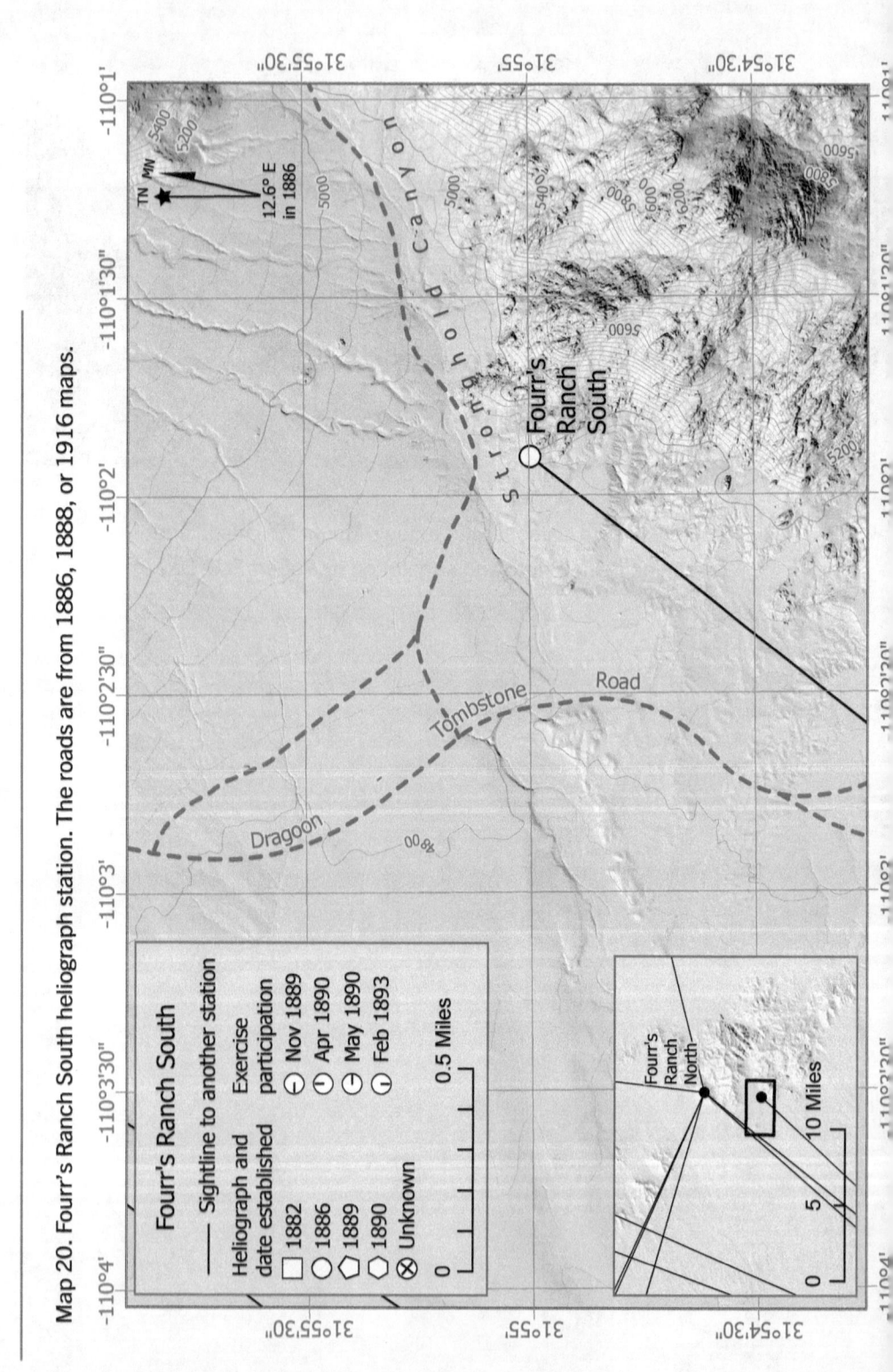

Map 20. Fourr's Ranch South heliograph station. The roads are from 1886, 1888, or 1916 maps.

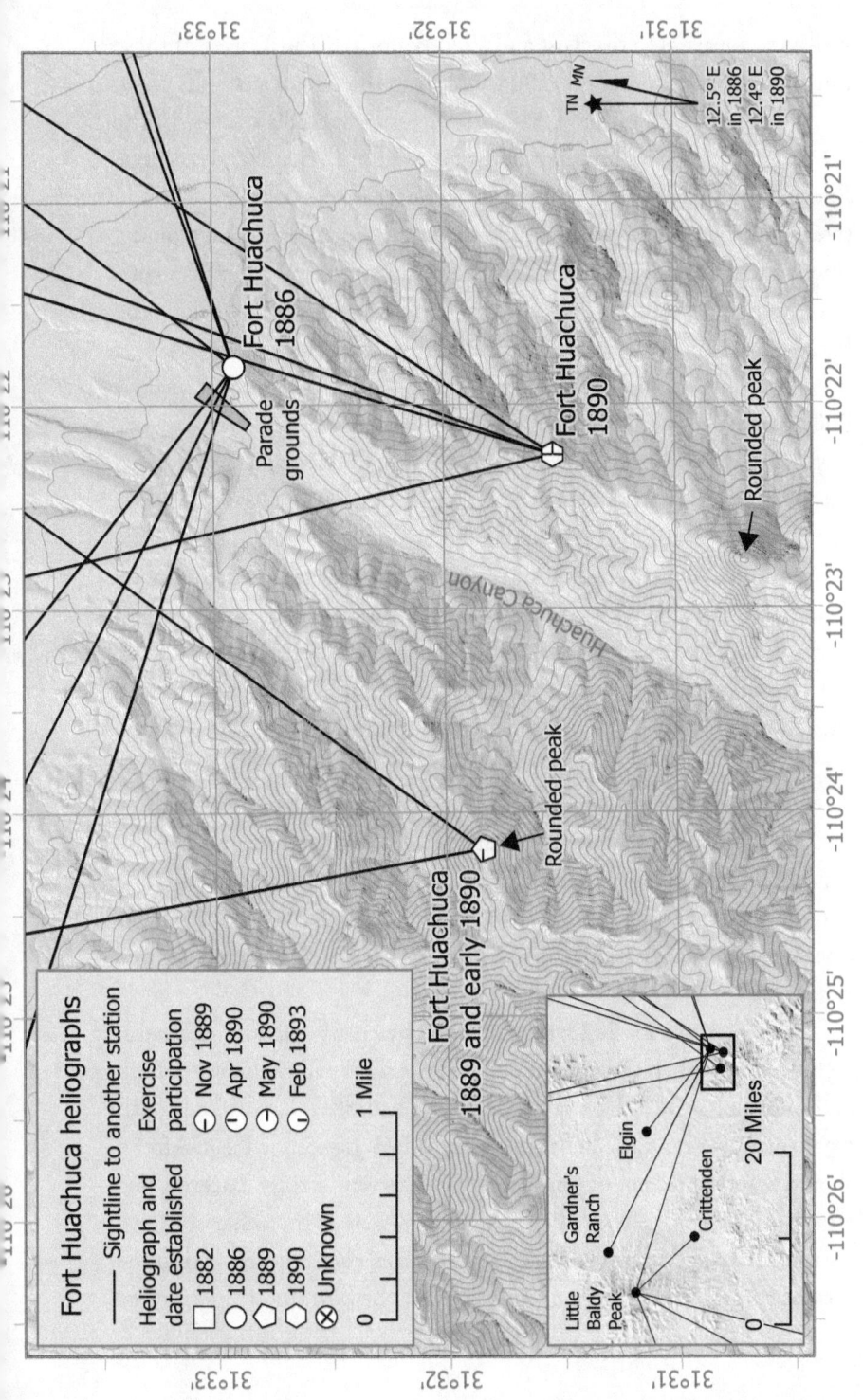

Map 21. Fort Huachuca heliograph stations.

well as some areas to the west and slightly north. (The nearest of these northwestern areas is about 700 yards from the parade grounds.)

Fuller made a rather detailed sketch of the 1886 station.[26] The sketch includes distant locations of the Fourr's Ranch South and Antelope Springs stations to the north and east. Closer in, the confluence of two streams, or washes, is visible. At the station, Fuller sketched a tent, a tree, benches, the heliograph and the posts supporting it, and a rock outcrop just adjacent to the heliograph. The sketch does not show the parade grounds or any other fabricated feature of the fort.

Based on the view of Fourr's Ranch South and Antelope Springs, Fuller's point of view in the sketch appears to be toward the northeast and east. Since the parade grounds are omitted, it can be reasonably deduced that his perspective is east of the parade grounds. The likely location, offering both visibility and suitable elevation, is on the hill east of the parade grounds, just south of the current-day water tank. Furthermore, a close examination of the Spencer map shows the heliograph station to the east of the fort.

In Fuller's tabular reports, he gives an establishment date for the Huachuca station of May 23, though he mentions connecting with the fort on May 21 and again on May 22 (from Antelope Springs). After leaving Antelope Springs, he went to Tombstone, where he remained until May 25, when he received orders from Miles to move to Fort Huachuca and extend the line farther west, likely to support Lawton's battalion at Calabasas. Fuller arrived at Huachuca on May 25, leaving it unclear whom at Fort Huachuca he had been signaling from Antelope Springs.[27]

On February 11, 2023, accompanied by Fort Huachuca's museum technician, Stephen Gregory, I followed the dirt road to the water tank and then along the hill, or ridge, heading south.[28] As we walked south from the water tank, we kept an eye on today's Josephine Peak. After a short distance, it fell out of view behind an intervening ridge. Backtracking a bit until Josephine was clearly in view, as well as the other stations, we found a rock outcrop—the only substantial rock outcrop within the area of common visibility on this ridge. At the bottom of the ridge to the east is a reservoir where two visible washes meet. Based on this, the elevation, and common visibility, this station is mapped at this outcrop.

If a heliograph station on Fort Huachuca were to connect to the expanded 1890 system, it required visibility to both Colorado Peak and the Pinaleño Mountains. To achieve this, it needed to be situated at a higher elevation than the 1886 station. In November 1889, as part of that month's exercise, Lieutenant Frank H. Albright, an infantryman, who went on to serve in the Philippines and then in World War I as a brigade commander,[29] established a station south of and above the fort, enabling a connection to the broader heliograph system.[30]

In March 1890, Lieutenant Gale, stationed on Colorado Peak, successfully connected with Albright's heliograph station at Fort Huachuca. Gale noted that the Huachuca station was located on a rounded peak, standing 2,000 feet above the fort, and commented that the station seemed higher than necessary.[31] Two prominent peaks match Gale's description. The first is east of Huachuca Canyon, about 3 miles south of the parade grounds, with an elevation of 7,500 feet (2,400 feet above the fort's 5,100-foot elevation). The second is a rounded peak on the west side of Huachuca Canyon, standing just over 7,200 feet (2,100 feet above the fort). Gale recorded the direction to the peak as 157°, which, after accounting for the 1890 magnetic declination of 12.4°, becomes 169.4° true.[32] The measured direction from Colorado Peak to the peak east of Huachuca Canyon is 168.3°; to the peak on the west it's 169.9°. Both rounded peaks were visible and suitable for connecting to the required stations in 1889 and 1890.

Of the two options, the peak on the west side of Huachuca Canyon seems to be a better fit, as it is closer to the elevation gain mentioned by Gale, aligns more closely with the direction reported from Colorado Peak, and offers a line of sight to the parade grounds—unlike the eastern peak, which lacks this visibility. As a result, the station has been mapped on the rounded peak to the west of Huachuca Canyon. In either case, Gale recommended to Huachuca's signal officer that they consider locating the station lower and closer to the fort. The signal officer responded that he would search for a lower one.[33]

CHAPTER 4

The station was indeed relocated to a lower position. In his report on the May exercise, Second Lieutenant William H. Hart of the Fourth Cavalry, Fort Huachuca's ASO and a future quartermaster general, described establishing the new station on April 24. He wrote that it was "about two miles from the post, and easily reached by a trail leading up a long ridge. Elevation about 6,800 feet; above post, 1,600 feet."[34] He further noted that the station had a clear line of sight to the parade grounds.

Visibility analysis from three stations (Colorado Peak, Fourr's Ranch North, and the station on the Pinaleño Mountains) as well as the parade grounds shows several places south of the fort where the station was visible to all. One of these places is 2 miles from the parade grounds, along the long ridgeline to the east of Huachuca Canyon and atop a small peak with an elevation of 6,700 feet (1,600 feet above the fort). Moreover, the direction given on the Volkmar map between Colorado Peak and the 1890 Fort Huachuca station is 156.2° magnetic, which, after accounting for the 1890 magnetic declination of 12.4° east, becomes 168.6° true. The measured GIS direction between Colorado Peak and this peak east of Huachuca Canyon is 167.4°. Considering Lieutenant Hart's description and the results of the visibility analysis, this station is mapped atop the small peak 2 miles up the long ridge from the fort.

In November, prior to establishing the Fort Huachuca station, Hart was tasked to find a suitable heliograph station in the Dragoon Mountains for the upcoming May 1890 exercise. After setting up camp at Fourr's Spring and after some difficulty locating a suitable station location, he found a spot "upon a spur of the mountains running N. E. from Fourr's" on November 16, 1889.[35]

A small mountain range runs northeast from Fourr's Ranch (the ranch is still there), extending to the next drainage of Jordan Canyon, where Dragoon Springs is located. Midway along the ridge, a spur juts off at a right angle to the northwest. From this spur, Hart was able to see several other mountains and features; these were the highest point of Mount Graham, Fort Grant, Fort Huachuca, Bowie Peak, a peak beyond the Santa Ritas, the highest point of the Catalinas, the Dragoon railroad station, and Tombstone.[36] Of these, the positions are known for all except the peak beyond the Santa Ritas, for which we don't have enough specifics. However, it is reasonable to assume he could see the Santa Ritas, so Baldy Peak is included (today's Mount Wrightson, the highest point of the Santa Ritas).[37] These known locations allow for overlapping visibility diagrams, which helps pinpoint a probable location for the heliograph station. The diagrams converge on a northeast spur near the center of the small range, as well as an area along the top of the main Dragoon Range, close to today's Mount Glenn.

A distribution analysis of the pattern of points at the ends of azimuth-distance lines from other heliograph stations helps narrow down the two potential locations (see the explanation in chapter 1). The tabular data from the Volkmar map provides magnetic azimuths and distances from Fourr's Ranch North to Colorado Peak, Fort Huachuca, and Bowie Peak.[38] Hart also provided a magnetic azimuth of 163° at 4 miles for the Dragoon rail station. To establish a true north azimuth line and distance, a magnetic declination is applied to each location in a GIS. For example, the listed azimuth and distance from Fort Huachuca to this Fourr's Ranch station are 24° and 37 miles. By adding Fort Huachuca's magnetic declination (in 1889) of 12.4° east, a true azimuth from Fort Huachuca to Fourr's Ranch North is determined to be 36.4°. Combining this azimuth and distance yields both a direction and magnitude, forming a vector. A point is then plotted at the end of this vector, the terminal point. Repeating this process for all the listed azimuths and distances results in a collection of terminal points that cluster on the ridge northeast of Fourr's Ranch.

The calculated mean center of those terminal points, the average location, plots just to the southwest of the spur atop the small range (see Map 22). Moreover, if a circle with a radius of the standard distance from this mean center were drawn, it would not overlap the other area near Mount Glenn, making that area less likely the location for the station (see the insert in Map 22).

Based on the physical description by Lieutenant Hart, the topography, the overlapping visibility diagrams, and the mean center analysis, the heliograph station is mapped on the spur in the center of the small range northeast of Fourr's Ranch.

On May 20, 1886, Fuller left White's Ranch for Antelope Springs, arriving on May 21. He established a station and connected to Fort Huachuca *through* the heliograph station at Emma Monk.[39] To communicate with the fort to his west, Fuller signaled east to Emma Monk, whose heliograph operator relayed the message back over Fuller's position to Fort Huachuca. This flexibility of the heliograph system is impressive, and it is helpful for analysis because it suggests that this location was visible to Emma Monk and not Fort Huachuca.

The visibility diagram showing where Emma Monk is visible and where Fort Huachuca is not includes large areas of east-facing slopes and flatter areas near Antelope Springs (which is still there), both east of a line running north to south through present-day Signal Hill.[40] There is an east-facing slope just to the west and north of Antelope Springs where this condition is also true; however, with most of these east-facing slopes, the tops are visible from both Huachuca and Emma Monk, so it is doubtful Fuller put the station on this or any of the east-facing slopes. Since Fuller was approaching from the east (the Spencer map shows a road leading to the heliograph station from almost due east), he likely set up this first heliograph station along his route of march, east of Antelope Springs and southeast of Signal Hill. Unfortunately, a better fix to this station's location is impossible given the lack of available data.

Recognizing the complexity of establishing a connection back through Emma Monk to reach Fort Huachuca, on May 22, Fuller made the decision to relocate the station, "so as to communicate directly into the post of Huachuca."[41] Fuller specified that the new station was situated on a "small peak at an elevation of 4,950 feet above sea level."[42] However, he also mentions an elevation of 4,750 feet for the same hill in his tabular report.[43] Additionally, Fuller reports the spring as being roughly half a mile away, though his tabular report shows the water source as 0.75 miles distant. Furthermore, he provides an elevation for the camp (presumably at the base of the small peak) of 4,525 feet in his tabular report. Consequently, the local relief of the hill upon

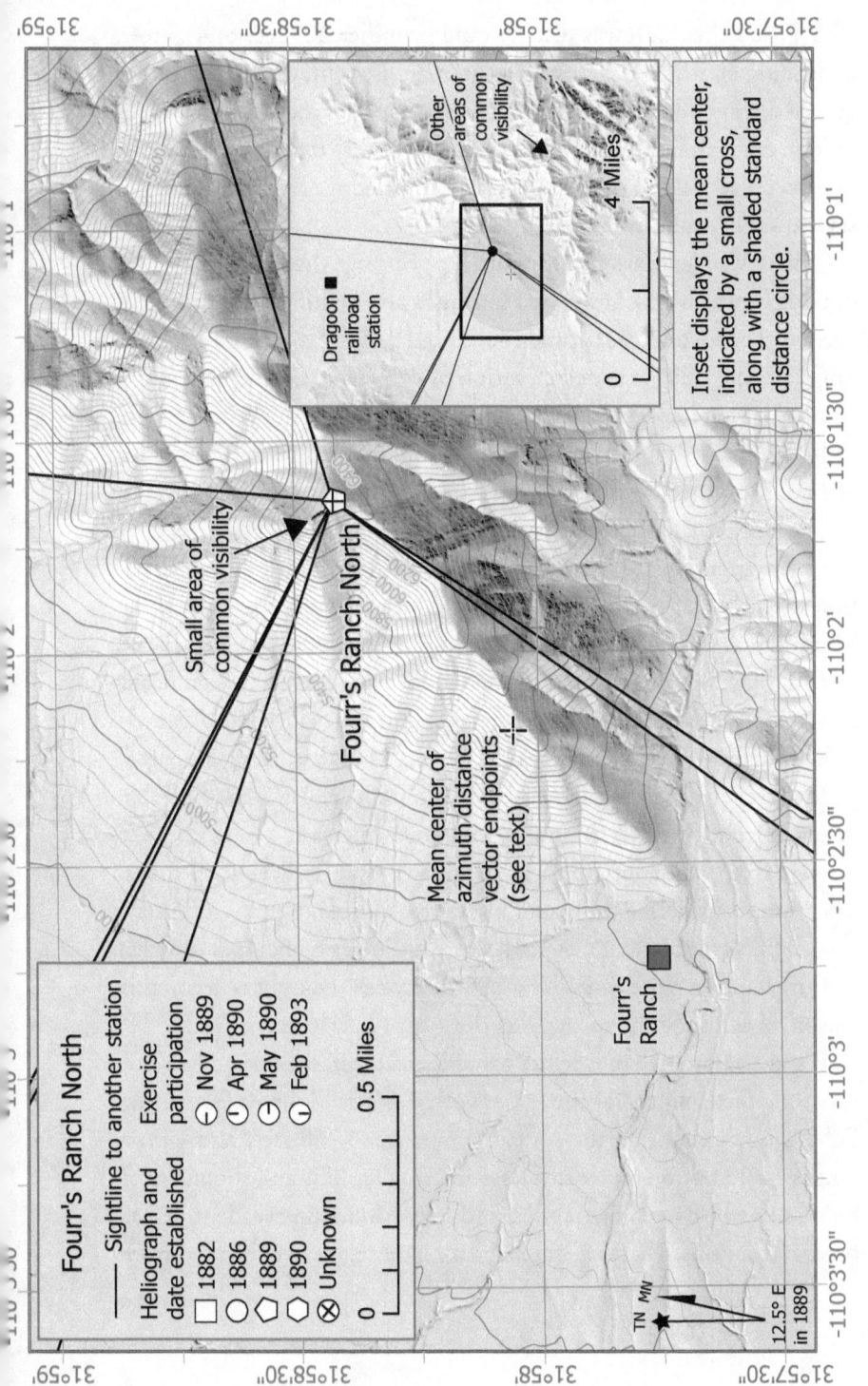

Map 22. Fourr's Ranch North heliograph station.

which the heliograph was placed could be either 225 feet or 425 feet, depending on which elevation measurement is accurate.[44] Lastly, Fuller gives a distance of 31 miles from the Antelope Springs station and 21 miles from Emma Monk. Range rings from those two distant stations intersect just to the east of and adjacent to Antelope Springs. There are several small peaks in this area.[45]

If all visible locations near Antelope Springs that are within sight of both Emma Monk and Fort Huachuca are identified and this set is narrowed down to those between 4,850 and 5,050 feet (using a 100-foot range around Fuller's stated elevation of 4,950 feet for the heliograph station), several options remain: an area just off the summit of Signal Hill, the north shoulder of Outlaw Mountain, the northern shoulder of Hay Mountain, a small hill to the northwest of Antelope Springs, and another small hill to the east of Antelope Springs.

The approximate local relief of these four sites is as follows: Signal Hill with a relief of 300 feet, Outlaw Mountain with 420 feet, Hay Mountain at 600 feet, and the small hills at less than 200 feet. Based on this, Hay Mountain can be dismissed for two reasons: its significant local relief and its size, spanning 2 miles end to end, which likely exceeds Fuller's definition of a small peak. The small hills are also eliminated, as their relief is too small. Outlaw Mountain lies at a considerable distance from the intersection of the range rings and is slightly over 3 miles from Antelope Springs, far greater than Fuller's 0.75 miles. Therefore, considering the distances, it is reasonable to rule out Outlaw Mountain as the station's potential site. This leaves Signal Hill—which has an elevation close to Fuller's 4,950 feet, has the right relief, and is reasonably close to the spring—as the most likely location.[46]

On the other hand, if the same logical criteria are applied but this time focusing on locations between 4,650 and 4,850 feet (using Fuller's other elevation for Antelope Springs, 4,750 feet, also with a range of 100 feet on either side), several smaller hills and shoulders of hills situated between Signal Hill and Hay Mountain are identified as potential options. These sites are all reasonably close to 0.75 miles from

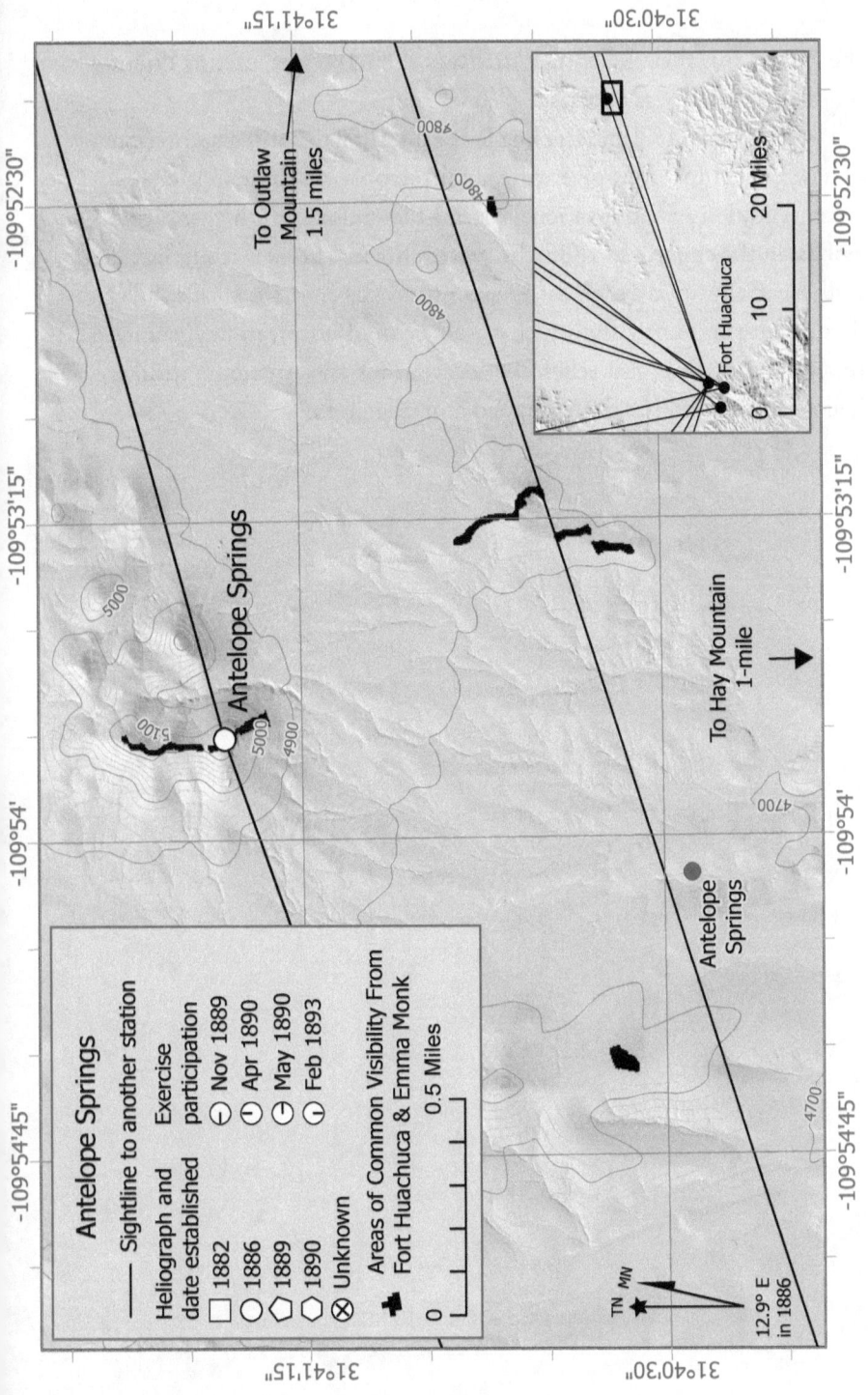

Map 23. Antelope Springs heliograph station.

the spring, but their relief measures less than 180 feet, casting doubt on their suitability as choices.

Both Signal Hill and the smaller peaks fulfill the primary criterion of visibility. However, given the dual purpose of these stations—communication and observation—placing the station at the higher Signal Hill is, in the context of 1886, the better choice. The view from Signal Hill, especially toward the west, is superior to that from the smaller hills. Taking into account visibility, elevation, span of observation, distance to the spring, and local relief, the location for the Antelope Springs heliograph station has been mapped atop Signal Hill.

Chapter 5

Fort Cummings

FORT CUMMINGS, NEW MEXICO, FOUNDED IN 1863, WAS THE EXPANSION of a stage station along the Butterfield stage route. The nearby Cook's Spring was the impetus for the stage station's location.[1] The fort was named after Major Joseph Cummings, who was killed by Navajos in 1863.[2] Dravo reported that the heliograph station at the fort was established on May 16, 1886, and "located on a hill 3° E. of S. [or 177° magnetic] from Fort." He also called the "ascent of hill easy" and said that operators should camp back at the fort each evening.[3]

After accounting for the 1886 magnetic declination of 12.1° east, Dravo's magnetic azimuth of 177° is 189.1° true. A line plotted at 189.1° from the center of the fort (some of which is still there) crosses the summit of a nearby 5,500-foot hill only 1.3 miles distant as well as another hill, elevation 5,600 feet, a short distance beyond.[4] From either of these hills, the four 1886 connections are visible (Henely, Lockhart's Well, Lake Valley to the north, and Deming).[5] The additional stations required for the 1890 exercise—Rincon, New Mexico (referred to here as Rincon NM to distinguish it from stations in the Rincon Mountains of Arizona), and San Andreas—are visible as well. From the more southerly location (the hill beyond), the short-lived station at Heatley's Well could be seen, in addition to all the other 1886 and 1890 stations.

Based on Dravo's description, visibility, and the requirement to connect to Heatley's Well in 1886, the heliograph station at Fort Cummings is mapped at the summit of the hill farther to the south, at an elevation of 5,600 feet. However, after June 11, 1886, when the connection to Heatley's Well was no longer required, the station could have easily been relocated to the hilltop closer to the fort.

Map 24. Chapter 5 heliographs and connections.

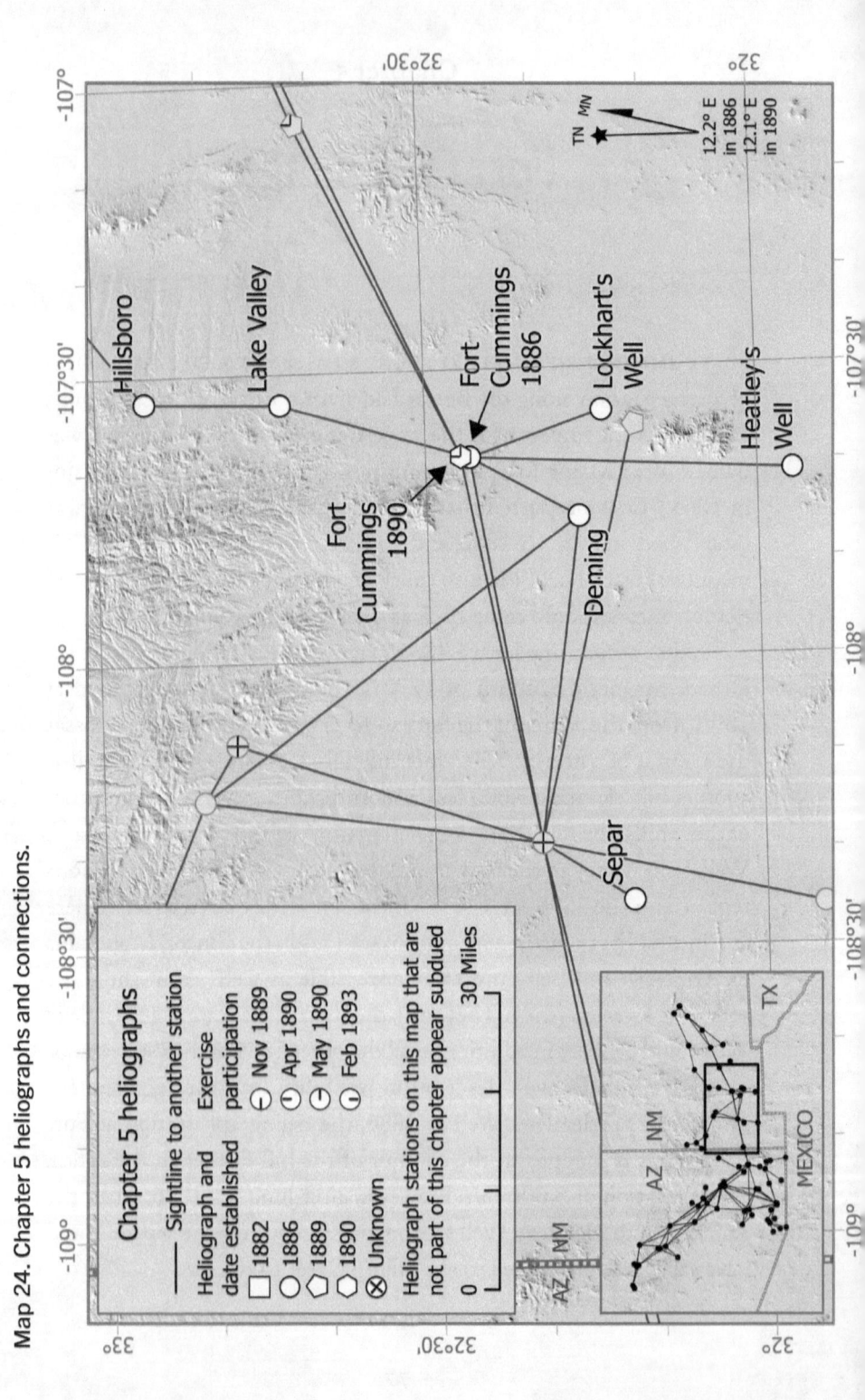

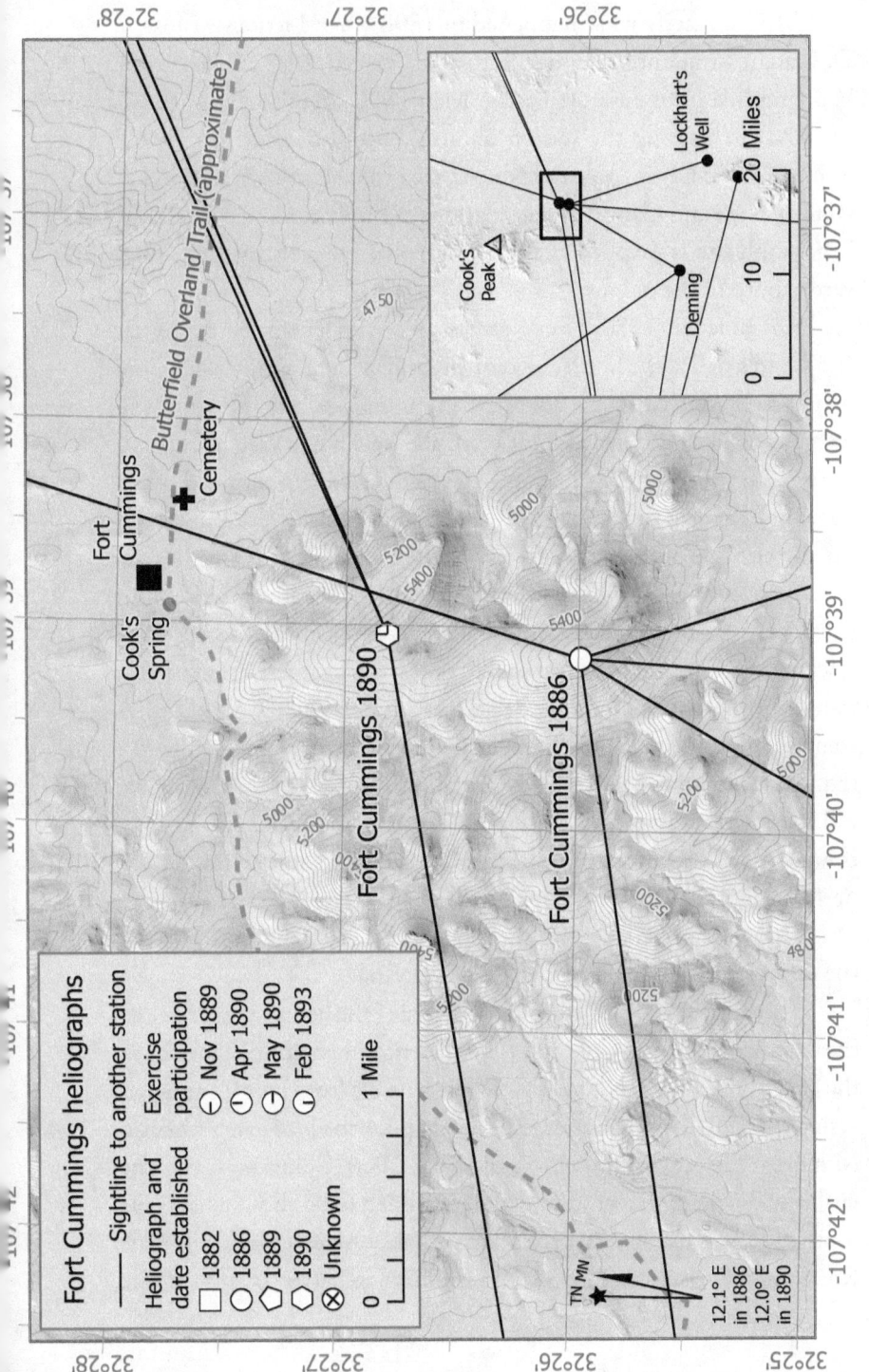

Map 25. Fort Cummings heliograph stations.

CHAPTER 5

This is exactly what happened in 1890 when Lieutenant Joseph D. Leitch, an infantryman and future major general, established the heliograph station supporting the May 1890 exercise on April 27, 1890.[6] Leitch set up the station "about 2,000 yards south of the old military fort" and was able to connect with the required San Andreas, Rincon NM, and Camp Henely stations.[7] The Fort Cummings 1890 heliograph site is mapped at the topographic crest of this hill, 2,000 yards south of the old fort.

On June 11, 1886, Dravo established the heliograph station at "Lockhart's well at the north end of the Florida Mountains."[8] Aside from noting the station's distance from Fort Cummings—15 miles—this is all he mentioned regarding the location of the Lockhart's Well heliograph station. The Spencer map, however, shows the station positioned near a detachment of cavalry along the western bank of the Rio Mimbres, on the eastern side of the northern end of the Florida Mountains.

Almost four years later, in December 1889, Lieutenant Hovey, in an effort to reestablish this station and after some difficulty locating the Lockhart Ranch, as it had been sold to the Coors brothers, established a camp "on the ground occupied in 1886."[9] His trouble finding the old camp stemmed from two factors: the change in ranch ownership and the fact that the location was on the plain rather than atop an easily identifiable hill.[10] Hovey established a heliograph station there and successfully connected with Fort Cummings. He exclaimed that "communication was perfect and it is evident that during the time referred to by Lieutenant Dravo, the station had been in the camp."[11] Hovey went on to say that the camp was 3 miles from the base of the range.

These, then, are the factors in determining the heliograph's location: the distance from Fort Cummings, the Spencer map, the distance from the base of the range, and the station being visible from Fort Cummings. Using the 1901 US Geological Survey map for the location of the Rio Mimbres, the range ring of 15 miles from Fort Cummings, and the visibility diagram, the location was determined to be about a half mile south of today's County Road 549 and a quarter mile east of Lewis Flats Road. As there is no feature, such as a hilltop, to better fix this station, its location is speculative and at the mercy of the distance given from Fort Cummings, which could be inaccurate. The distance from the base of the range (measured at 2 miles) was not used; instead the camp is on the same side of the river, as shown on the Spencer map.

While Dravo was involved as a witness in a court-martial and prior to May 30, Captain Chaffee, the commander of the district around Fort Cummings, had his men establish an outpost with a heliograph station in the vicinity of Heatley's Well. Heatley's Well, associated with Heatley's Ranch, was located north of Columbus, New Mexico, near today's Highway 11, at a gap called Seventysix Draw, between the Florida and Tres Hermanas Mountains.[12] Upon Dravo's arrival at this station on May 30, 1886, he found it was successfully connecting with Fort Cummings, 34 miles to the north.[13] Based on visibility diagrams, the heliograph station was likely in Greasewood Hills, about a mile to the southwest. The station and detachment were moved to Lockhart's Well to the northeast a few days later. The station is mapped in Greasewood Hills, on the nearest hill that was able to connect with Fort Cummings.

Lake Valley, a silver mining community north of Fort Cummings by almost 20 miles, was a boom town in 1886. Silver was found in 1878, spurring development of mines and the town. In 1882 the discovery of a large amount of silver at the Bridal Chamber Mine drove the population to four thousand inhabitants. In 1884 the railroad line was extended from the Santa Fe railroad to support mining and the population.[14] The railroad station was located near the intersection of Keil and Railroad Avenues (if Keil had extended all the way to Railroad).[15]

On June 15, 1886, Second Lieutenant Evan Malbone Johnson Jr. of the Tenth Infantry, who later commanded an infantry division in World War I,[16] established the heliograph station at Lake Valley.[17] Dravo described this heliograph's location as "on a peak 1 mile 12 1/2° N. of E. from R. R. station."[18] A direction of 12 1/2° north of east is 77.5° magnetic, which, after accounting for the 1886 magnetic declination of 12.2° east, becomes 89.7° true. A distance of 1 mile almost due east from the railroad station stands today's Monument Peak at just under 5,700 feet.[19] This is where the Lake Valley heliograph station is mapped.

Map 26. Heatley's Well heliograph station.

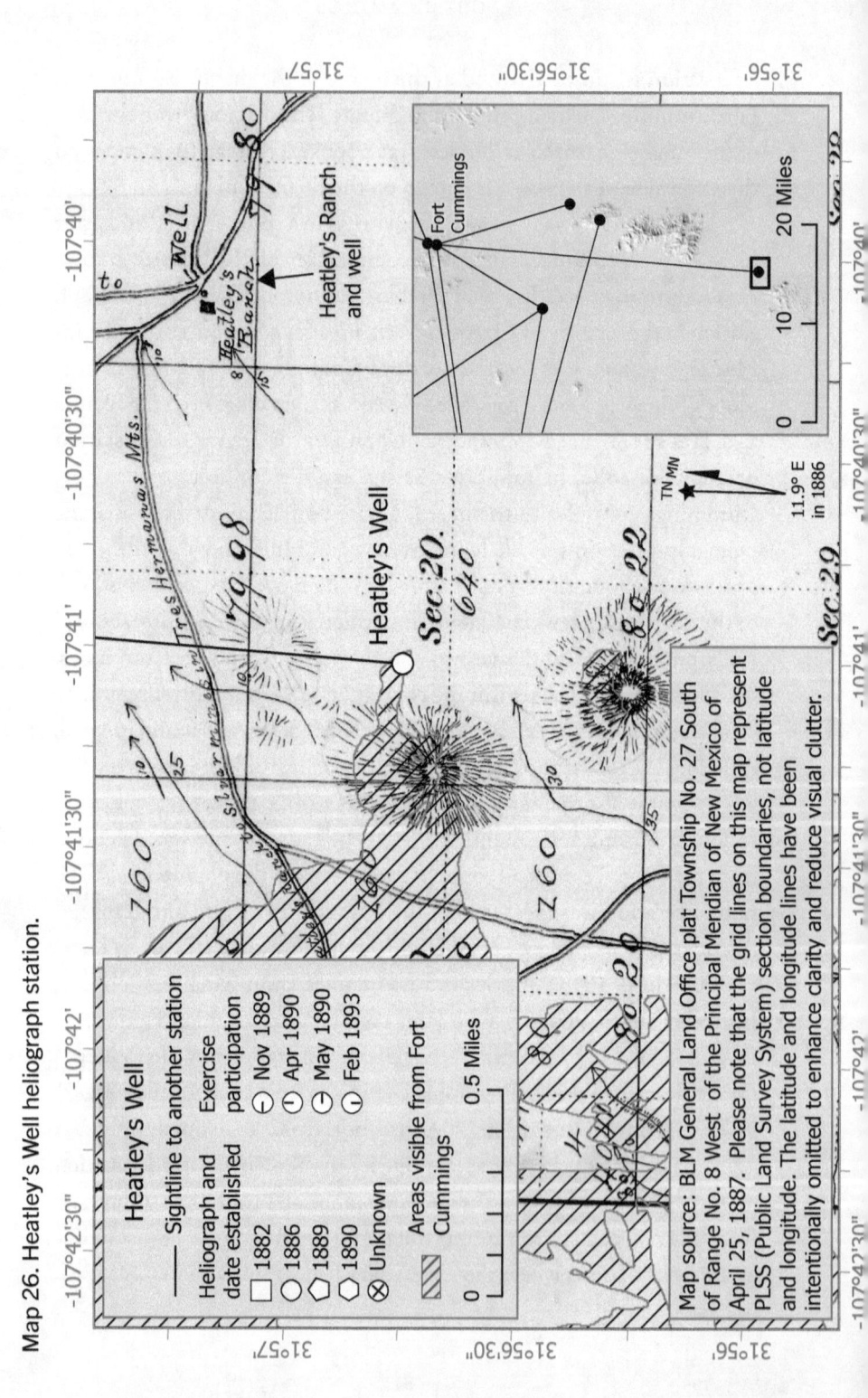

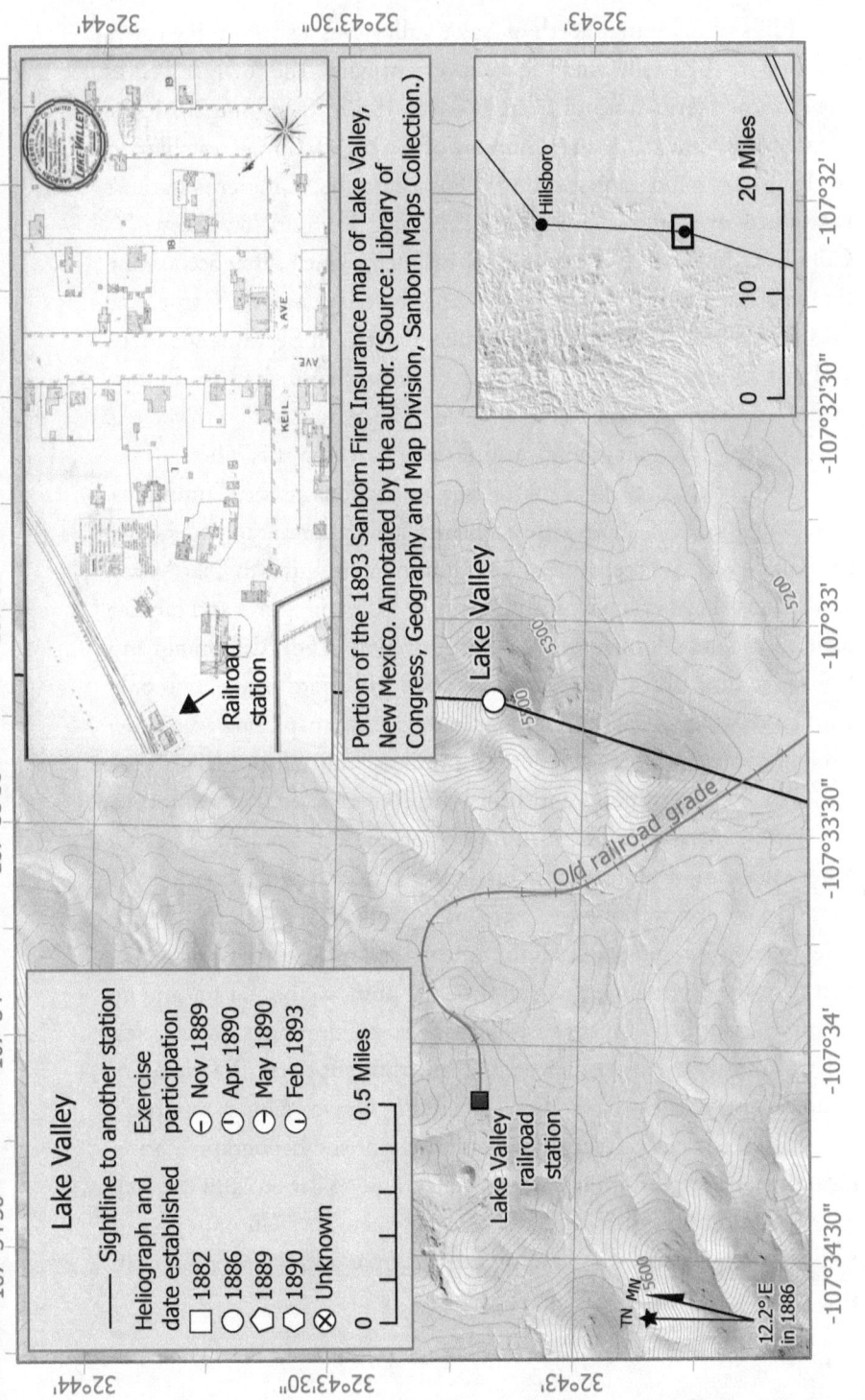

Map 27. Lake Valley heliograph station.

Hillsboro, located north of Lake Valley, was also a mining town, founded in 1877 following the discovery of gold. The town served as the seat for Sierra County from 1884 to 1936.[20] On June 20, 1886, Lieutenant Quincy O'Maher Gillmore of the Eighth Cavalry established the Hillsboro heliograph station.[21] Dravo described Gillmore's station's location as "on a hill ... 1 mile 22 1/2° N. of E. from smelter in town."[22] Gillmore's 22.5° north of east is 67.5° magnetic, which, after accounting for the 1886 magnetic declination of 12.3° east, becomes 79.8° true. The GLO map from 1882 shows the smelter between Sections 9 and 16 and just south of the line in between.[23] From the smelter, 1 mile east at 79.8° true is very close to today's Hill 5723, where a radio tower was once sited.[24] The heliograph station is mapped atop this hill.

While the locations of the heliograph stations at Fort Cummings, Lake Valley, and Hillsboro are established, it is important to understand the visibility among these three. Fort Cummings is visible to Lake Valley, and Lake Valley is visible to Hillsboro. In his two reports and tabular table, Dravo did not mention a connection between Fort Cummings and Hillsboro directly. His tabular report had Hillsboro connecting only with Lake Valley. Both the Volkmar and Spencer maps, however, show a direct connection between Fort Cummings and Hillsboro.

For Fort Cummings and Hillsboro to be visible to one another, the stations needed to be much higher, or one or both had to be at a different location altogether. This is unlikely, as the descriptions of those heliograph stations are clear and logically consistent. In all three cases, Dravo offered a start point and an azimuth to the station. In two cases, he gave a mileage from the start point to the station, and he gave a description of an easy ascent from Fort Cummings. In all cases, this results in an easily identifiable hill that supports the visibility requirements to the other stations laid out in Dravo's tabular reports.[25]

Given the clear descriptions by Dravo of the heliograph station locations at Fort Cummings, Lake Valley, and Hillsboro and the lack of text indicating a direct connection between Fort Cummings and Hillsboro, the connection lines on the Volkmar and Spencer maps, in this case, appear to be inaccurate.[26]

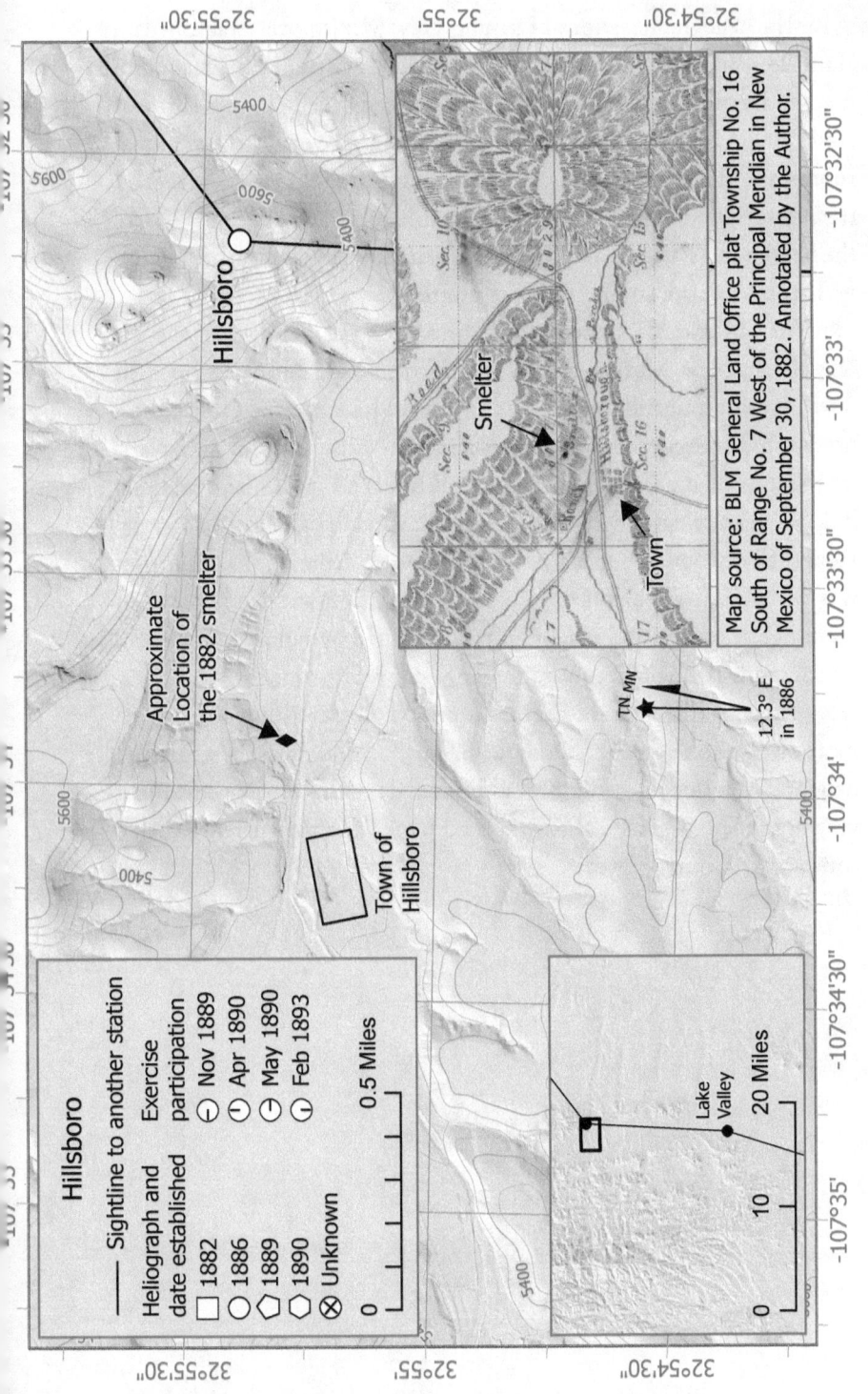

Map 28. Hillsboro, New Mexico, heliograph station.

CHAPTER 5

The heliograph station at Deming, New Mexico, was established on "June 28, 1886 at the R. R. Depot" by Dravo.[27] A detailed fire insurance map of Deming from June 1886 places the railroad station north of Rail Road Avenue between Silver and Platinum Avenues.[28] Individual rooms are identified within the station, and the telegraph office is at the western end of the building. Near this telegraph station is where the heliograph station is mapped. Dravo mentioned that the heliograph was atop a rail car, so the station is plotted just a bit to the north.[29]

Just 11 miles southwest of Camp Henely's heliograph station was Separ, New Mexico. Separ, at the crossroads of an old Spanish north–south trail and the Southern Pacific Railroad, was a rail station used by local ranchers for cattle transfers and by the army for logistics.[30] The station had access to the telegraph line, and it was here where a link between telegraph and heliograph was made on May 9, 1886, by Robert Sherwood.[31] Separ's sole heliograph-to-heliograph connection was with Camp Henely. On July 28 a connection was made between Deming's telegraph and heliograph (established by Fuller on June 28), which made the heliograph station at Separ redundant.[32]

There is little available information to help determine the precise location of the telegraph station at Separ in 1886 (and therefore the heliograph station). Lacking any additional data, this station is mapped very near the crossroads of the railroad tracks and the current north–south road through Separ.

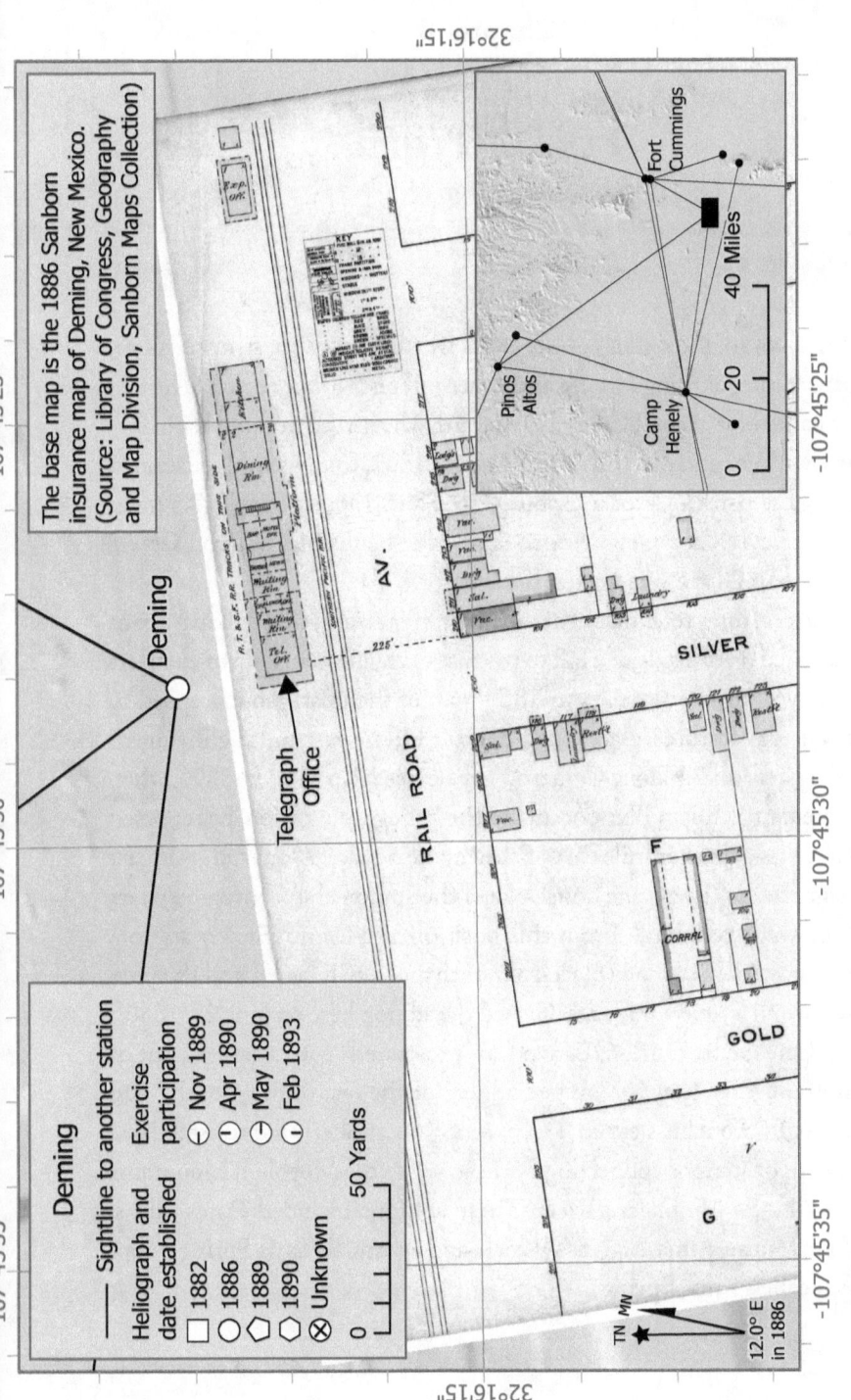

Map 29. Deming heliograph station. The telegraph office was in a building near the intersection of Rail Road and Silver Avenue. (Source: Library of Congress, Geography and Map Division, Sanborn Maps Collection.)

Chapter 6

Fort Bayard

FORT BAYARD, ESTABLISHED IN 1866 IN NEW MEXICO, SERVED AS AN outpost safeguarding mining and other interests in the region. The fort was named for general George D. Bayard, who served in the West before the Civil War and died on December 14, 1862, from wounds sustained during the Battle of Fredericksburg.[1] The fort is approximately 7.5 miles east of Silver City, New Mexico, along the banks of Cameron Creek. Lieutenant Dravo was posted here.

According to Dravo, the heliograph station at Fort Bayard was positioned on the "side of a hill to the east of engine house, three-quarters of the way up the slope."[2] An 1884 map of the post, georeferenced to present-day features, places the Engineer House east and slightly north of the parade field, along Cameron Creek (see Map 31).[3] In 1890, when Lieutenant William Black occupied the heliograph station, he reported that it was a quarter mile east of the engine house.[4] A quarter mile east (magnetic) of the engine house places the station about three-quarters of the way up the hill. From this position, the Camp Henely station, the Pinos Altos station (placed a month and a half later), and the post itself are all visible. Dravo established the station here on June 28, 1886.[5]

On August 6, 1886, Dravo was "prostrated" with a severe bout of remittent fever, leaving him bedridden for the rest of the month.[6] On August 14, from his sickbed, Dravo sent Robert Sherwood to establish a network of stations connecting Fort Bayard to the Mogollon Mountains along the San Francisco River. These stations included Pinos Altos, White House, Alma, Siggins's Ranch, and eventually Lyda Spring (more often called Mule Spring).

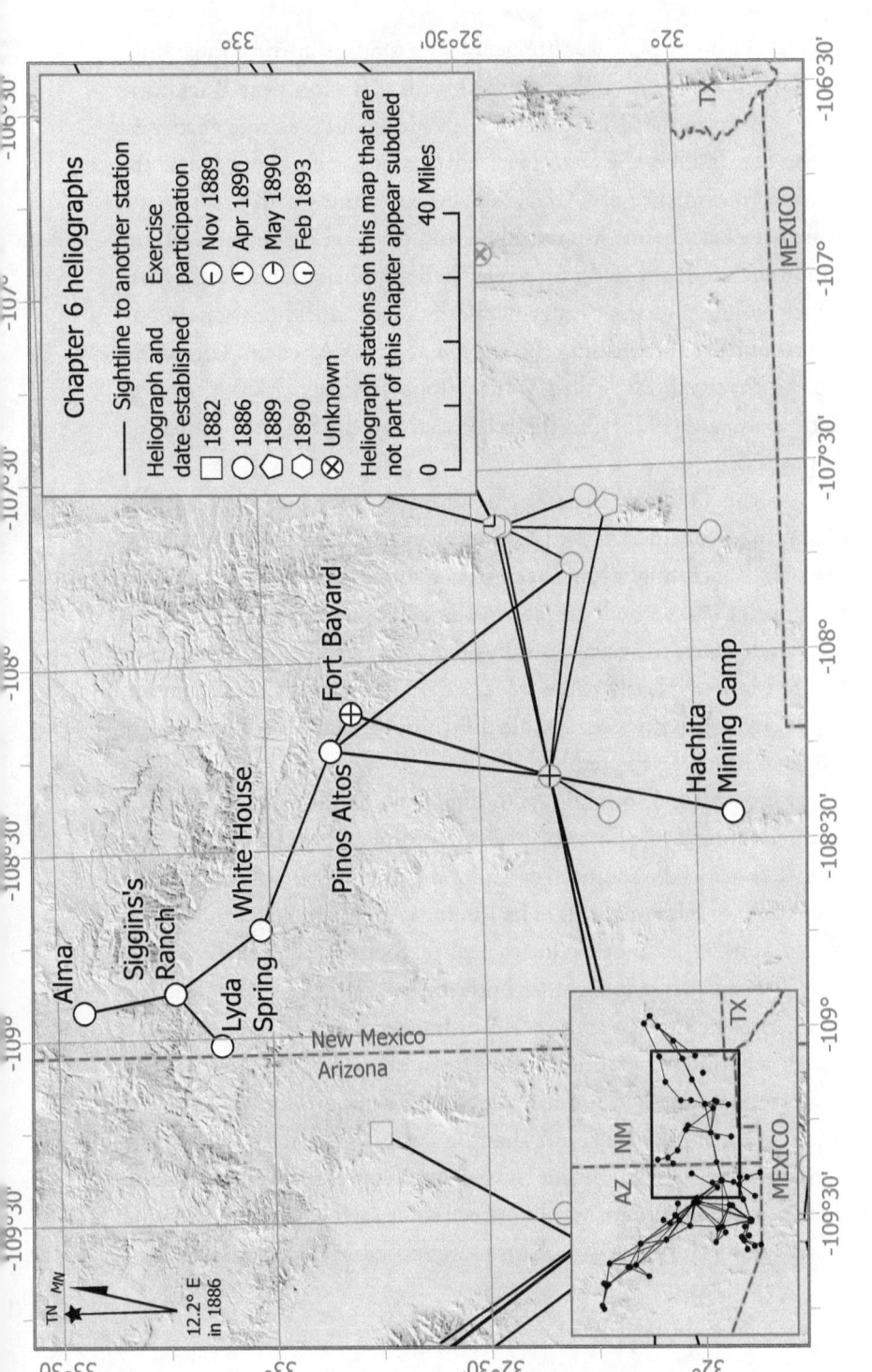

Map 30. Chapter 6 heliographs and connections.

CHAPTER 6

Sherwood's plan was to establish a station in the Pinos Altos Mountains to connect Fort Bayard with a station near Buckhorn, New Mexico, in the Duck Creek Valley. He would then ride northwest to the San Francisco River, travel north along the river to locate the cavalry detachment near Alma, New Mexico, and set up a heliograph near that detachment. Afterward he would retrace his route down the San Francisco River to set up a relay heliograph station near Siggins's Ranch, connecting the cavalry near Alma with Buckhorn. Sherwood moved quickly, establishing the station at Pinos Altos on August 14; the Buckhorn station, called White House, on August 20; the Alma station on August 23; and the relay station near Siggins's Ranch on August 26.[7]

At this juncture, Lyda Spring did not factor into his planning. Had it been considered, the positioning of the intermediate station at Siggins's Ranch might have been better suited to reach the area near Lyda Spring. As it stood, the intermediate station only needed to bridge the gap between the cavalry near Alma and the heliograph station at White House.[8] Nearly three weeks after the Siggins's Ranch station was set up, and with considerable difficulty, the station at Lyda Spring was established on September 15, 1886.[9]

Pinos Altos Mountain, standing at an elevation of 8,100 feet, is situated north of Silver City, New Mexico, and to the west of Fort Bayard. It marks the southern extent of the Pinos Altos Range, with the town of Pinos Altos just east of its summit. Approximately 1 mile south of this summit lies another, lower peak at about 7,600 feet elevation.

The Spencer map includes both the town of Pinos Altos and the Pinos Altos heliograph station. According to this map, the heliograph station is positioned southwest of the town, suggesting that a lower peak south of Pinos Altos Mountain served as the station's location. A close examination of the map shows that the station is situated about a mile from the town, a measurement consistent with present-day data for the lower peak, though there may have been some cartographic adjustments to account for the map's scale. This more southerly peak offers excellent visibility both to the valley in the northwest, where White House is located, and to Fort Bayard. Its lower elevation and closer proximity to the valley made it logistically easier to support from Fort Bayard.

The Volkmar map includes an estimated azimuth from Fort Bayard to Pinos Altos of 285° magnetic, which, after applying the 1886 magnetic declination of 12.4° east, becomes 297.4° true. The GIS-measured azimuth to the more southerly, lower peak is 299.8° true, while the azimuth to the other likely peak, present-day Pinos Altos Mountain, is 306.3° true. Given that the direction aligns more closely with the lower peak and the relative position shown on the Spencer map, the heliograph station is mapped on the lower peak.

In 1880 Tom Lyons and his business partner, Angus Campbell, bought the Nogales (White House) Ranch. They left their mining enterprise in Silver City and took over the headquarters of the ranch (the White House) at Buckhorn. The pair created one of the largest ranches in the United States, controlling one million acres of land.[10] In 1890 the Lyons & Campbell (LC) Ranch moved its headquarters to Gila, New Mexico, about 10 miles southeast.[11] The White House headquarters in Buckhorn burned to the ground that same year.

On August 20, 1886, Sherwood established a heliograph station at White House and "opened communication with Pinos Altos."[12] Visibility analysis from Pinos Altos to the area around Buckhorn reveals that most of the area is visible, suggesting that the station could have been located on the plain near Duck Creek. However, since this was an intermediate station needed to connect farther north into the San Francisco Valley, it was more likely that the heliograph station was positioned atop one of the two nearby hills just west of Buckhorn, separated by what is now Buckhorn Creek. Of these two hilltops, the higher one is where the heliograph station is mapped.

Approximately 4 miles north of Alma, New Mexico, along a large bend in the San Francisco River, is where the Alma heliograph station is mapped. The Spencer map shows the heliograph station about 4 miles north of Alma on the east side of the river, and both the Spencer

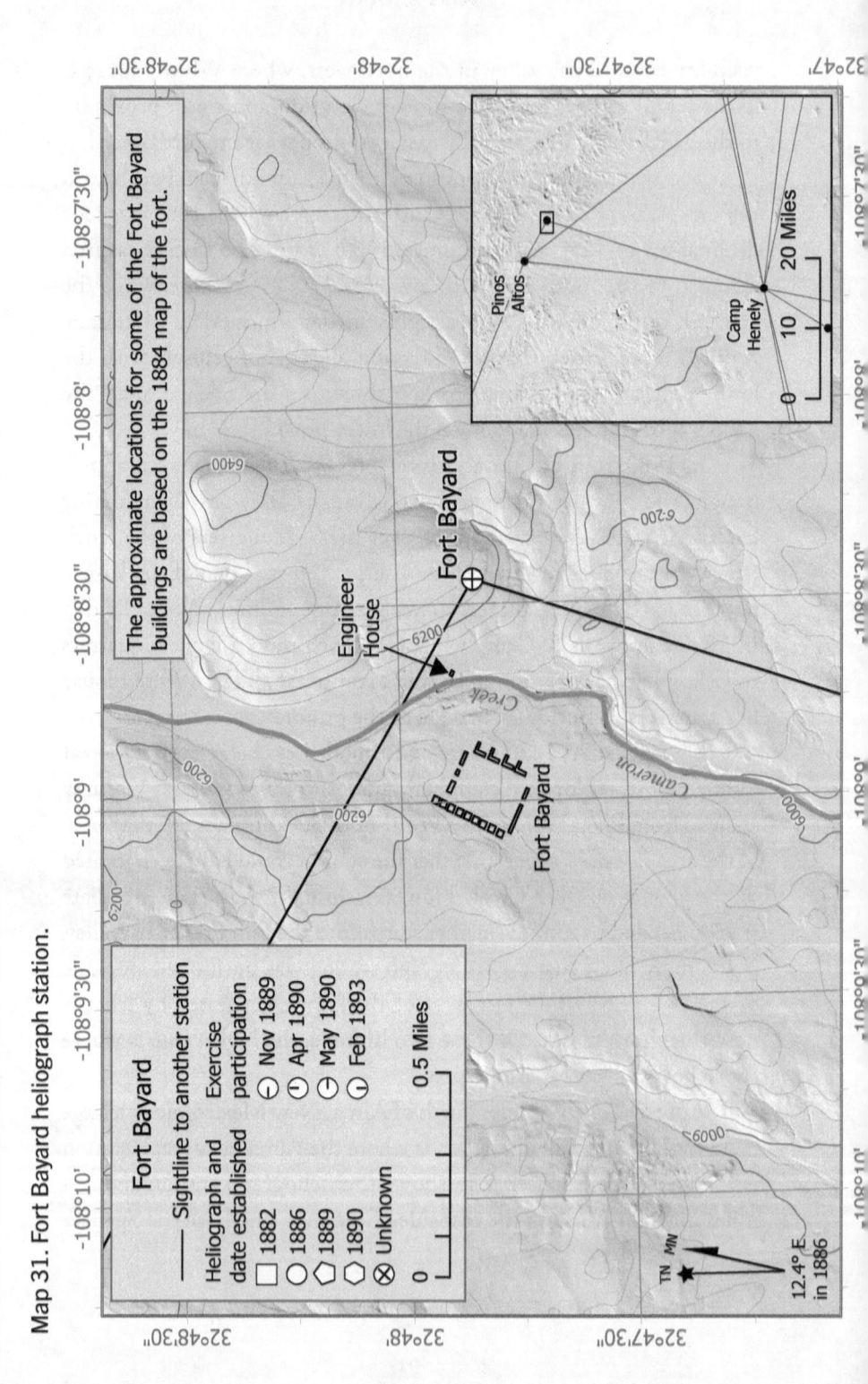

Map 31. Fort Bayard heliograph station.

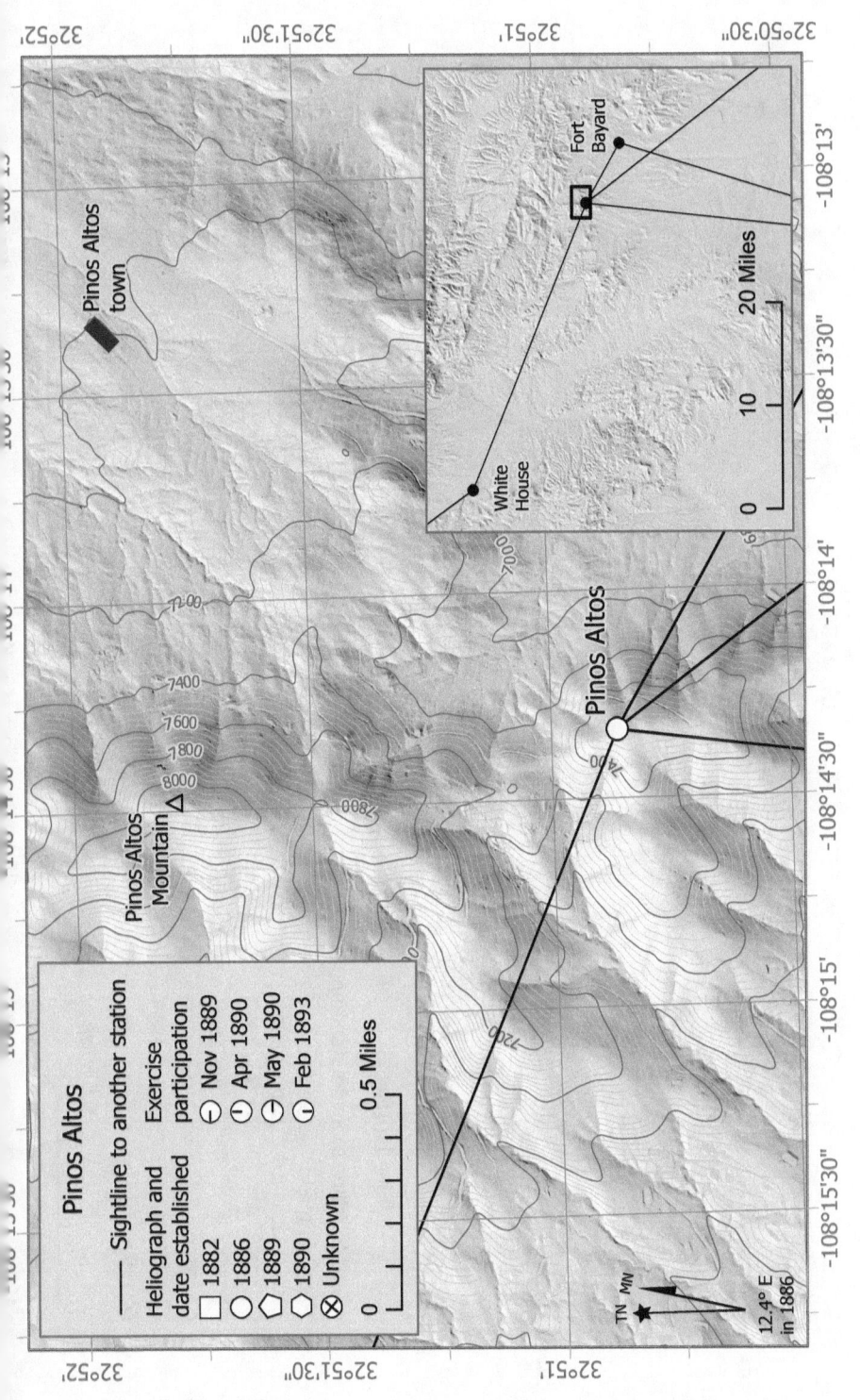

Map 32. Pinos Altos, New Mexico, heliograph station.

Map 33. White House, New Mexico, heliograph station.

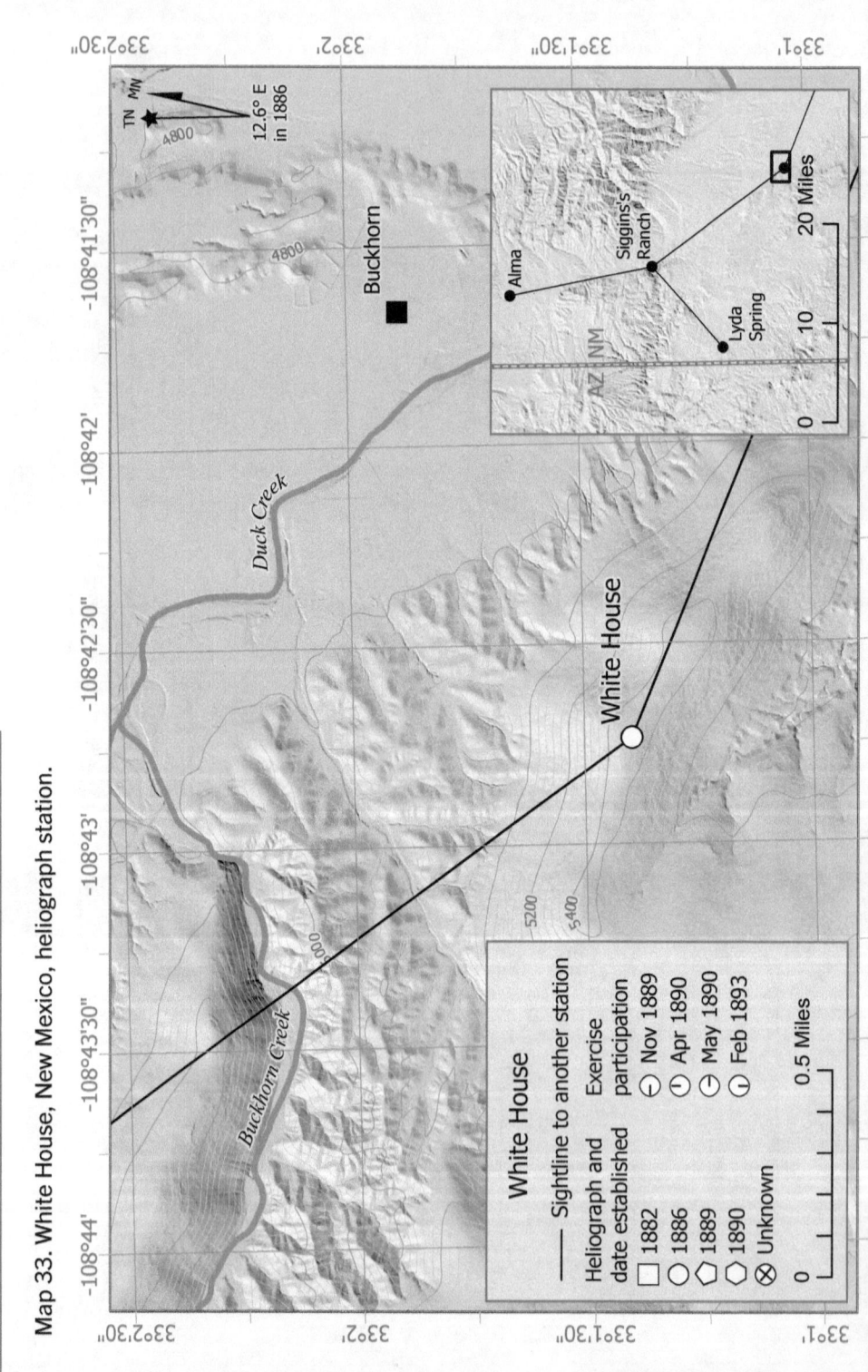

and Volkmar maps place the station east of a large bend in the river. Additionally, the 1883 US Army map of southeastern New Mexico depicts a convergence of five roads and trails near this area, making it a logical location for an army presence.[13] A location atop a spur along the eastern bank of the river would provide a good view to the south. Considering the period maps and the distance from Alma, the heliograph station has been mapped atop this spur.

The Siggins's Ranch heliograph station was a relay between the stations at White House and Alma. Sherwood set up this station on August 26, 1886.[14] Isaac Siggins was one of the first white ranchers in the area, and his ranch was located along the Big Dry Wash, about 2.5 miles upstream from its confluence with the San Francisco River. His ranch house was at or near the site of the current SI Ranch.[15] This area has seemingly been occupied intermittently for thousands of years by Indigenous peoples and others.[16]

As Siggins's Ranch was situated near a valley floor, elevation was required for the heliograph to relay signals between Alma and White House. Among the nearby peaks, two offer views to both stations. The first is present-day Sundial Mountain, and the second is a peak south of Sundial Mountain and east of present-day Outlaw Mountain, marked as "6033" on older US Geological Survey topographic maps (see Map 35).

Sherwood chose Sundial Mountain for the location of the Siggins's Ranch heliograph station. While this location effectively connected the stations at Alma and White House, it was less suitable for communication with the Mule Creek Valley station to the southwest. The summit of Sundial Mountain is partially obstructed by the taller Outlaw Mountain, limiting the view in that direction. Peak 6033 is visible from much of the Mule Spring Valley, including Lyda Spring (Mule Spring), the Lyda Ranch, and several other ranch houses in the area, so it would have been easy for Sherwood to connect from Mule Spring Valley had he chosen Peak 6033.

While Sherwood was attempting to place a heliograph station near Lyda Spring, he encountered "great trouble . . . in selecting a point near the camp at Mule Springs, from which connection could be made

Map 34. Alma, New Mexico, heliograph station.

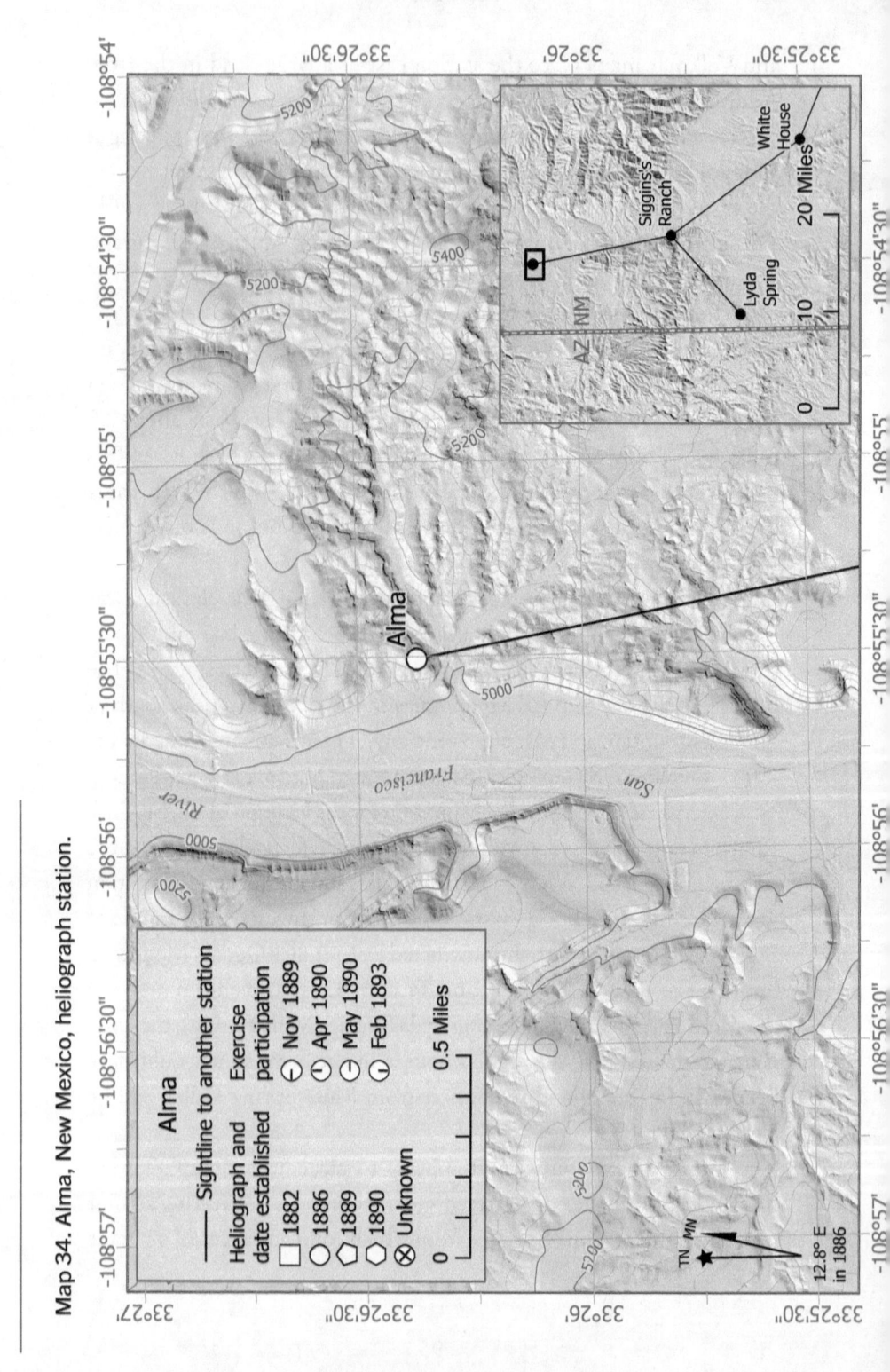

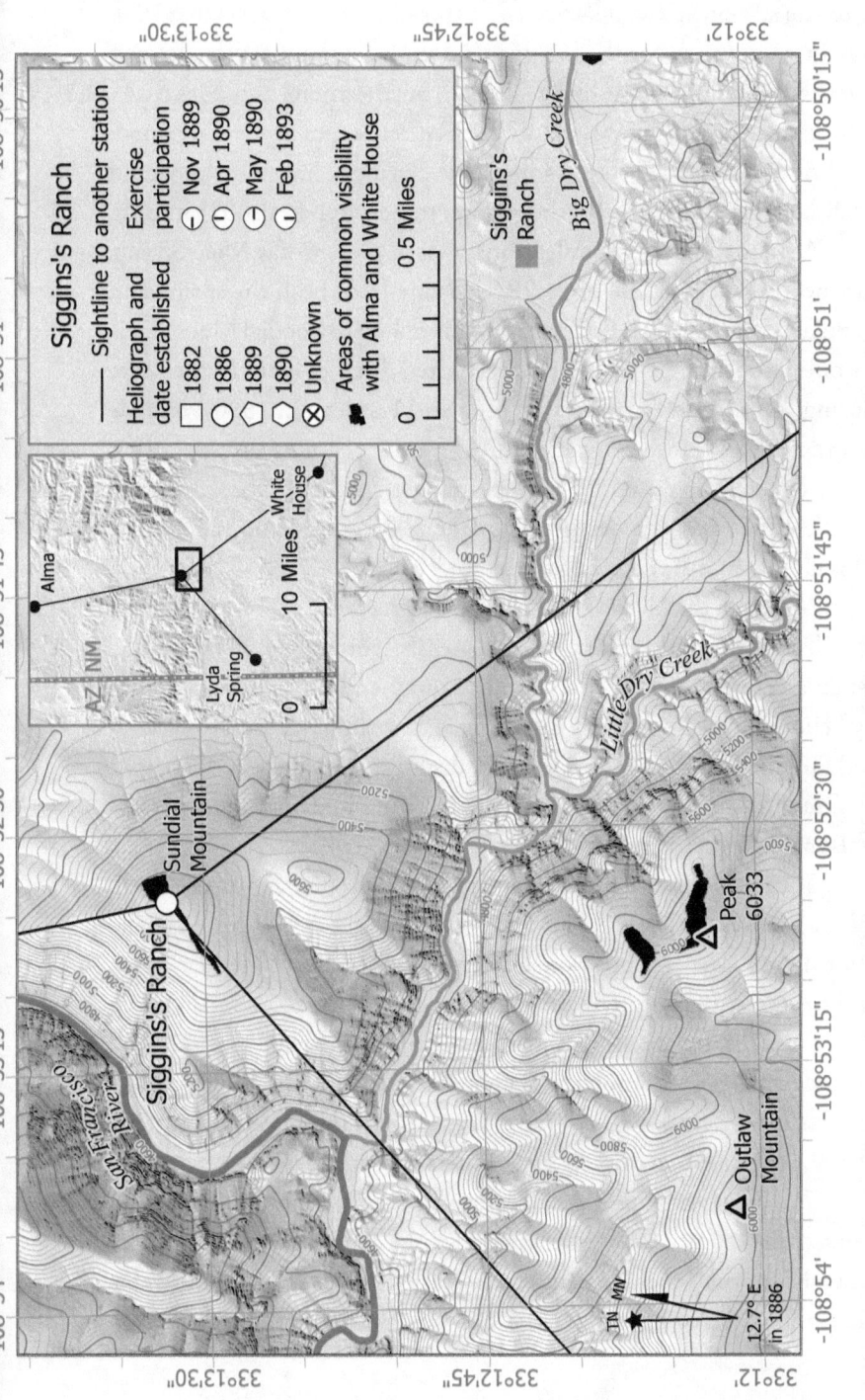

Map 35. Siggins's Ranch, New Mexico, heliograph station. Dark areas along the stream banks and other very steep slopes are shadows and thus locations not visible to Alma and White House.

CHAPTER 6

with the station at Siggins's ranch."[17] This difficulty suggests that Peak 6033 was not used for the Siggins's Ranch heliograph station, leaving Sundial Mountain as the likely location. Furthermore, it implies that a station at Lyda Spring was an unforeseen requirement for Sherwood. Had he anticipated the need for a Lyda Spring connection, he would likely have placed the Siggins's heliograph station atop Peak 6033 instead.

A hilltop located 2.7 miles northwest of present-day Mule Spring, just north of today's Highway 78, is visible from both the spring and Sundial Mountain. The distance from this hilltop to Sundial Mountain is 11 miles, which aligns with Dravo's reported distance between the Mule Spring station and Siggins's station (also 11 miles). In 1890 Volkmar recorded an unverified magnetic bearing from the Siggins's Ranch station to the Mule Spring station as 214.5° magnetic, which, after accounting for the 1886 magnetic declination of 12.7° east, becomes 227.2° true. This closely matches the GIS-measured angle of 227.3°.

Although the hilltop is 2.7 miles from the camp near Mule Spring, this distance is mitigated by a clear line of sight, making it feasible to establish communication between the camp and the station, whether by heliograph or possibly by flags. Given its visibility to both the camp and Sundial Mountain, along with the distances and directions reported by Dravo and Volkmar, it is plausible that the Mule Spring heliograph station was located on this hill.

The locations of the heliograph stations at Alma, Siggins's Ranch, White House, and Mule Spring are identified based on the best available information. They are positioned to ensure visibility among each other and to closely match the reported distances and directions. However, their exact locations remain a matter of conjecture. Neither of Dravo's documents—his report on the system and his later descriptive notes—provide more detailed descriptions of these locations, likely because he was not the one who initially established the stations.

The Hachita Mining Camp, in the Empire mining district, is marked on the 1913 GLO map as simply "Hachita."[18] The 1886 Spencer map shows the Hachita heliograph station, situated with a cavalry detachment, on or just to the east of a north–south road, depicted as

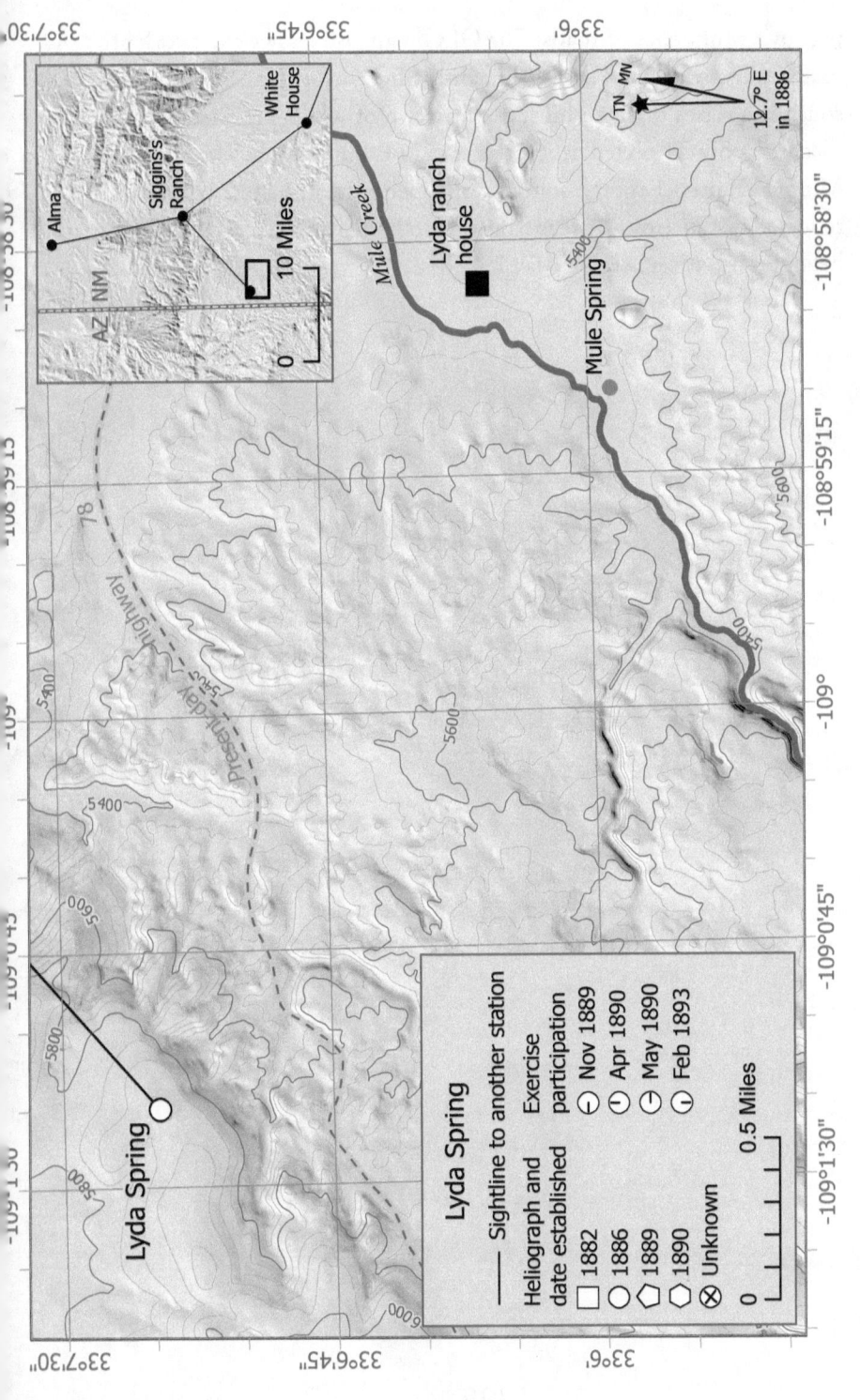

Map 36. Lyda Spring (Mule Spring) heliograph station.

CHAPTER 6

part of a confluence of roads. The GLO map also shows a network of roads to the south and east of Hachita. Slightly farther east and a bit south is a series of small hills, the first two of which are visible from Hachita's only connected heliograph station, Camp Henely. The Hachita Mining Camp heliograph station is mapped here, just short of Dravo's reported distance of 31 miles from Camp Henely. Dravo established a heliograph station here on May 29, 1886.[19]

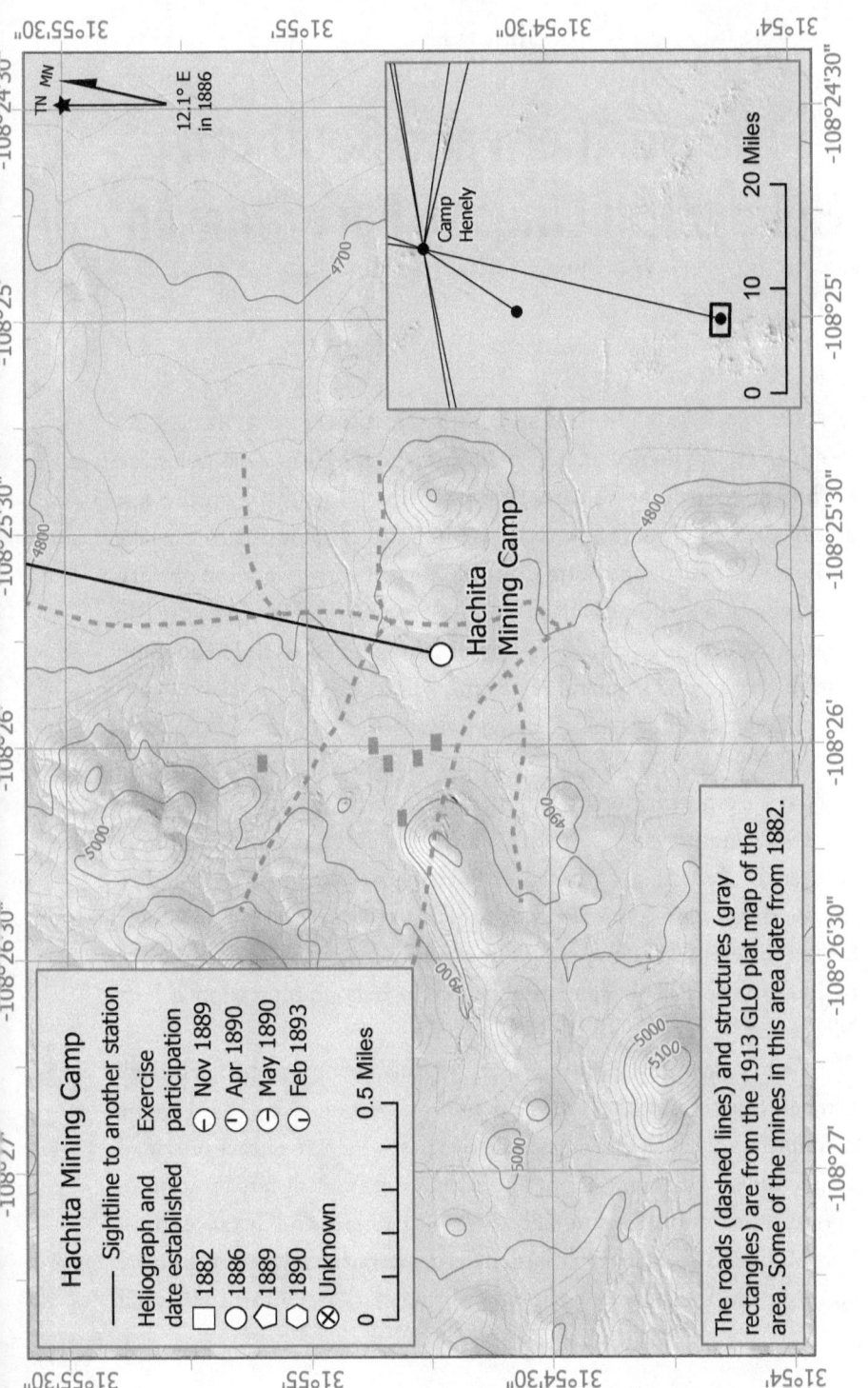

Map 37. Hachita Mining Camp heliograph station.

Chapter 7

The Northern and Southern Heliograph Lines to Fort Stanton

FORT MCRAE, ESTABLISHED IN 1862 AND ABANDONED AS A PERMANENT fort in 1875, was named after Captain Alexander McRae, who was killed in the 1862 Battle of Valverde, New Mexico.[1] The fort is on the east side of the Rio Grande and today is often underwater due to the lake created by the Elephant Butte Dam, built in the early twentieth century. The fort sits in a valley (McRae Canyon) just over 2 miles from the Rio Grande. Valleys are poor locations for heliographs, so the heliograph station was located some miles distant, where it could be seen by both the Hillsboro and Dripping Spring stations.

This and the rest of this 1886 line of heliograph stations extending eastward from Hillsboro were installed in the latter half of November 1886, more than two months after the surrender of Geronimo. Lieutenant Pershing, stationed at Fort Bayard, had just returned to the fort from more than two weeks of mounted patrol to find orders for him to establish a heliograph line to the east.[2] He left Fort Bayard on November 10, 1886, and established the heliograph station at Fort McRae on November 16, 1886.[3]

While Pershing set up the Fort McRae, Dripping Spring, Nogal, and Fort Stanton stations, the locations of these stations are based on the descriptions given in Dravo's descriptive notes.[4] Dravo's notes provided a direction and distance of 38° and 31 miles from Hillsboro station to Fort McRae station. Additionally, he gave directions and distances from Fort McRae to its heliograph station (as it was not collocated at the fort) and from Fort McRae to Dripping Spring station. Unfortunately, all

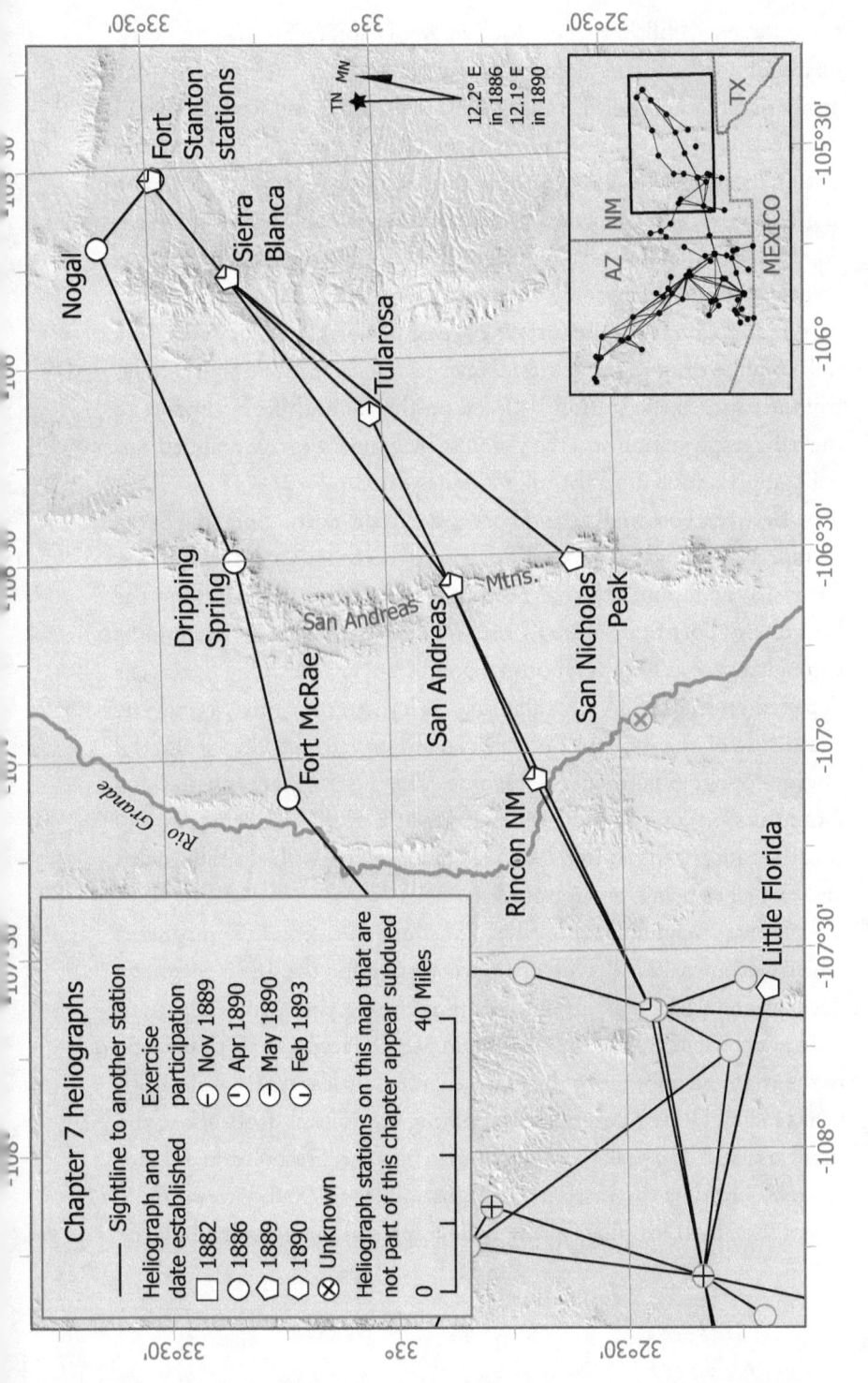

Map 38. Chapter 7 heliograph stations.

CHAPTER 7

three stations (Hillsboro, Fort McRae, and Dripping Spring) are nearly on a common line. This alignment, coupled with the potential for small errors in direction over distances of 30 to 40 miles, renders any sort of intersection of those lines useless as an aid to determine Fort McRae's station location. However, the direction from the fort to its heliograph station (only a short distance away) provides valuable information.[5]

Visibility analysis from Hillsboro, Dripping Spring, and the fort reveals locations of common visibility.[6] The most likely location, based on the direction from the fort of 25° east of north, or today's 37°, is a hill 3.6 miles distant. (Note that Dravo gave a distance of only 1.5 miles from the fort to the station.)[7] Based on the azimuth from the fort to the heliograph station and the visibility diagrams, we have mapped this heliograph station atop the hill 3.6 miles distant.

Between Fort McRae and Fort Craig to the north, along the Santa Fe railroad line, was Lava Station, where there was a telegraph office. From Lava Station, the telegraph line extended eastward along the Fort Craig–Fort Stanton road, crossing the San Andreas Mountains through a pass at today's Thoroughgood Canyon. Dripping Spring, as depicted on the 1883 GLO plat map, is located near this pass on the south side of the road and is still marked on current US Geological Survey topographic maps.[8] Dravo described the heliograph station's location as "on top of a hill 3 miles 25° S. of E. from the spring," noting that the route to the station led up a gradual slope, with the hill's south and east sides being "precipitous."[9]

Dravo's magnetic azimuth of 25° south of east is 115° magnetic, which becomes 127.2° true after accounting for the 1886 magnetic declination of 12.2° east. From Dripping Spring, this azimuth leads to a ridge known as Silver Top Mountain, which features two peaks, one to the west and another to the east, separated by less than a mile. Both peaks match Dravo's description of having precipitous drop-offs to the south and east. The peak more closely matching the direction and distance given by Dravo is the eastern one at just under 8,100 feet elevation.[10] Given this, the Dripping Spring heliograph station is mapped on the

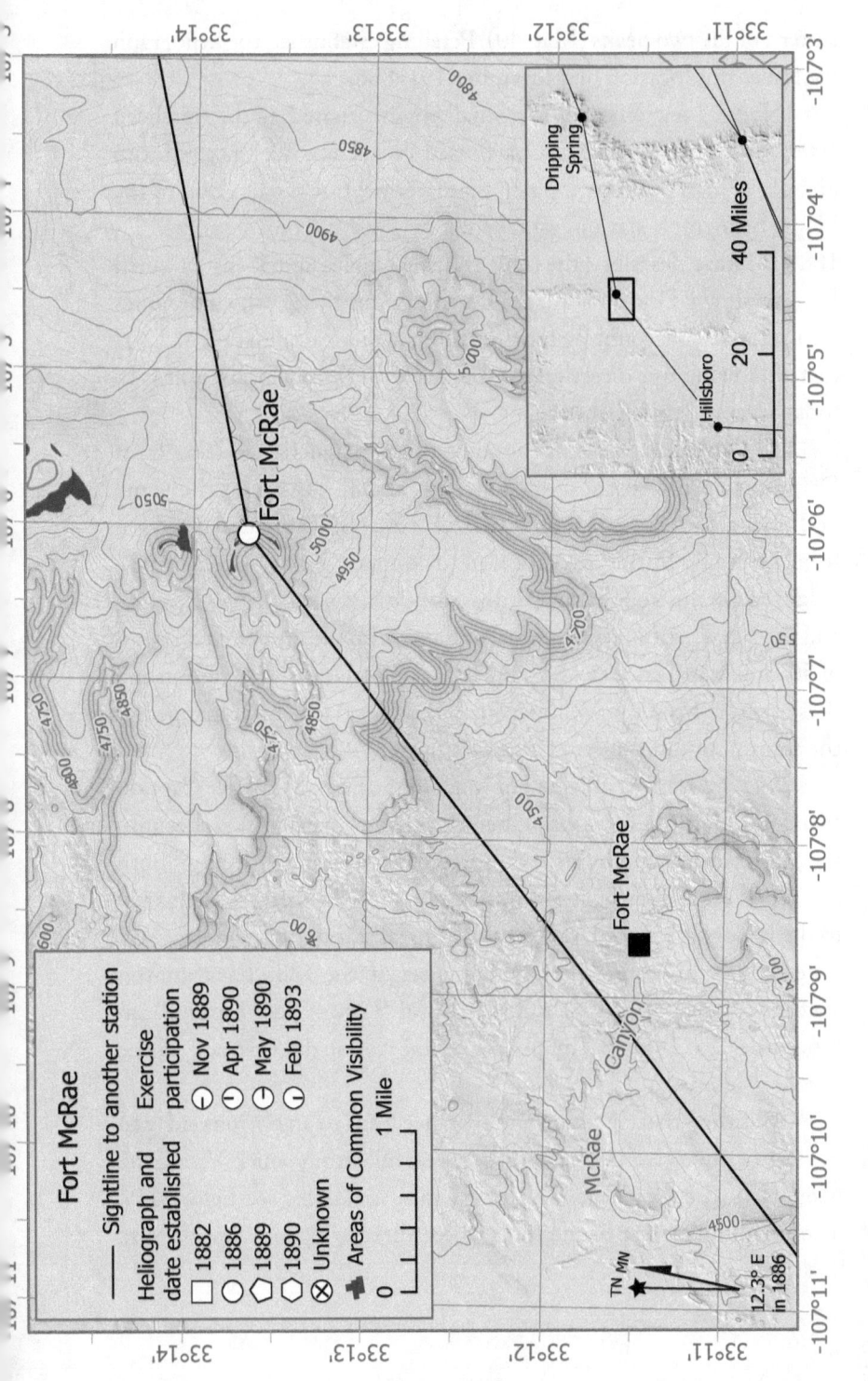

Map 39. Fort McRae heliograph station.

eastern of the two peaks (Map 40). Pershing established the heliograph station at this location on November 19, 1886.

Nogal, New Mexico, is a small town situated at the northern extent of the Sierra Blanca. Dravo used Nogal as a reference point to identify a "high prominent peak" that later became the location of the Nogal heliograph station, established by Pershing on November 21, 1886.[11] Dravo described the peak as being 5 miles and 5° east of north from the town. Dravo's 5° east of north is 5° magnetic, which becomes 17.1° true after accounting for the 1886 magnetic declination of 12.1° east. Following this direction for 5 miles leads us to a location near to today's Vera Cruz Mountain.

The Volkmar map lists the (unverified) magnetic direction from Dripping Spring to the Nogal station as 55°. Applying the same declination of 12.1° east results in 67.1° true. The measured direction from Dripping Spring to Vera Cruz Mountain is very close, at 68°. Additionally, the Volkmar map gives 51 miles between Dripping Spring and the Nogal station, which closely aligns with the measured distance of 50.1 miles to Vera Cruz Mountain. Based on Dravo's description and these vectors from the Volkmar map, the Nogal station is mapped at the summit of today's Vera Cruz Mountain.

Fort Stanton, established in May 1855 and named after Captain Henry W. Stanton, who was killed during a pursuit of Apaches into the Sacramento Mountains,[12] is situated at the eastern extent of both the 1886 and 1890 heliograph systems. In 1886 the fort was connected to the heliograph system through the Nogal station to the northwest. To establish this connection, Pershing set up the 1886 Fort Stanton station "on edge of mesa 1 mile 23° E. of S. from post hospital" on November 23, 1886.[13] The post hospital was at the east end of the parade grounds.[14]

Following that direction and distance leads to an obvious ridge to the south. On top of that ridge are several hills from which Vera Cruz Mountain (the Nogal heliograph station) is visible (see Figure 11). The small hill closest to the fort (where there is a good view of Vera

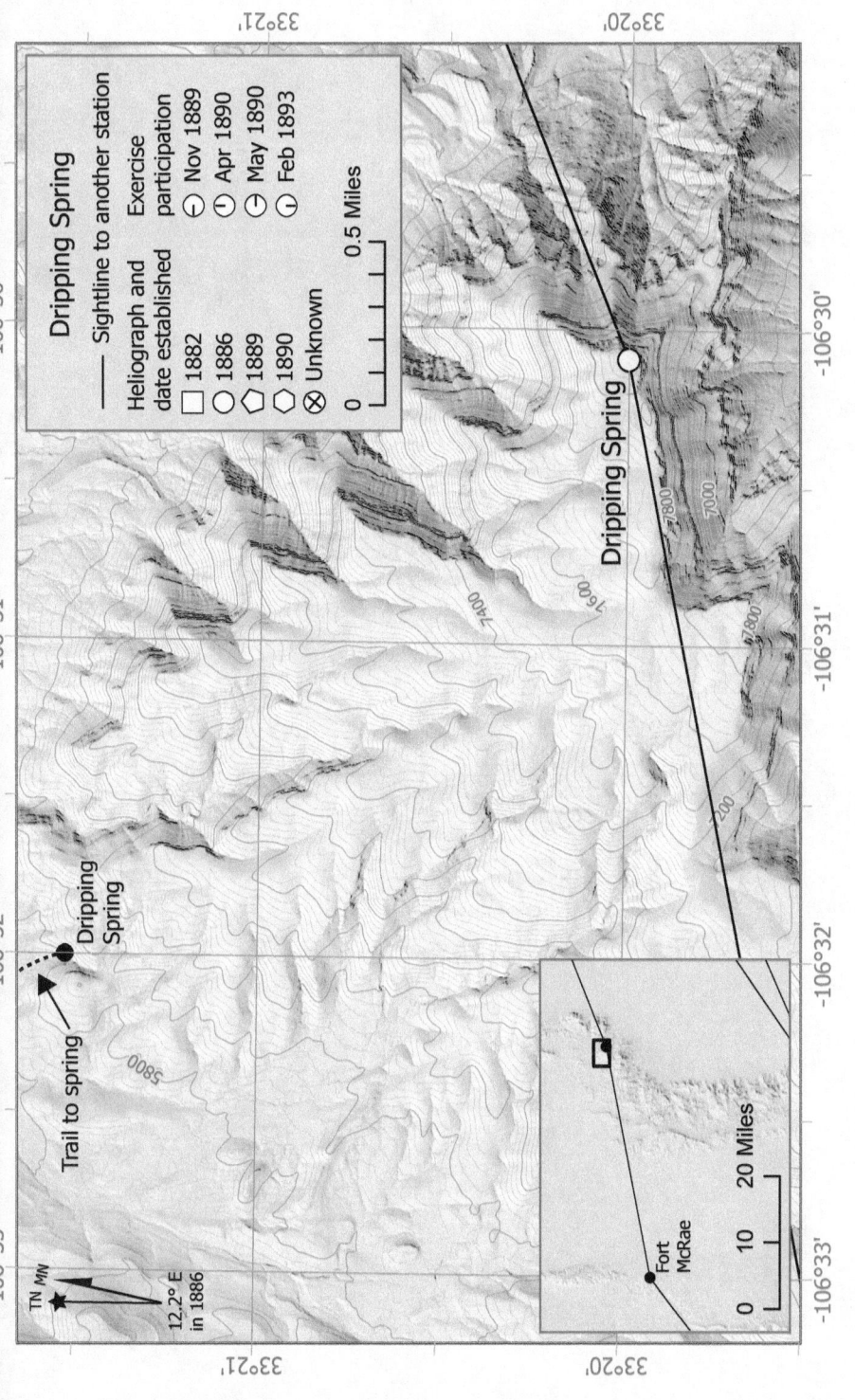

Map 40. Dripping Spring heliograph station.

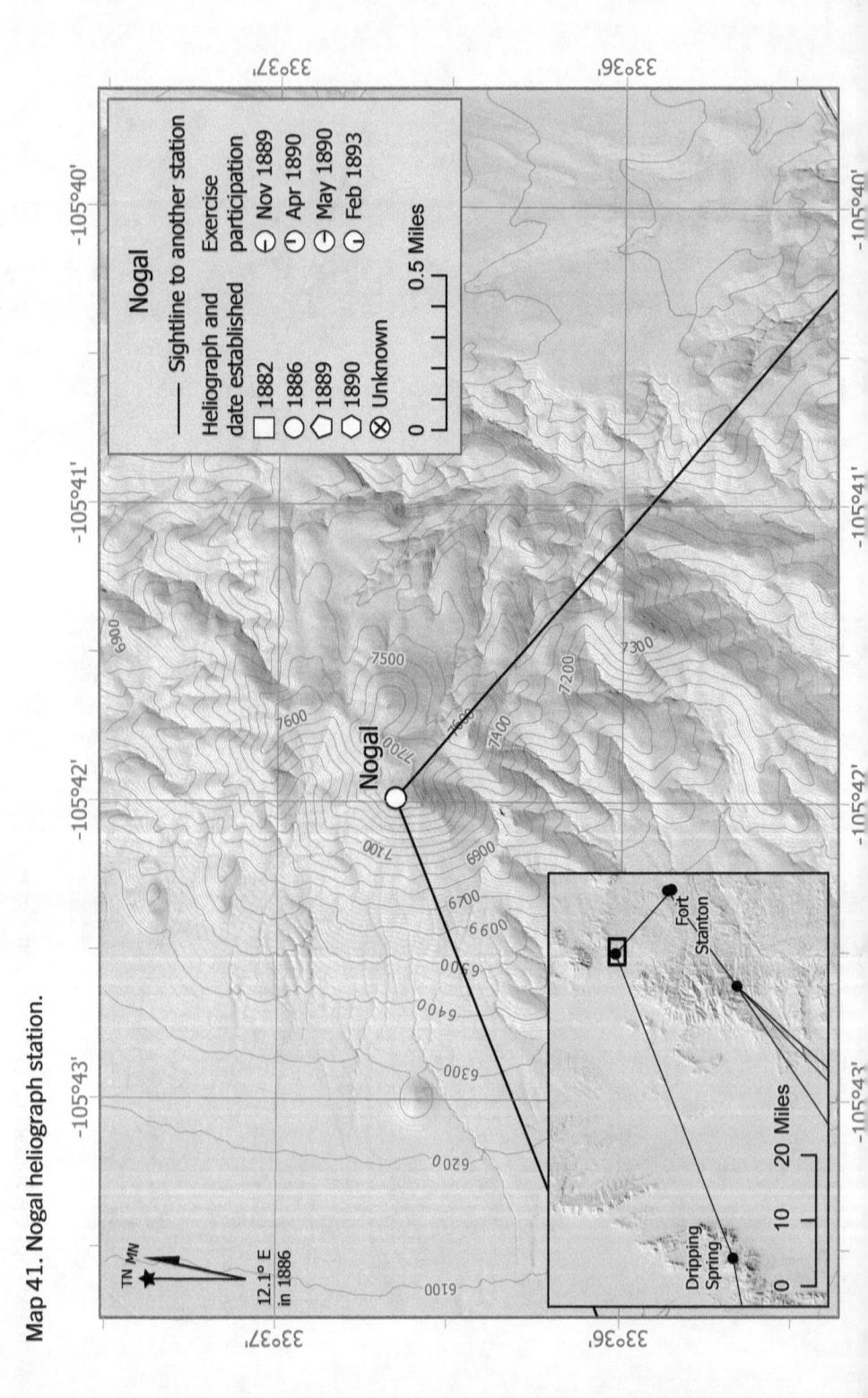

Map 41. Nogal heliograph station.

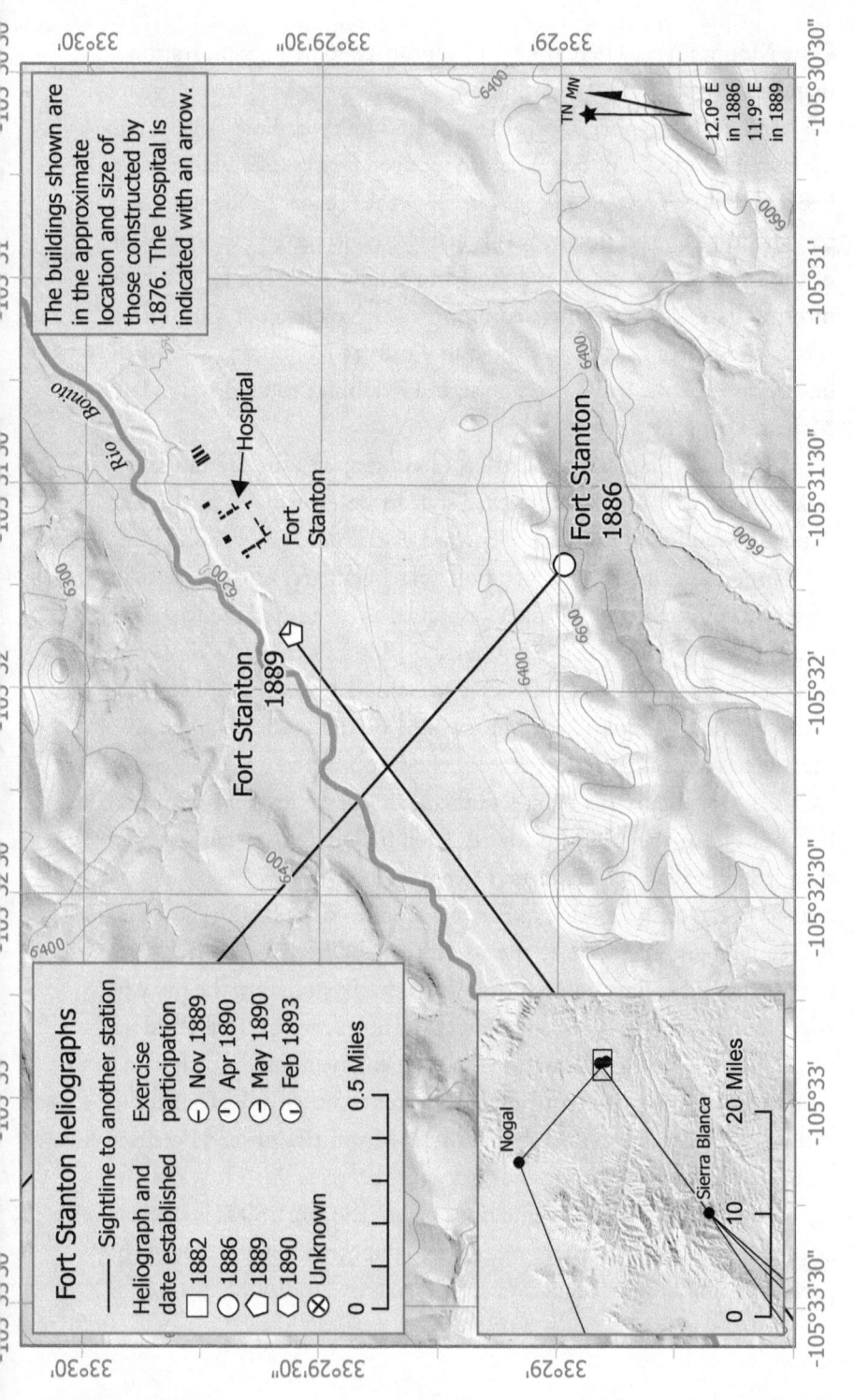

Map 42. Fort Stanton heliograph stations.

Cruz Mountain) is where the Fort Stanton 1886 heliograph station is mapped.[15]

In late November 1889, Lieutenant Hovey of Fort Bayard and Lieutenant Paddock of Fort Stanton received orders to establish a new heliograph line connecting Fort Stanton to the broader heliographic system anchored at Fort Bowie.[16] This line was constructed in anticipation of the May 1890 exercise and aimed to reduce the need for multiple intermediate stations when communicating from Fort Cummings across the Sierra Blanca and the San Andreas Mountains. This new line bypassed Lake Valley, Hillsboro, and Pershing's route through Fort McRae, Dripping Spring, and Nogal.

Paddock's plan was to position a heliograph in the Sierra Blanca, linking Fort Stanton with another station he envisioned in the San Andreas Mountains atop San Nicholas Peak. From San Nicholas Peak, he planned to connect with a station near Lockhart's Well or perhaps higher up in the Little Florida Mountains, as he needed the flashes to continue west. (This was prudent, as the 1886 Lockhart's Well station was on the plain and was not visible to San Nicholas Peak.) From the Little Florida Mountains, flashes would be directed toward Camp Henely to connect with the greater heliograph system.

To coordinate this plan, Paddock informed Hovey by mail that he would be at the 1886 Lockhart's Well heliograph station site near the Little Florida Mountains on December 17.[17] Before this, Paddock needed to establish a station at Fort Stanton to connect with the Sierra Blanca heliograph station to the southwest, a different direction than the northward connection required in 1886. Fortunately, the peak he selected in the Sierra Blanca was directly visible from the fort. Paddock positioned the heliograph station near the "engine house," which held a steam engine likely used for pumping water or other industrial applications, located approximately 0.25 miles from the post.[18] Pershing occupied this station during the May 1890 exercise.[19]

The location of the engine house from the late 1800s is unclear. Some records place the more recent (1941) power systems east of the parade grounds, near the laundress's quarters. Additionally, there is

Figure 11. View from the heliograph station atop the ridgeline about a mile south of Fort Stanton. Fort Stanton is on the right of the picture in the valley. Photo and annotations by Robert E. C. Davis.

Figure 12. View from the 1889 Fort Stanton heliograph station to the Sierra Blanca heliograph station. Photo and annotation by Robert E. C. Davis.

mention of a power system installed in 1899.[20] Through multiple field visits, I observed that it is difficult to see the Sierra Blanca heliograph station from east of the parade grounds, as buildings and trees in that area obstruct the view to the station, which sits low on the western horizon. The visibility analysis from the heliograph station in the Sierra Blanca shows that the peak is more easily seen from the side of the post closer to the Rio Bonito. Paddock recorded the direction to this station as south 40° west, or 220° magnetic, which becomes 231.9° true after applying the 1889 magnetic declination of 11.9°. He estimated the distance to this peak from Fort Stanton at 21 miles.

Based on Paddock's description, visibility from the connected peak, and my field experience, the heliograph station is mapped on the west side of the post, approximately 0.25 miles from the parade grounds' center. It is likely that Paddock established this station as he departed for the Sierra Blanca in early December 1889.

On December 4, 1889, Paddock left Fort Stanton with a dozen mounted soldiers and a small supply train, heading west into the Sierra Blanca.[21] Following Carriso (or Carrizo) Creek to its head, he made the ascent to a peak, which he noted was southeast of Sierra Blanca Peak (which he called Blanco). He established the Sierra Blanca station on this peak near the headwaters of Carriso Creek. The GIS azimuth from Fort Stanton to this peak is 232.9°, which is very close to Paddock's 231.9° true.[22] This peak's straight-line distance to Fort Stanton is measured at 18.3 miles, reasonably matching Paddock's estimate of 21 miles.

Lieutenant Andre W. Brewster, who manned this station during the May 1890 exercise, described it as a bald peak situated between the "South Fork of the Rio Ruidoso and the Carrizo Creeks," estimating the elevation to be 9,000 feet.[23] Brewster's description is consistent with Paddock's. Based on these descriptions, visibility, and azimuths, the Sierra Blanca heliograph station is mapped on this unnamed peak. Brewster later received a Medal of Honor during the Boxer Rebellion, became Pershing's inspector general during World War I, and retired as a major general.[24]

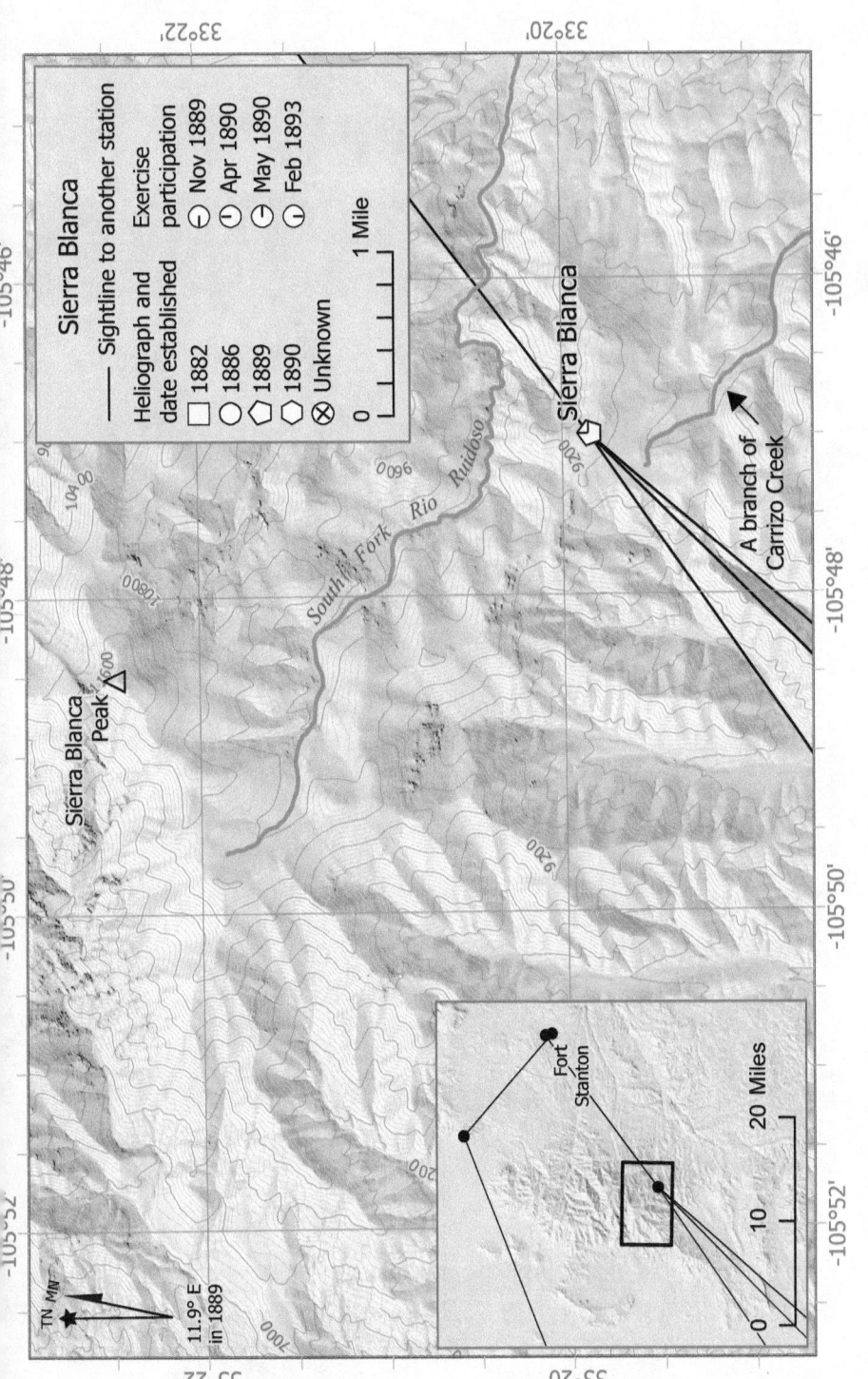

Map 43. Sierra Blanca heliograph station.

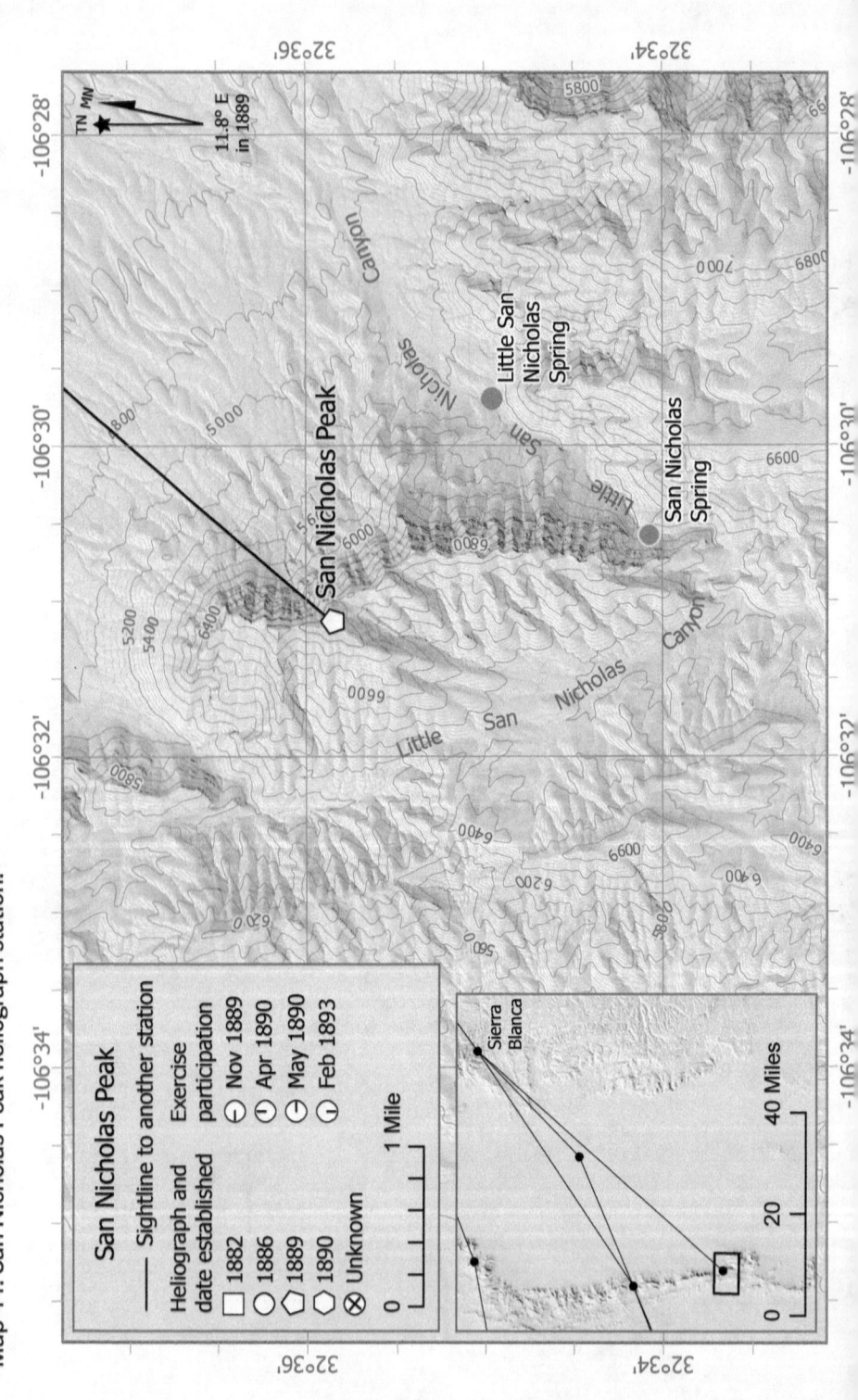

Map 44. San Nicholas Peak heliograph station.

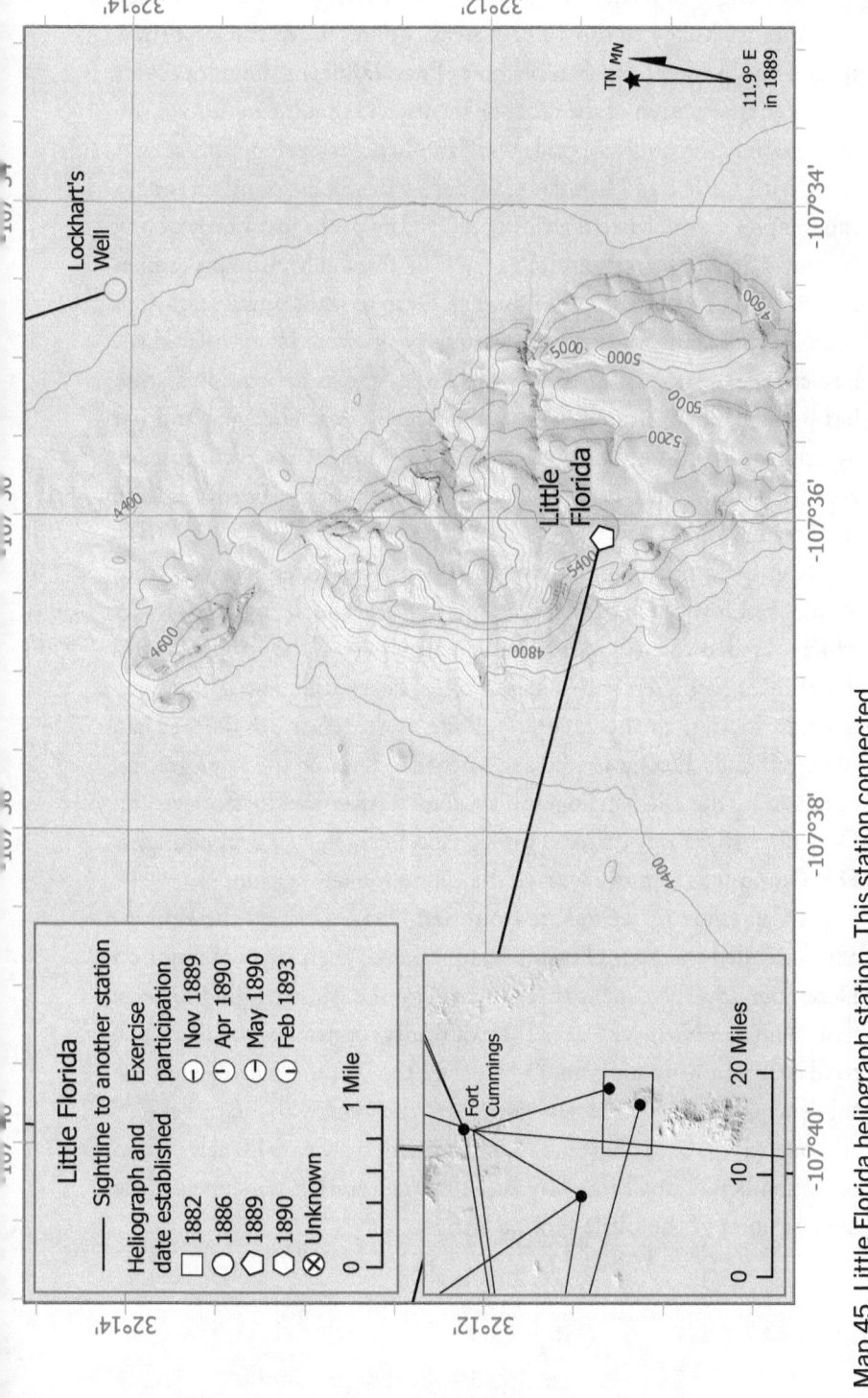

Map 45. Little Florida heliograph station. This station connected briefly in late December 1889 to Camp Henely to the west.

Leaving a detachment in the Sierra Blanca, Paddock crossed the Tularosa Valley to the San Andreas Mountains to the west (now spelled San Andres) and San Nicholas Peak. While he did not provide a detailed description of the peak, he mentioned that the mountain was in the vicinity of, and accessible via, San Nicholas Spring. This spring, along with Little San Nicholas Canyon,[25] can still be found on today's topographic maps, labeled as "Nicolas."[26] The peaks just north of San Nicholas Spring correspond to today's Big Brushy Mountains (sometimes called Black Brushy Mountains). Despite some possible hilltops south of the spring, the higher northern peaks would have offered the best chance of success. Therefore the San Nicholas heliograph station has been mapped on present-day Big Brushy Peak. Leaving another detachment on this peak, Paddock traveled to Lockhart's Well, possibly via Fort Selden, which housed only a detachment of the Twenty-Fourth Infantry in residence.[27]

On December 14, in accordance with the letter received from Paddock, Hovey left Fort Bayard with all available signal soldiers and proceeded to Lockhart's Well. Along the way, he dropped off a detachment of soldiers with a heliograph at Fort Cummings. After some difficulty locating the Lockhart's Well site, as the ranches in the area had changed hands, Hovey arrived and pitched a camp "on the same ground occupied by the 1886 heliograph station."[28] After Paddock arrived on December 18, Hovey directed the signal soldiers he had dropped off at Fort Cummings to move west to the Camp Henely station.

In an effort to see flashes from San Nicholas Peak, the lieutenants ascended the Little Florida Mountains to their highest point on December 19. They spent three unsuccessful days struggling to see a flash from San Nicholas Peak. Their difficulty connecting was likely due to adverse weather conditions.[29] However, they communicated with the soldiers now at Camp Henely, 45 miles to the west, with relative ease.[30] Although it was used only once to connect with the Camp Henely station on December 21, the Little Florida heliograph station is mapped at the highest point of the Little Florida Range.

On December 21, Paddock still could not establish communication with the station on San Nicholas Peak. Logistics were pressuring him—he was running low on rations, and his detachment nearly 70 miles away on San Nicholas Peak had been out of contact for several days. That evening, back at camp, the lieutenants conferred and chose to abandon the attempt to connect Little Florida to San Nicholas Peak. They decided to pack up and move to Fort Cummings, where they tried to formulate another plan, but without success. Paddock and Hovey ultimately decided to abandon the effort and regroup later in the season.[31]

Paddock, unwilling to abandon the effort easily, developed another plan soon after leaving Fort Cummings. He headed northeast toward Rincon, New Mexico. There, on Christmas Day 1889, he found a hill from which Fort Cummings and a likely mountain in the San Andreas were visible, where he could set up another heliograph station. He sent a letter to Hovey at Fort Bayard, letting him know he had left a detachment at Rincon and asking him to connect with it from Fort Cummings while he struck east toward the San Andreas Mountains to establish a new station there.[32]

Paddock described the Rincon station as "on a hill N. 12 1/2° W. from R. R. depôt at Rincon, N. M., and about 1 mile distant."[33] Paddock's N 12 1/2° W is 347.5° magnetic, which becomes 359.5° true after accounting for the 1889 declination of 12.0° east. From the collection of buildings shown on the 1919 GLO map, 1 mile north at an azimuth of 353.7° stands a hilltop with excellent views of both the Fort Cummings and San Andreas heliograph stations. This hilltop corresponds with Hovey's later report, in which he noted that the hill is "on the center of a long, low ridge of hills easily distinguished by being the only ones of that character in view."[34] This hilltop is where the Rincon NM heliograph station is mapped.

Paddock left Rincon for the San Andreas Mountains the next day, December 26, informing Hovey that he would reach the range in about three days. He also requested that Hovey attempt communication from Fort Cummings to both Rincon NM and the San Andreas, starting on

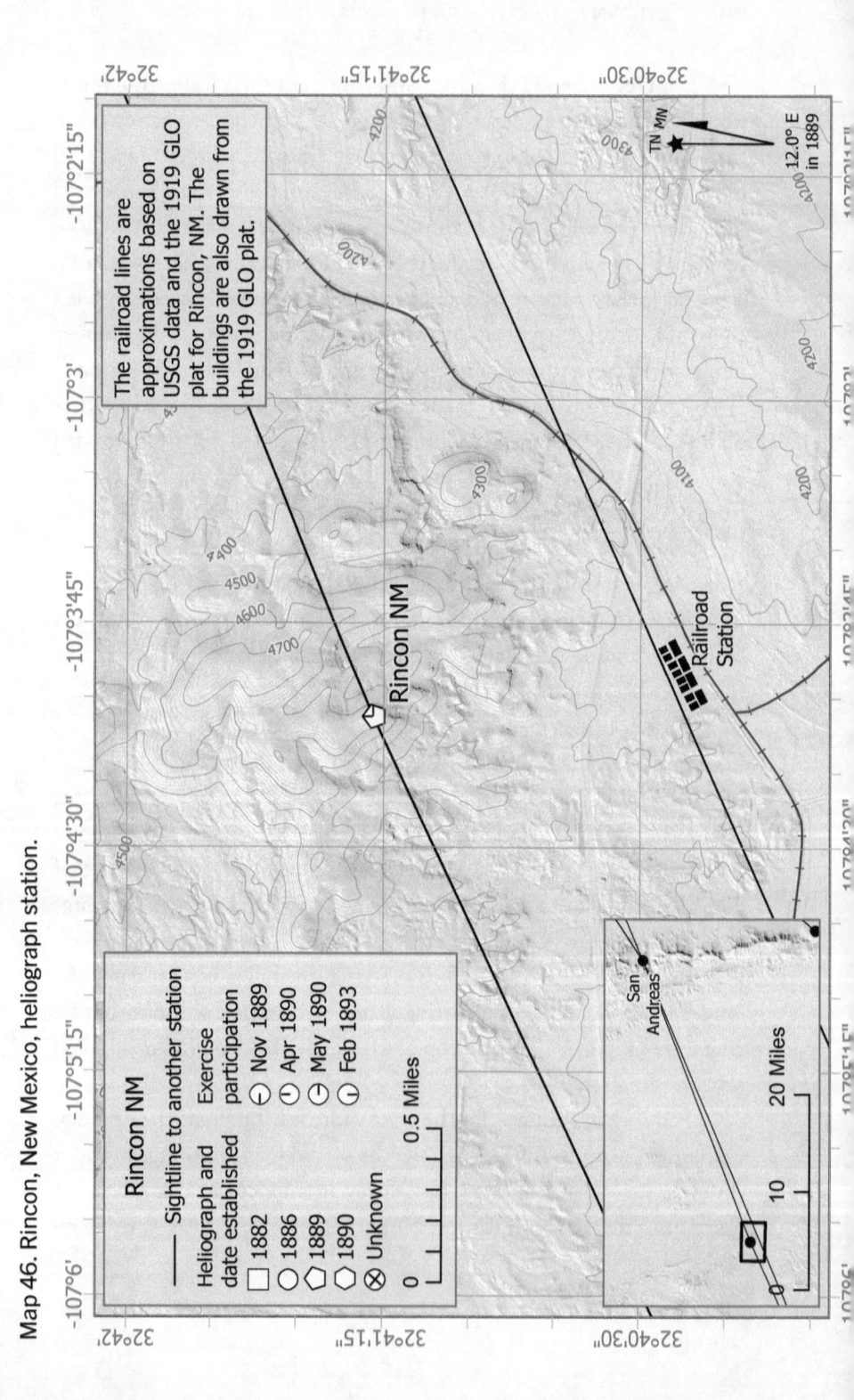

Map 46. Rincon, New Mexico, heliograph station.

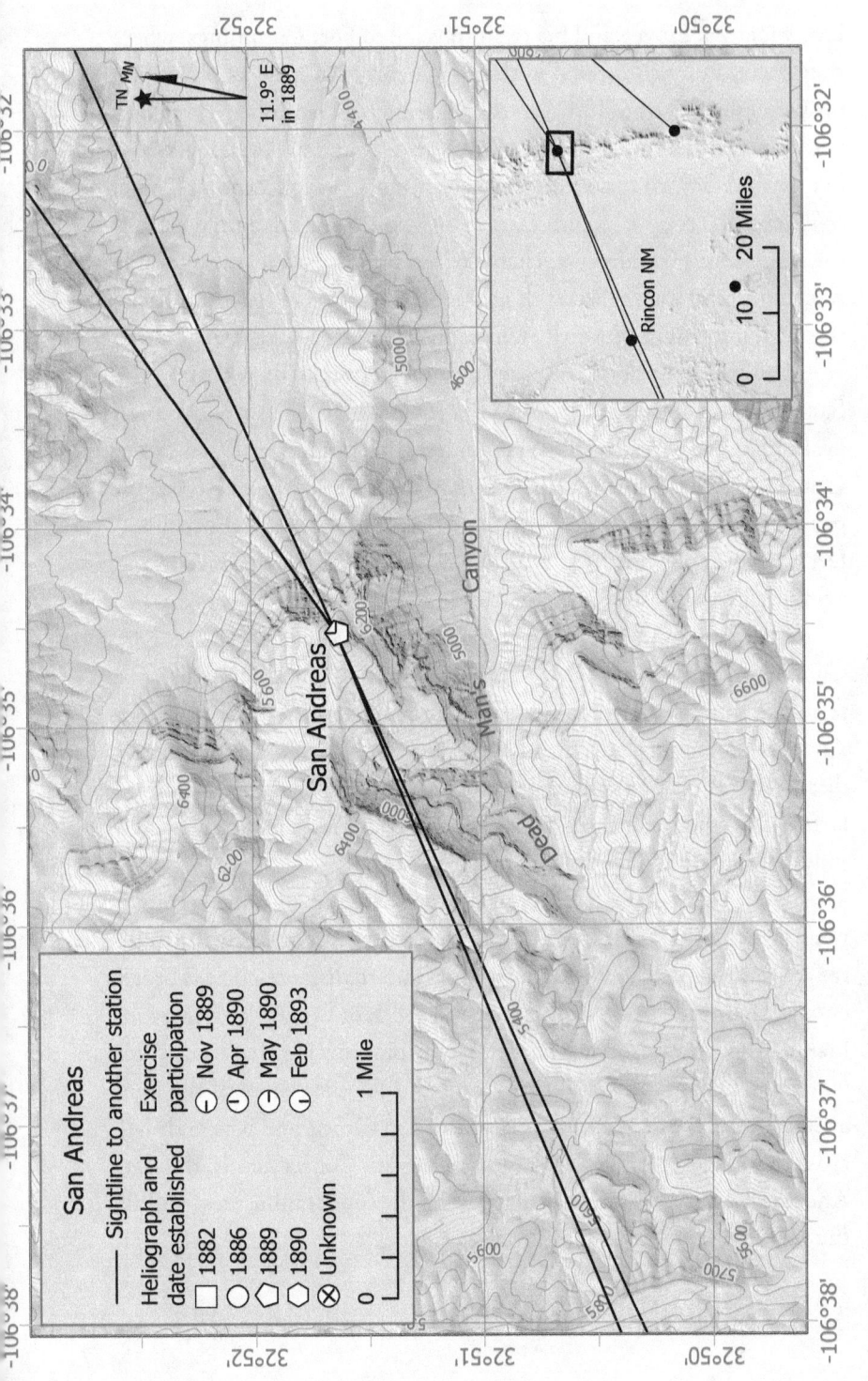

Map 47. San Andreas heliograph station.

December 29. Hovey and his team traveled to Fort Cummings, where they successfully connected with the Rincon NM heliograph station on December 29 and 30.[35] On the morning of December 31, Hovey became ill with a "severe attack of rheumatism and malarial fever" and left for Fort Bayard to seek medical attention, leaving Corporal Daniel Williams of Troop L, Tenth Cavalry—who was on the initial rolls of the black Tenth Cavalry—in charge of the station.[36] Williams was able to communicate with Paddock in the San Andreas either that day or the next, informing Hovey by telegraph on January 1, 1890.[37]

Paddock described the San Andreas station as "on a peak in the San Andreas Mountains, in Dead man's Cañon."[38] There are two high peaks along the crest of the San Andreas Mountains at Dead Man's Canyon, one to the north and another to the south of the canyon. Both peaks are visible from all the connected stations; both are above 6,400 feet. However, the northern peak aligns more closely with the reported azimuths from those connected stations.

There are four connected stations (Sierra Blanca, Tularosa, Fort Cummings, and Rincon NM). Of these four, the azimuths and distance terminal points from three cluster around the central ridge of the San Andreas Mountains, near Dead Man's Canyon. Tularosa's azimuth and distance terminal point, however, plots in the valley to the west, 9 miles farther than the other three. (Had Paddock reported the distance as 27 miles instead of 37, it would be much closer.)[39]

The weighted mean center of the four azimuth-distance terminal points (including Tularosa) plots approximately 2.5 miles west of the range's central ridge. Unfortunately, the standard distance circle is broad enough to include peaks both north and south of Dead Man's Canyon, making the mean center less effective in pinpointing the exact peak where the heliograph station was located. However, three of the four azimuth-distance lines (from Tularosa, Sierra Blanca, and Rincon NM) cross the range north of Dead Man's Canyon. Consequently, the San Andreas heliograph station is mapped on the topographic crest within the visible area north of Dead Man's Canyon.[40]

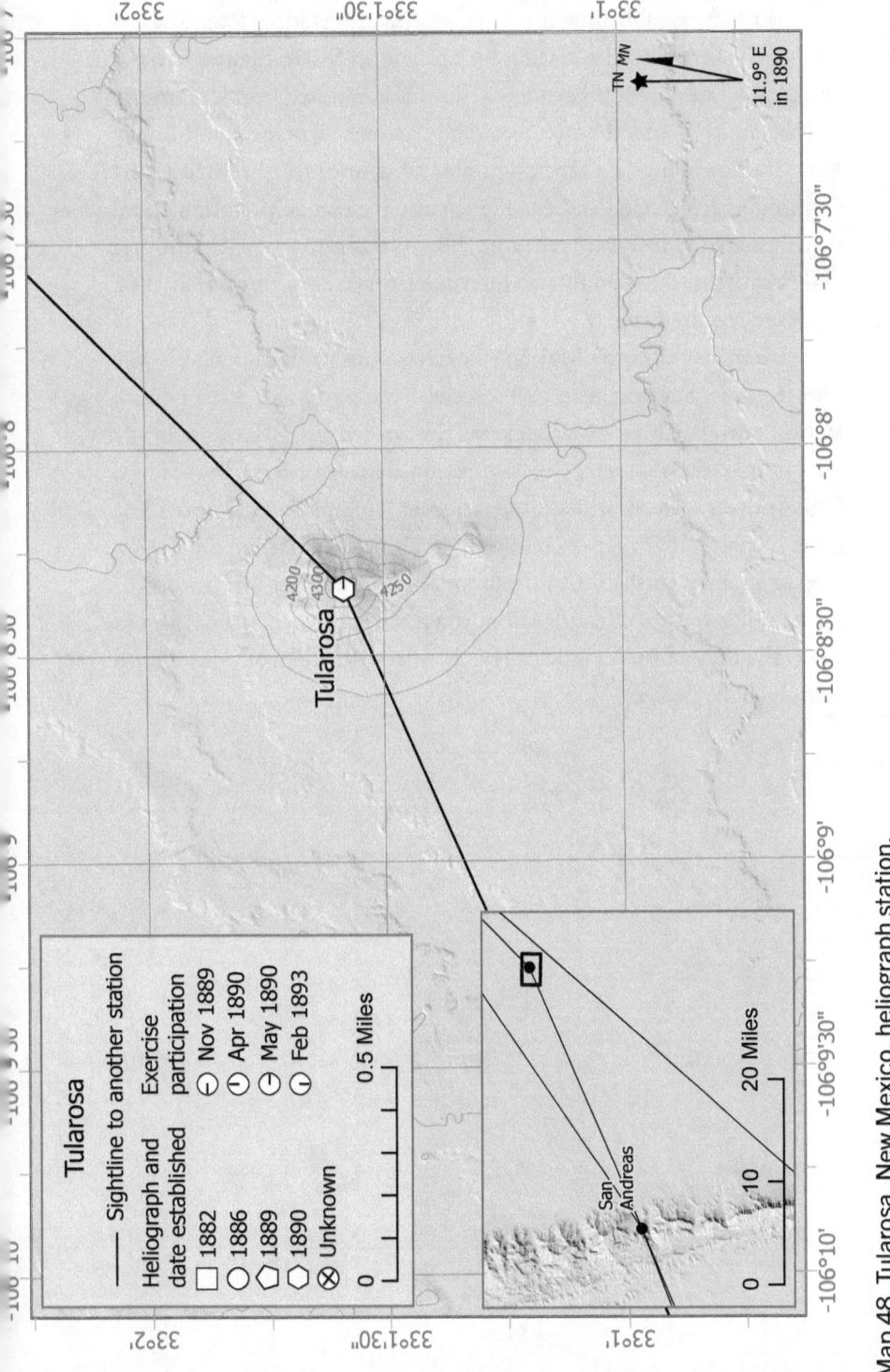

Map 48. Tularosa, New Mexico, heliograph station.

CHAPTER 7

With the establishment of the San Andreas station, Paddock was able to connect with the station he had left on Sierra Blanca, thereby completing the new southern line. This line connected Fort Cummings to Rincon NM and then to San Andreas, Sierra Blanca, and finally Fort Stanton. Since Paddock was able to connect directly with Fort Cummings from San Andreas, he felt the Rincon NM station could be eliminated.[41] If it were removed, he could connect Fort Cummings and Fort Stanton with only two intermediate stations, whereas in 1886 five were required.

Later, just after the May 1890 exercise and while making his way back to Fort Stanton, Paddock received orders to site a heliograph station near Tularosa, New Mexico. The concern was that dust in the area frequently obscured the connection between Sierra Blanca and San Andreas. An intermediate station at Tularosa would provide a more reliable path in dusty conditions. Paddock established the station atop an easily identifiable "butte about seven miles west of Tularosa."[42] Seven miles west of the railroad in town stands present-day Tularosa Peak, the only butte for miles. This is where the Tularosa heliograph station is mapped.[43]

Chapter 8

The Sierra Ancha

OLDER US GEOLOGICAL SURVEY TOPOGRAPHIC MAPS, THE SPENCER map, and GLO maps place Camp Reno along a trail or road following Reno Creek, a short distance west of Tonto Creek.[1] Heading west, the trail crosses Reno Pass and descends into the Sycamore Creek Valley, passing by Sunflower Ranch on its way south to Fort McDowell.

Lieutenant Glassford described Mount Reno as "above old Fort Reno" and "in direct line of trail from [Fort McDowell] to the Tonto Basin."[2] In April 1890 Lieutenant Overton spent four hours ascending Mount Reno from the west or northwest side, following streams and mining trails from Sunflower Ranch. From the summit, he signaled back to Fort McDowell on April 4, 1890.[3]

Lieutenant Edmond Wittenmyer, who was on loan to Fort McDowell from the San Diego Barracks specifically for the 1889 and 1890 heliograph exercises, described the distance from the Sunflower Ranch to Mount Reno as 8 miles over very rough trails.[4] Wittenmyer later commanded the Seventh Infantry Division in World War I as a major general.

Lieutenant Frank DeWitt Ramsey, an infantryman and the acting signal officer from Fort Verde, who later served in Cuba, the Philippines, and China, measured a back azimuth of 347.5°, or a fore azimuth of 167.5° magnetic, to the station at Mount Reno from Baker's Butte. Ramsey's 167.5° magnetic becomes 181.1° when the 1890 magnetic declination of 13.6° is applied. This is very close to the GIS-measured azimuth of 182.9° to today's Mount Ord.[5] The

Volkmar map and table give a direction of 169° magnetic (182.6° true) and a distance of 38 miles from Baker's Butte to Mount Reno. The GIS-measured distance is 37.7 miles.[6]

All these descriptions and measurements point to present-day Mount Ord in the Mazatzal Mountains. However, Mount Ord—and not Mount Reno—has been consistently listed as such on topographic and other maps extending back to 1880. Additionally, these same maps consistently show an "Ord Peak" about 100 miles east of the Mazatzal Mountains. To avoid confusion between the two, the army may have referred to Mount Ord in the Mazatzal Mountains as Mount Reno, based on its proximity to Camp Reno. Regardless of the differing names, the Mount Reno heliograph station is mapped at the topographic crest of what is today called Mount Ord in the Mazatzal Mountains.

Lieutenant Leonard Wood spent two days in late August 1888 on Lookout Peak in the Sierra Ancha Range, northeast of Phoenix, connecting with an unnamed heliograph operator on Mount Turnbull on the second day, August 29.[7] He described the area as "lofty mountains, covered in pine and spruce, well-supplied with water."[8] From Lookout Peak he had a clear view of Baker's Butte, Mount Turnbull, Triplets near San Carlos, Mount Graham, the Pinal Mountains, and Mount Thomas (today's Bald Mountain in the White Mountains).[9]

A year after Wood was at Lookout Peak, in September 1889, Lieutenant Overton departed from Fort McDowell, heading toward the Tonto Basin. He was scouting locations for heliograph stations. Overton's journey took him over Reno Pass, along the Tonto River to the Salt River, and then upstream to "Salamis Canyon" (likely present-day Salome Canyon). After traveling 17 miles up Salamis Canyon, he stopped for the night.

The following day, he continued for another 8 miles upstream in one of the tributaries to Salome Creek (either Reynolds or Workman Creek), arriving at the base of Baker Mountain. Overton ascended a peak that was "much higher, than anything adjacent," pointing out that

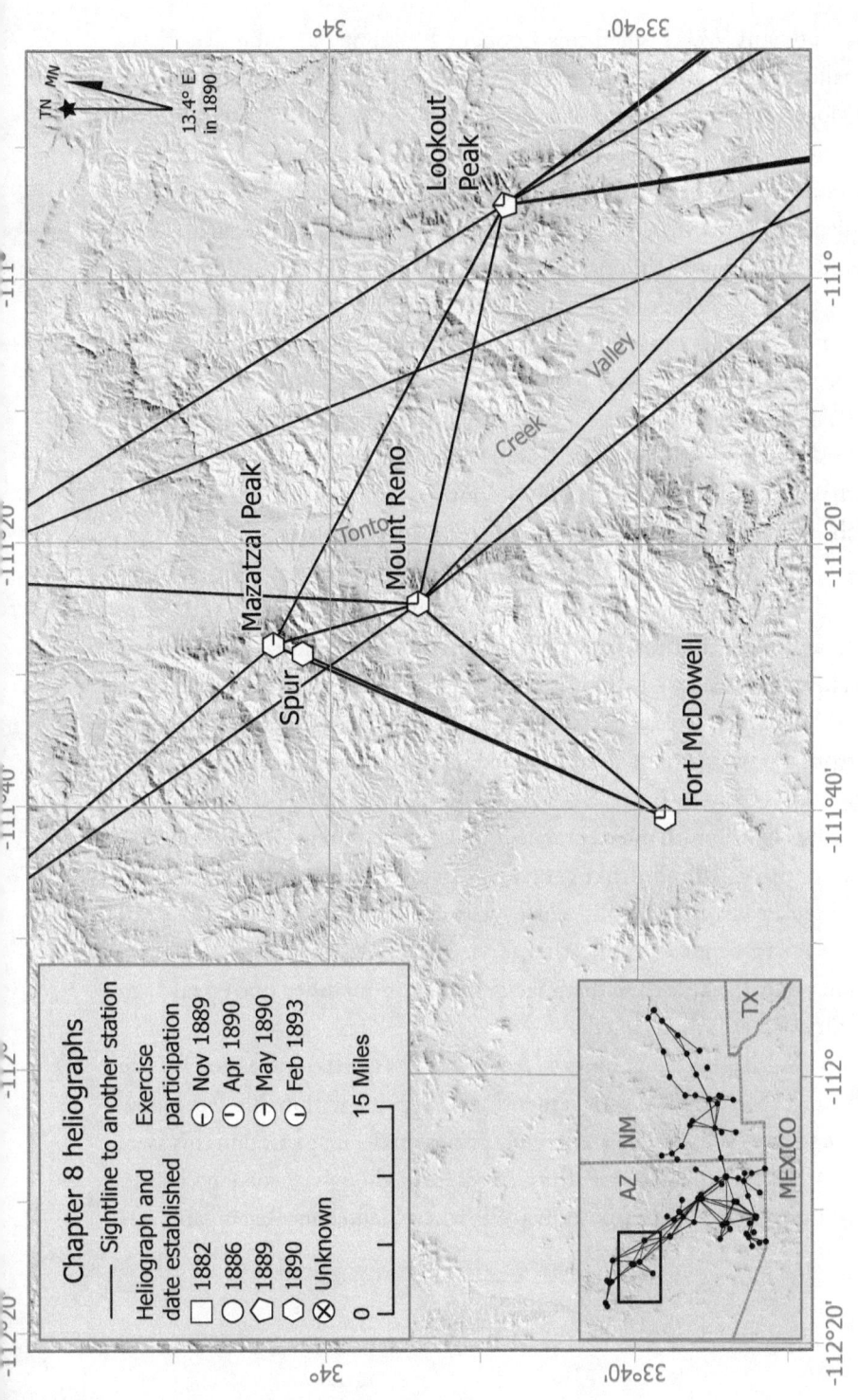

Map 49. Chapter 8 heliograph stations and connections.

CHAPTER 8

Lieutenant Wood called this Lookout Peak but clarifying that it was called Baker's Butte (distinct from Baker's Butte to the north on the Mogollon Rim).[10] From this peak, Overton was able to see Picacho Colorado, the Apache Mountains, Triplets, Mount Turnbull, Mount Reno, and Four Peaks.[11] In April and May this peak's heliograph station also connected with Saddle Mountain (Stanley Butte), Mazatzal Peak, and Baker's Butte (the one on the Mogollon Rim and east of Fort Verde).

There are four locations in this area where all those peaks can be seen.[12] These locations form a roughly north–south line about a mile long, anchored by present-day Aztec Peak at the southern end. Aztec Peak is the highest of the four (though only slightly, as Murphy Peak, less than a mile to the north, is only a few feet lower) and the tallest in the local area, including Baker Mountain, which is just over a mile to the northwest.[13]

The May 1890 exercise azimuth given between Mount Reno and Lookout Peak is 88.8° magnetic, which becomes 102.2° true after accounting for the 1890 magnetic declination of 13.4° east. The GIS-measured direction from Mount Reno to Aztek Peak is 102.6°, which is very close. Indeed, given the azimuths and distances from the peaks that would be reasonably known at this time (Mount Graham, Baker's Butte, Saddle Mountain, and Pinal Mountain) and using the azimuth and distance data provided on the 1890 Volkmar map, the weighted mean center is just over a mile away from both Aztec Peak and Murphy Peak, with Aztec being slightly closer to the mean center. (While this is visually helpful, the circle with a radius of the standard distance is large and includes both peaks, so statistically it means little.)

While this data points to Aztec Peak as the likely location for the heliograph station, Overton mentioned a spring within 300 feet below the peak.[14] While there are several springs on the maps around this area, all are farther than 300 feet from Aztec Peak, though reasonably close to Murphy Peak. Furthermore, in 1890, when Lieutenant Reichmann was

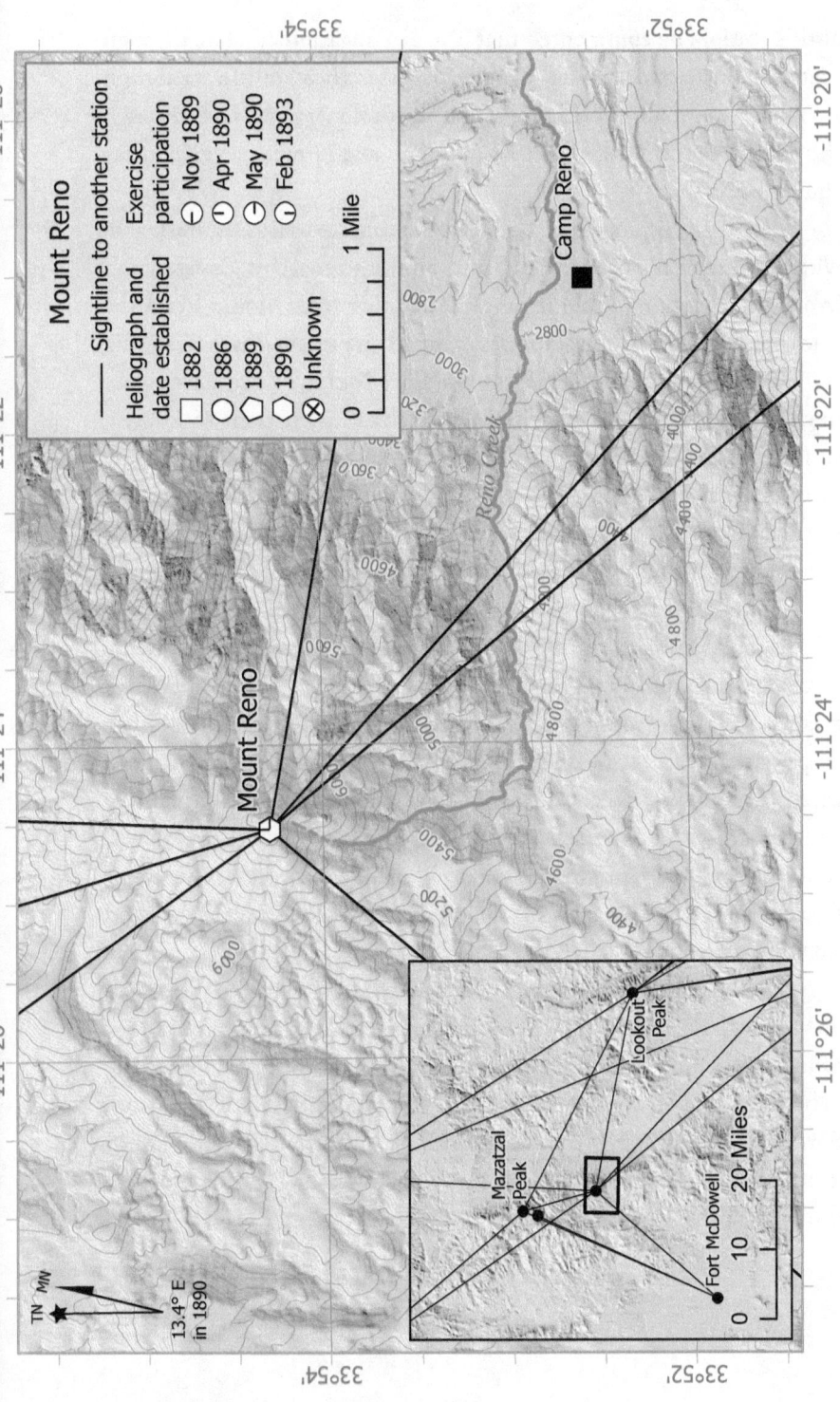

Map 50. Mount Reno heliograph station.

at this station, he commented that "the prolongation of a line through Natanes Butte and Picacho Colorado passes through the station on Lookout Mountain."[15] GIS shows that the prolongation passes through the country to the north of Aztec Peak by almost 11 miles at current-day Pine Mountain.[16]

However, despite the lack of nearby springs as well as the mismatch with Reichmann's observation, based on the strength of Overton and Wood's descriptions, visibility, and the azimuth from Mount Reno, the Lookout Peak heliograph station is mapped on present-day Aztec Peak.

On April 1, 1890, Overton left Fort McDowell to participate in the April heliograph exercise. He also had instructions to locate a summit that could link Fort McDowell, Lookout Peak, and Squaw Peak, thereby eliminating the need for both Mount Reno and Baker's Butte as stations.[17] He coordinated his efforts with the commanding officers at Fort McDowell, Fort Verde, and San Carlos, asking them to have heliographs available for communication on April 5.

It took him a day and a half in rain and snow to reach Mount Reno. Overton spent April 3 and 4 in touch with Fort McDowell and improving a camp built 200 feet below Reno's summit. On April 5, the weather was poor, and from the summit he was unable to reach any station except McDowell, which informed him that Lieutenant Wittenmyer had been dispatched to relieve him.[18] Overton was instructed not to return to McDowell but to "proceed to North Peak, or such other point for a station . . . with a view of signaling Squaw Peak, McDowell, and Lookout" directly.[19]

The North Peak listed on modern maps is just over 17 miles north of Mount Reno, line-of-sight. Visibility analysis shows that North Peak cannot be seen from Mount Reno due to the taller Mazatzal Peak lying in between (see Figure 13). Additionally, North Peak is not visible from Fort McDowell, though part of a ridge just under a mile southwest of it is. However, North Peak is visible from Lookout Peak, as much of the Mazatzal Range can be seen from there.

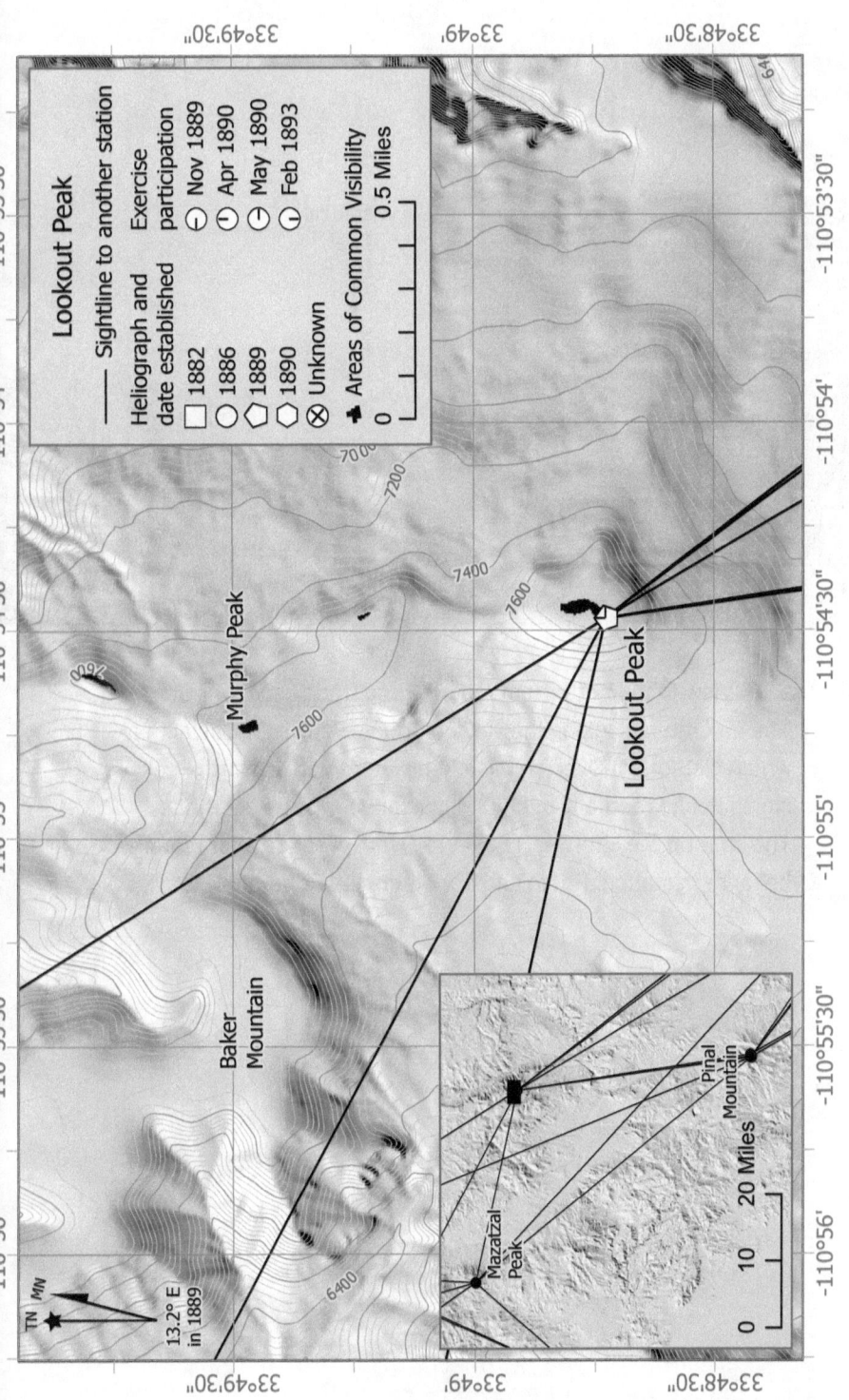

Map 51. Lookout Peak heliograph station.

Figure 13. View to the north from Mount Reno (today's Mount Ord). The mountain blocking the view to Squaw Peak is Mazatzal Peak, which Lieutenant Overton mistakenly identified as North Peak. North Peak is actually behind Mazatzal Peak and not visible. The spur on the left (west) is where Overton set up a temporary heliograph station. Photo and annotation by Robert E. C. Davis.

Nevertheless, Overton, likely frustrated with the maps at hand, notes that the field of view is "cut off" from the north to the northwest by North Peak and the Russet Hills. Overton was mistaken—Mazatzal Peak and a spur to its west were blocking his view, including the view of North Peak (see Figure 13).

From Mount Reno, Overton and Wittenmyer spent April 6–7 attempting to signal Baker's Butte without success. On the April 8, Overton concluded that Baker's Butte was invisible from Reno and left for North Peak that afternoon. On April 9, before ascending "North Peak proper," Overton positioned a heliograph on a spur somewhere between Mount Reno and North Peak.[20] From this vantage point he could see Fort McDowell, Squaw Peak, and Lookout Peak, and he exchanged messages with Fort McDowell.[21] Also on April 9, Wittenmyer was able to connect with Baker's Butte directly from Mount Reno.[22]

Since Overton does not mention Mount Reno as one of the visible peaks from the spur, several locations could satisfy the visibility criteria for Squaw Peak, Lookout Peak, and Fort McDowell. These include Mazatzal Peak, some ridges to its south and southwest, and the ridge just southwest of North Peak. Had he included Mount Reno as visible, only the areas south and southwest of Mazatzal Peak would meet the visibility criteria.

At this point, one might conclude that Overton was near North Peak, on the ridge about a mile to the southwest, where he could see Squaw Peak, Lookout Peak, and Fort McDowell. However, on the next day, April 10, Overton states that after four hours of climbing on hands and knees, he reached the summit of North Peak and connected only with Mount Reno by heliograph.[23] Since North Peak is not visible from Mount Reno, this suggests that what he referred to as North Peak was actually Mazatzal Peak and that the spur he mentioned was a ridge southwest of Mazatzal Peak (see Figure 13). On April 11, he went back to the spur and contacted Fort McDowell, receiving instructions to return to the post.[24]

CHAPTER 8

In late April Overton left Fort McDowell to participate in the May exercise, and on April 28, he "occupied Mazatzal Peak . . . a spur of North Peak proper, lying 3° west of south from it and just across the head of Deer Creek Cañon, and is distant . . . about five miles from North Peak."[25] Present-day Mazatzal Peak is located just under 6 miles south of North Peak and across the headwaters of Deer Creek when approached from the south.

During the May exercise, Lieutenant William A. Campbell, who later served in Cuba and the Philippines, noted, "2d Lieut. Clough Overton, 4th Cavalry, occupied . . . a spur of the Mazatzal range just south of North Peak, the bearing of which is north 10°30' east."[26] North 10.5° east is 10.5° magnetic, which becomes 24.0° true when adjusted for the 1890 magnetic declination of 13.5° east (at Fort McDowell). The GIS-measured direction between Fort McDowell and present-day Mazatzal Peak is 23.3° true, only a 0.7° difference.

Based on Overton's description of the location of Mazatzal Peak and Campbell's careful direction measurements, the Mazatzal Peak heliograph station is indeed on Mazatzal Peak. Furthermore, it is likely that the spur Overton connected from earlier in April was not the ridge southwest of North Peak but rather the obvious spur southwest of Mazatzal Peak. It is mapped as such.

The fort connected to this set of heliograph stations was Fort McDowell. This fort was very near the confluence of the Verde River and Sycamore Creek, about a half mile west of the Verde River, according to the 1902 GLO map.[27] This location is consistent with US Geological Survey topographic maps marked "Fort McDowell (Site)."[28] Overton, during the April exercise, described the heliograph's location only as "McDowell."[29] Lieutenant Campbell, of the Ninth Infantry, offered a bit more information during the May 1890 exercise, writing that the location of the heliograph was "just north of the garrison."[30] The heliograph station is mapped at the location marked "Fort McDowell (Site)" on the earlier US Geological Survey topographic maps.

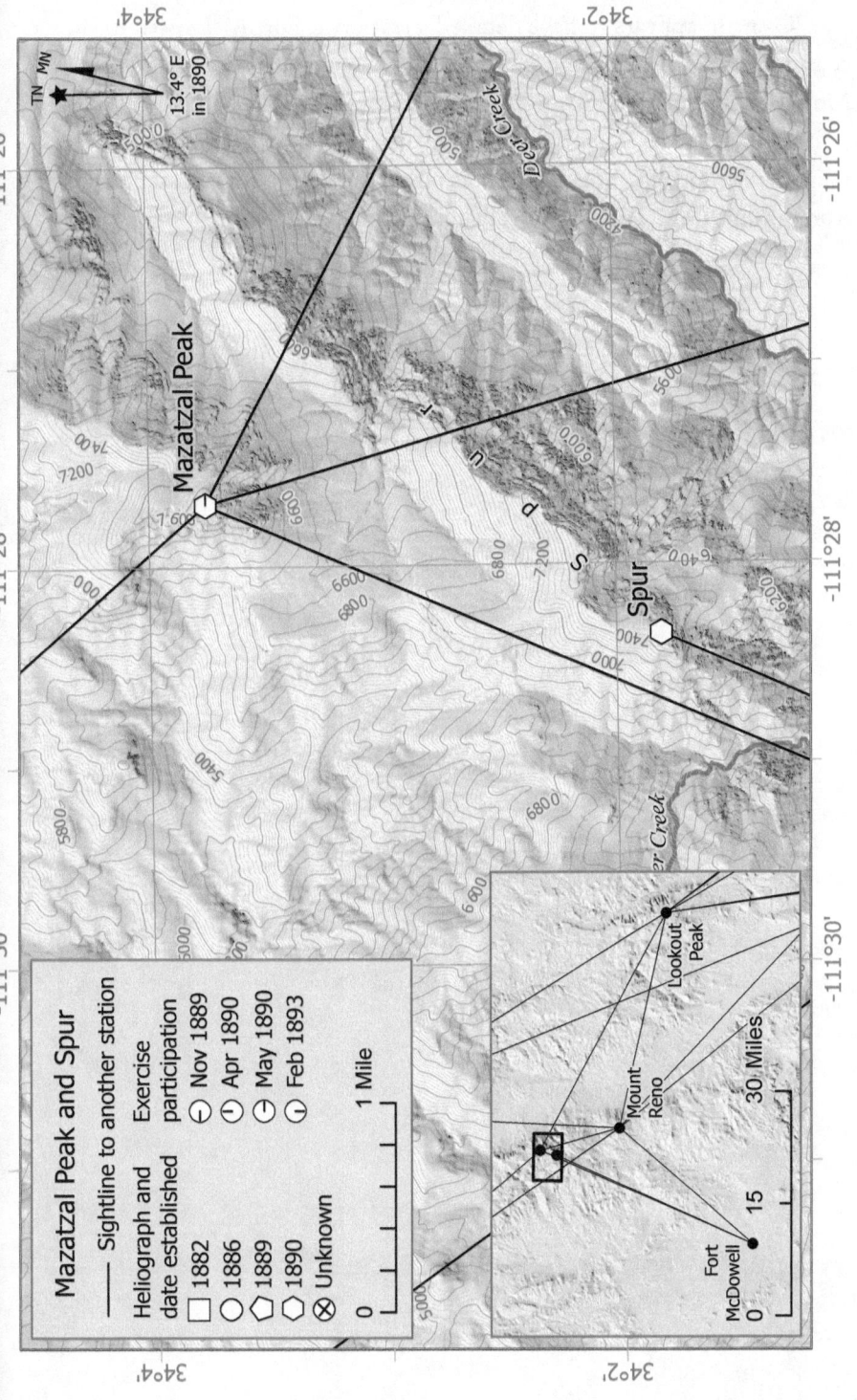

Map 52. Mazatzal and Spur heliograph stations.

CHAPTER 8

Overton appears to have detailed a station at Fort McDowell to communicate with him while he evaluated heliograph sites on and near Mount Reno during the April exercise. He left for Reno on April 1, so it's reasonable to think the Fort McDowell station was in place the same day. In any case, he communicated with the McDowell station from Mount Reno on April 3, 1890.[31]

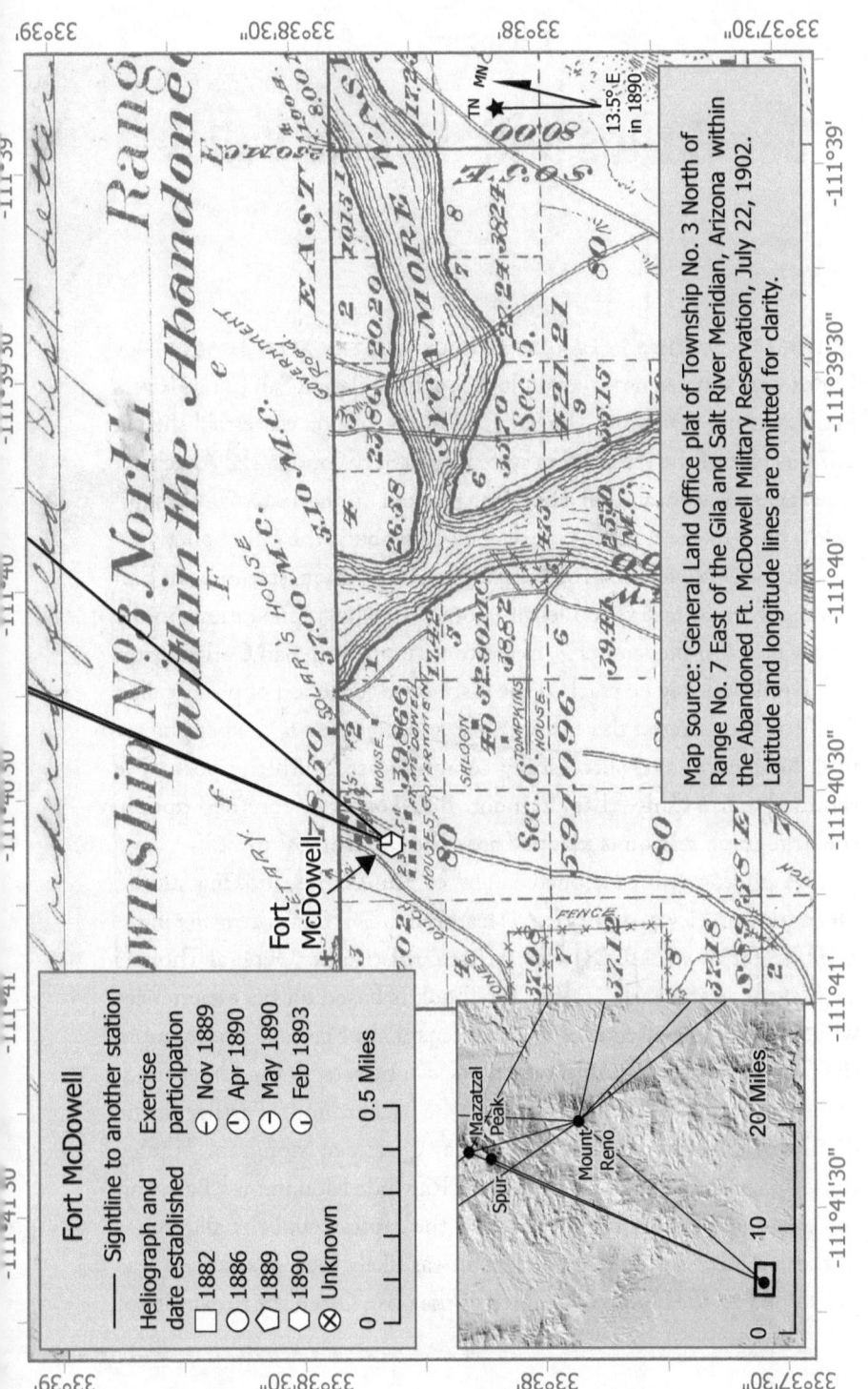

Map 53. Fort McDowell heliograph station.

Chapter 9

San Carlos and the Gila Valley

IN THE 1880S, THE SAN CARLOS AGENCY AND THE ARMY POST AT SAN Carlos were situated near the confluence of the Gila and San Carlos Rivers in the Arizona Territory.[1] The San Carlos Reservation was established in 1871 as part of the reservation system created to contain the Apaches.[2] The agency was located in a building marked on the 1916 GLO map.[3] A low bluff to the north of the agency overlooked the Gila Valley.

In early November 1889, Lieutenant Eggleston, stationed at San Carlos, received orders to identify potential heliograph stations in the Gila Valley. On November 7, Eggleston set out from San Carlos. After an 11-mile journey he reached the distinctive collection of peaks called Triplets. He selected the easternmost and highest peak, where he set up a heliograph and successfully communicated with the post.[4] He returned to San Carlos that afternoon. Based on Eggleston's description, the heliograph station is mapped atop the easternmost peak.

A photograph captioned "View of San Carlos, looking south," taken from an elevated position, is featured in *San Carlos Arizona in the Eighties: The Land of the Apache*, written by Brigadier General Thomas H. Slavens in the early 1940s. This book is based on his experiences while serving at San Carlos as a lieutenant. In Slavens's photograph, the post, consisting of numerous tents, can be seen in the foreground, while the agency buildings appear in the background.[5] Farther in the distance, the horizon reveals present-day Quartzite Mountain, Stanley Butte, Copper Reef Mountain, and the Rawhide Mountains.[6] Based on the photograph's elevated perspective, the caption, and the alignment of these mountains on the horizon, it was likely Slavens's camera was positioned on the bluff to the north of the post. Given the proximity of

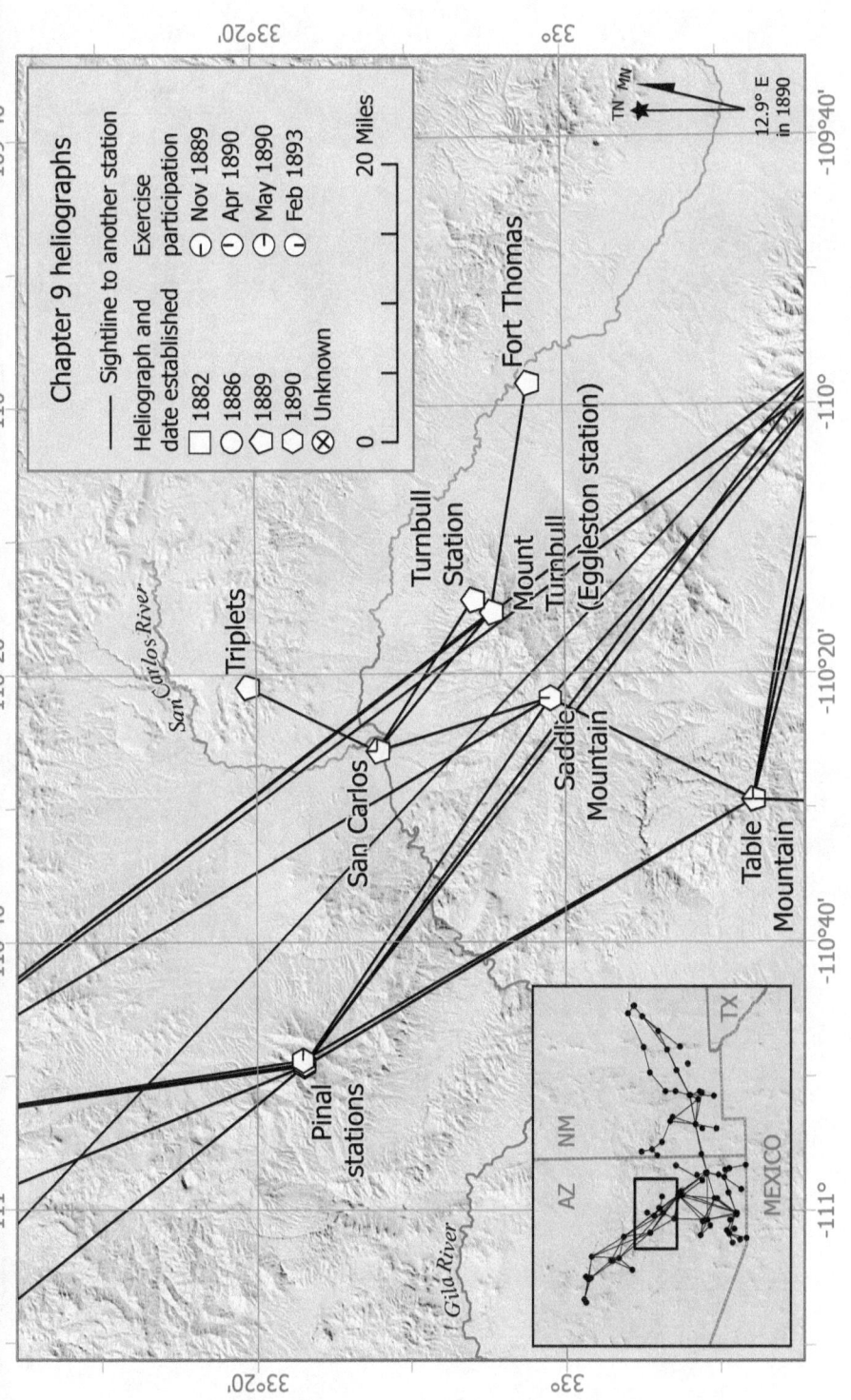

Map 54. Chapter 9 heliograph stations and connections.

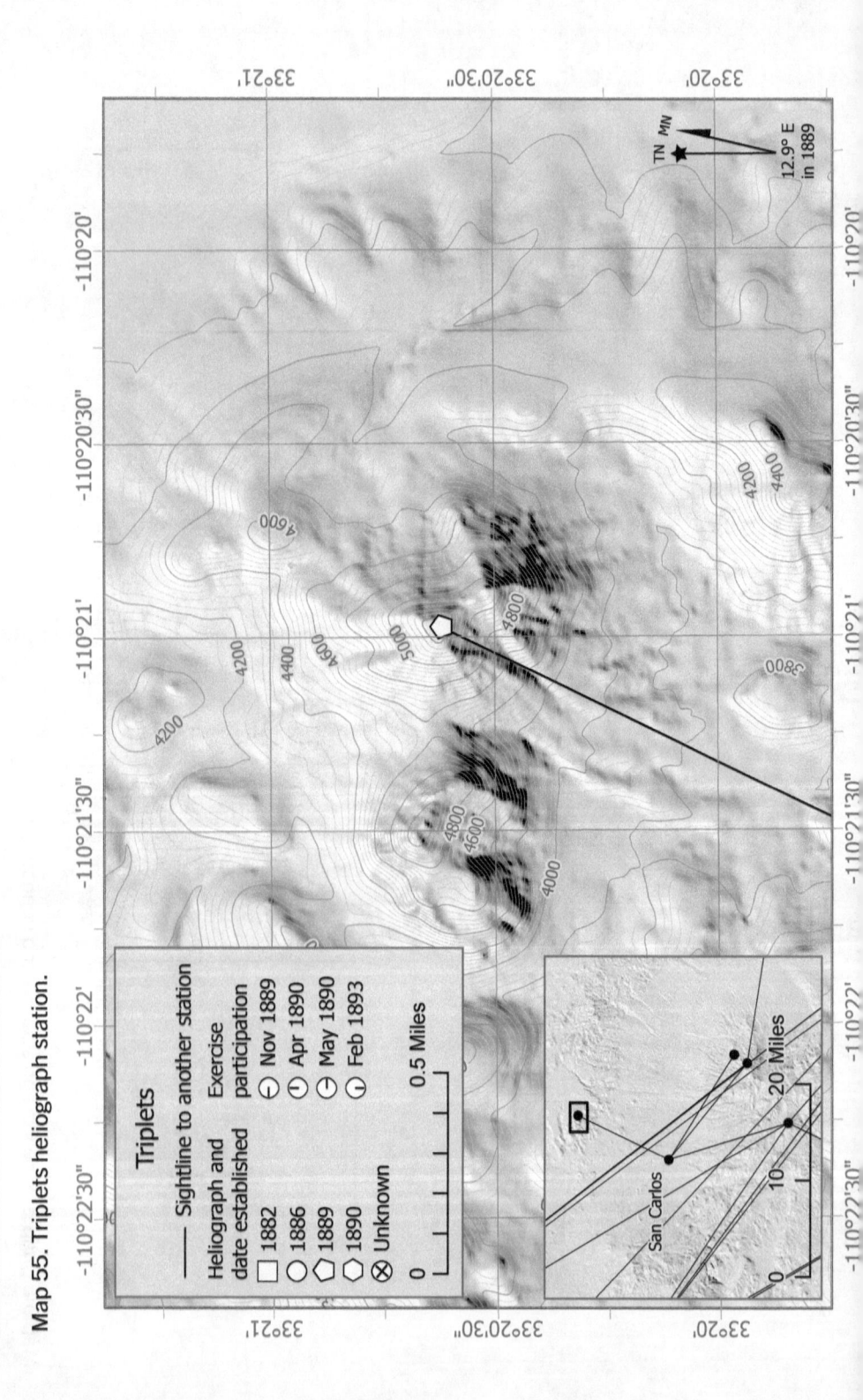

Map 55. Triplets heliograph station.

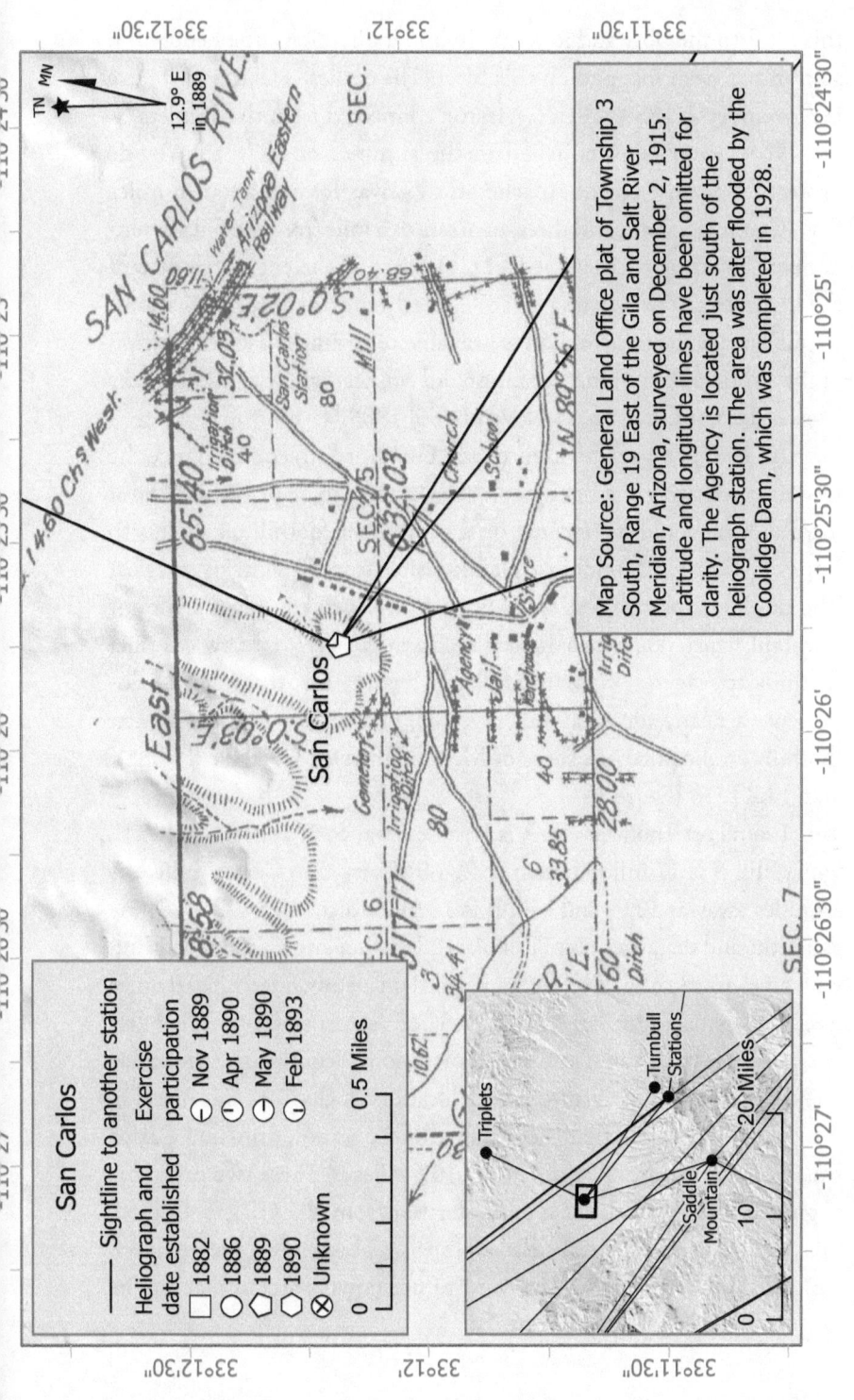

Map 56. San Carlos heliograph station.

this bluff to the post and its visibility in all directions, the heliograph station has been mapped on this bluff. The earliest account of its use is November 7, 1889, when Eggleston connected from the Triplets.[7]

Fort Thomas, established in the summer of 1876 to provide protection against recent Apache attacks, was located about 6 miles from Camp Goodwin and three-quarters of a mile from the Gila River.[8] A map drawn by Lieutenant E. D. Thomas of the engineer office of the Department of Arizona in 1877 clearly shows the fort's location.[9] While no specific information is available regarding the exact location of the heliograph station, the station did connect with Eggleston at the Mount Turnbull station on November 9, 1889.[10]

In 1887, as part of Lieutenant Glassford's reconnaissance, he recommended that if a heliograph station were to be placed at Mount Turnbull, it should be "located on a prominent foothill on the north slope of [Turnbull Mountain], plainly visible from San Carlos and Fort Thomas."[11] Looking west from Fort Thomas clearly reveals Mount Turnbull, which dominates the skyline. To the viewer's right, two distinct foothills appear to the north. Similarly, looking east from San Carlos presents a nearly identical, but reversed, view. These two prominent foothills on the northern slope of Mount Turnbull are labeled "Hill A" and "Hill B" in Figure 14.

From Fort Thomas, Hill A is 16 miles away at an azimuth of 286.1°, while Hill B is 17 miles distant at 284.9°. From San Carlos, Hill A is 14 miles away at 121°, and Hill B is 13 miles distant at 123.4°. These azimuths and distances place Hill A at a location known today as Hint, with an elevation of 5,306 feet, and Hill B on an unnamed hill 0.9 miles west and slightly south of Hill A, at 5,565 feet in elevation. Visibility analysis from both San Carlos and Fort Thomas confirms several areas on Mount Turnbull visible from both locations, including Hills A and B.

Glassford stated that the distance of the station from San Carlos was 13 miles and from Fort Thomas, 17 miles.[12] These two distances align precisely with the distances derived from the GIS for Hill B. Glassford also noted that a clear cool spring "pours into a rocky cañon" a mile to the southeast.[13] A canyon (of unknown water availability or

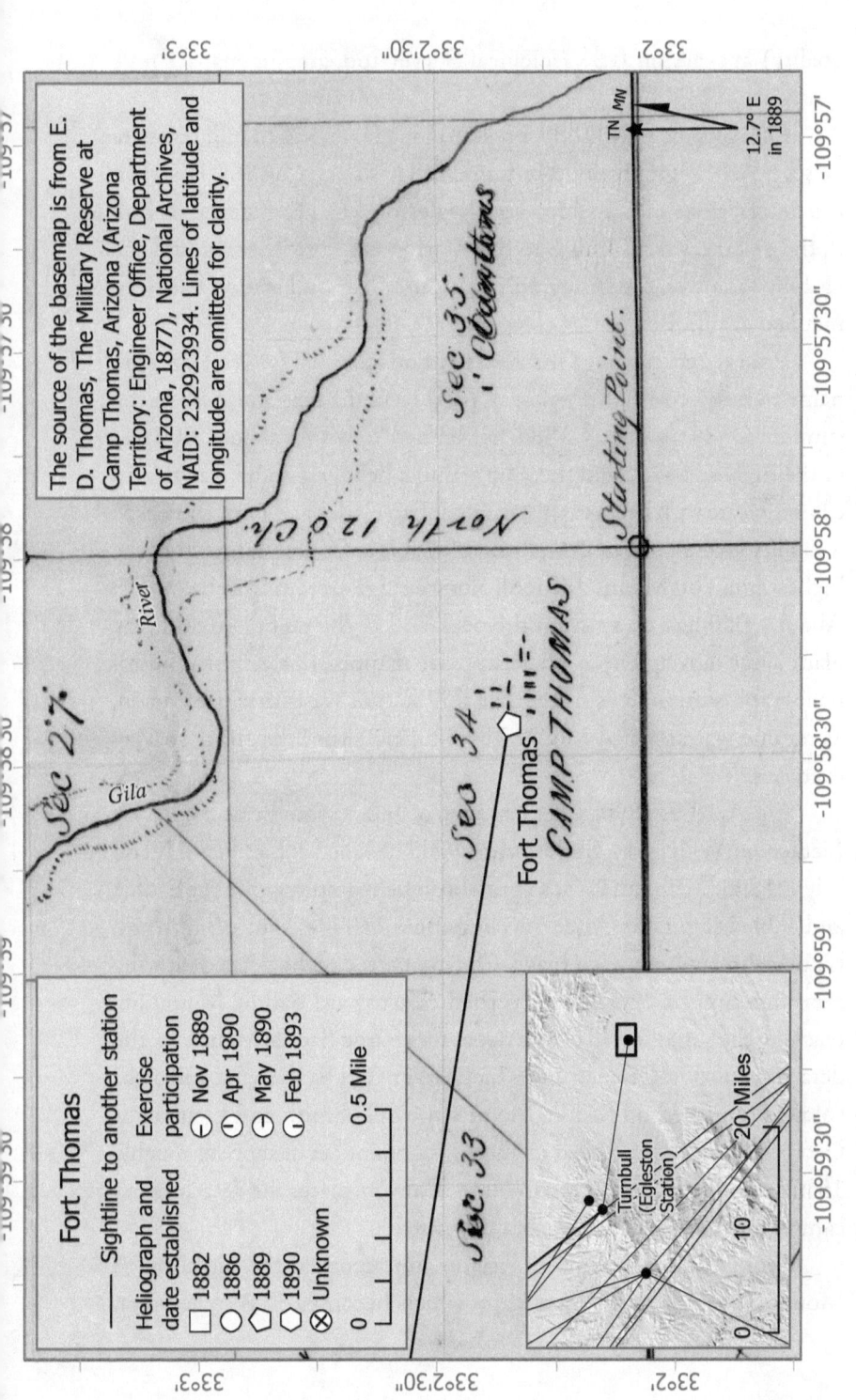

Map 57. Fort Thomas heliograph station.

quality) appears on US Geological Survey topographic maps 1 mile southeast of Hill B.

On November 17, 1889, Eggleston made his way to Hill B. There he set up a heliograph and communicated with San Carlos.[14] Based on the descriptions of Glassford and Eggleston, the prominence of Hill B, the distances from Hill B to San Carlos and Fort Thomas, and the visibility analysis, the heliograph at Mount Turnbull station has been mapped at Hill B.

Prior to reoccupying Glassford's station on Hill B (which Eggleston named Glassford Hill), Eggleston tried to find a station closer to the summit. On November 9, 1889, he climbed "to within about 1,000 feet of the highest point," and there he set up a heliograph and was able to communicate with heliograph stations at San Carlos and Fort Thomas.[15] Visibility analysis shows a thin band of visible areas descending from the highest point on Mount Turnbull along a ridge extending to the north. About 1,000 feet down from the peak and in the visibility band (the black areas in Map 54) is where we have mapped this very temporary heliograph station. It is likely that this station was never used again, as no one was assigned Mount Turnbull for either the April or May exercises.[16]

On April 6, 1890, a detachment of infantry and cavalry, led by Lieutenant William H. Smith, who would be killed at San Juan Hill on July 1, 1898,[17] left San Carlos to establish heliograph stations at Saddle and Table Mountains. After traveling about 15 miles south, Smith and his detachment camped in Hawk Canyon for the night.[18] The following morning, Smith's detachment turned east toward Saddle Mountain, reaching the summit well before noon. Once atop Saddle Mountain, the detachment split: German-born Lieutenant Carl Reichmann and eight soldiers remained on Saddle Mountain, while Smith and Lieutenant George E. Stockle proceeded to Table Mountain.[19] A distance of roughly 15 miles south of San Carlos, in Hawk Canyon, places the detachment's camp directly west of today's Stanley Butte.

From Saddle Mountain, Reichmann recorded the direction to Mount Turnbull as 44° magnetic, which becomes 56.8° true when

Figure 14. Photographs of Mount Turnbull. The top image shows the view from Fort Thomas (eastern perspective) while the bottom image shows the view from San Carlos (western perspective). "A" and "B" mark prominent foothills on the north-facing slope. Photographs and annotations by Robert E. C. Davis.

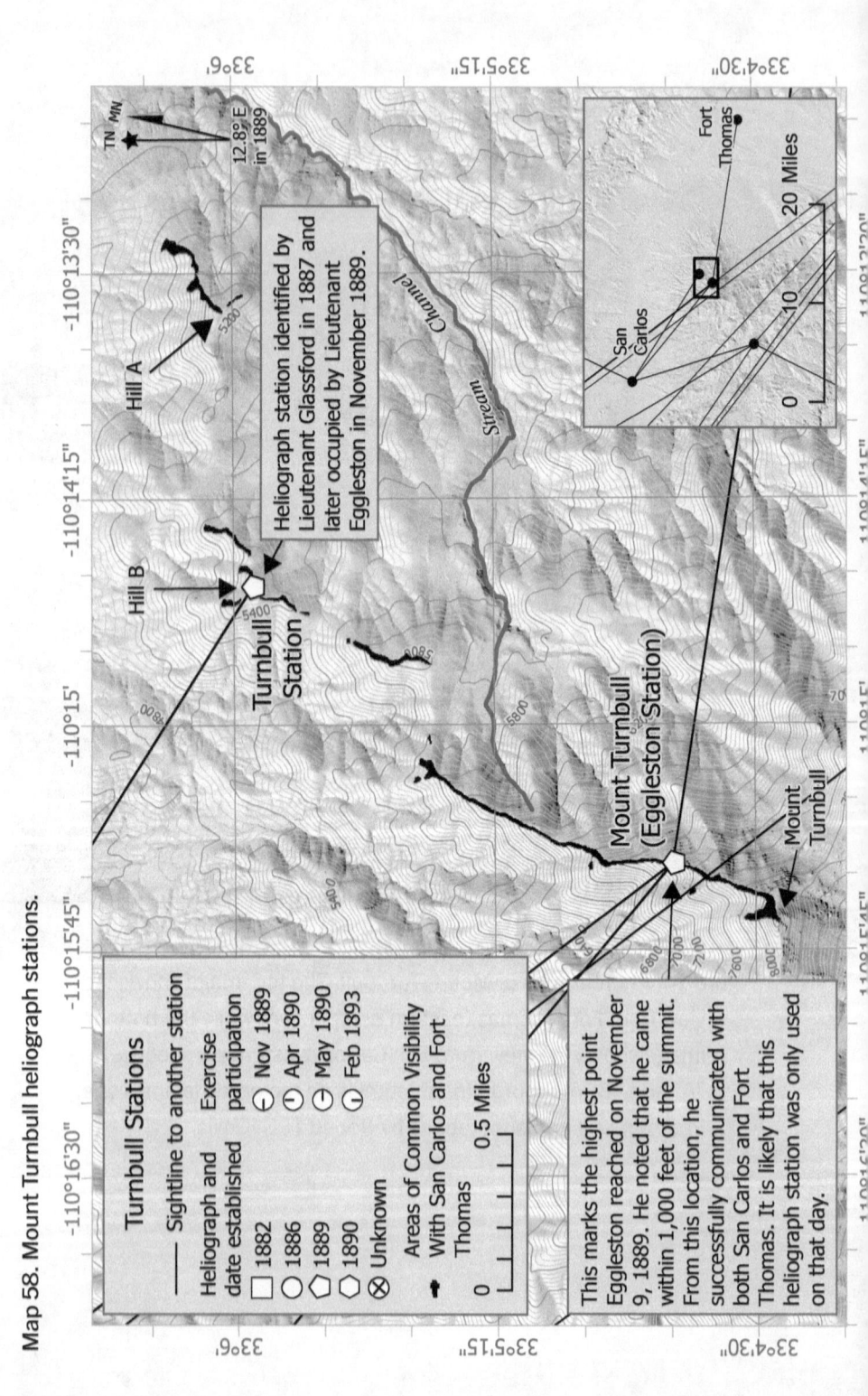

Map 58. Mount Turnbull heliograph stations.

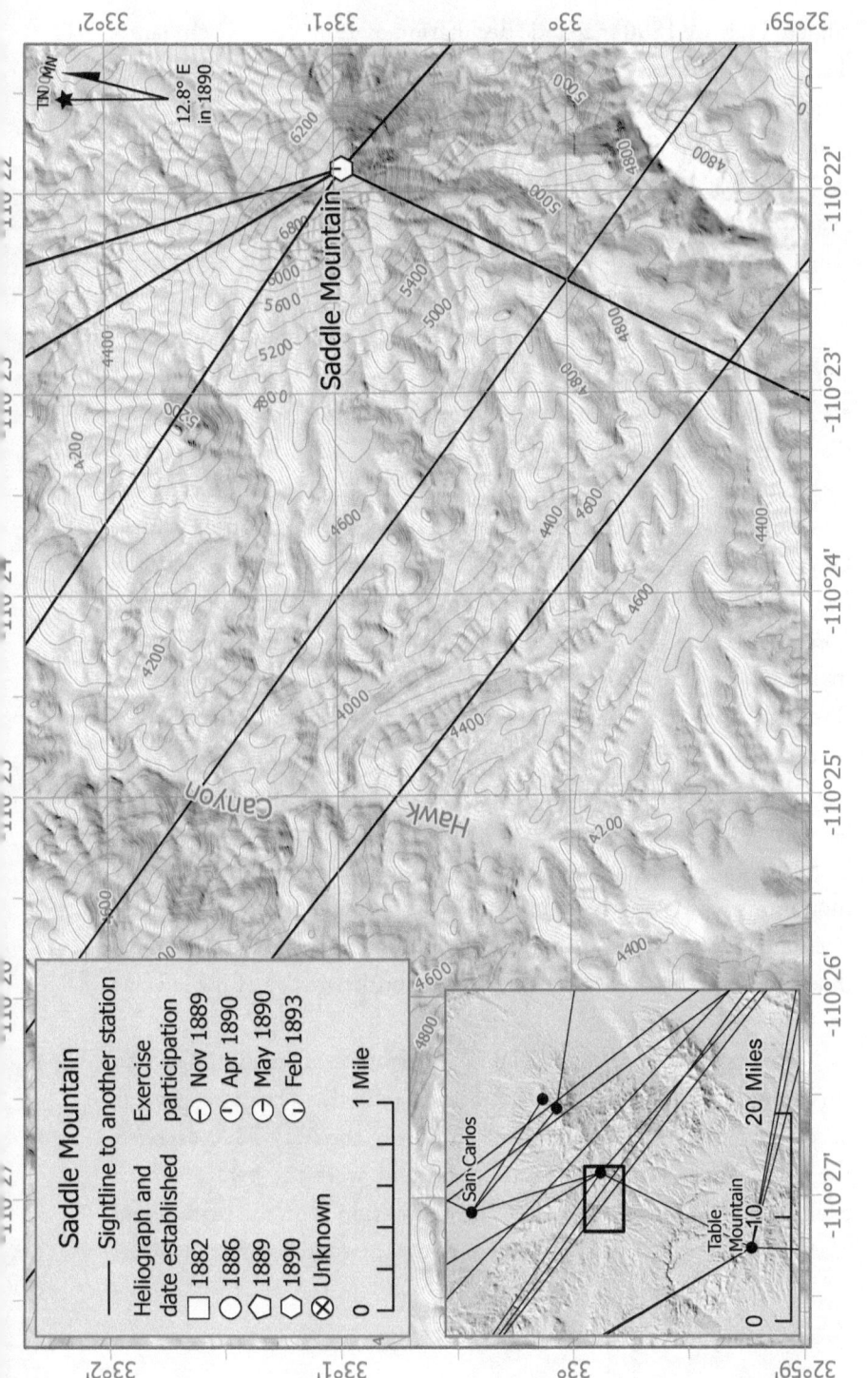

Map 59. Saddle Mountain heliograph station.

adjusted for the 1890 magnetic declination of 12.8° east. Reichmann's direction is very close to the GIS-measured direction from today's Stanley Butte of 56.2° true. Moreover, the GIS direction from Stanley Butte to San Carlos is 343.5°; Reichmann's recorded direction was very close at N 28° W, or 344.8° true in 1890.[20] Based on Reichmann's and Smith's descriptions and the close alignment with the measured directions to San Carlos and Mount Turnbull, the Saddle Mountain heliograph station is mapped atop the summit of what is now Stanley Butte.[21]

Before Smith's detachment left San Carlos, Lieutenant Dade, stationed at Fort Grant, received orders in early December 1889 to establish a heliograph station on Table Mountain. Dade, a cavalryman who later served in Cuba, the Philippines, and the Mexican Punitive Expedition,[22] reached the summit of Table Mountain on December 9 and successfully connected with the heliograph station at Mount Graham (later identified as the Alpina station in the Pinaleños). Dade's description placed the heliograph station at the southern end of a horseshoe-shaped ridge, which he described as "the point or nose of the mountain, which is also the toe of the shoe."[23] In April 1890, Lieutenant Smith, stationed at Table Mountain, confirmed the accuracy of Dade's report from December 1889.[24]

Additionally, Lieutenant Reichman, viewing Table Mountain from his perch at Saddle Mountain commented, "Table Mountain . . . cannot be mistaken. At the northern end of the Galiuro Range there are two mountains which for their shape, as seen from Saddle Mountain, no one could call anything else but Table Mountains. The bearing is almost due south and the station is nearly at the southern edge of the left one of the two mountains."[25]

Moreover, Lieutenant William T. Littlebrant, an infantryman who later served in Cuba and the Philippines and retired as a brigadier general, occupied the Table Mountain station during the May 1890 exercise. He simply stated that the Table Mountains were the "two northern peaks of the Galiuro Mountains, and the station is on the eastern end of the eastern peak."[26] The alignment of these two hills runs northwest

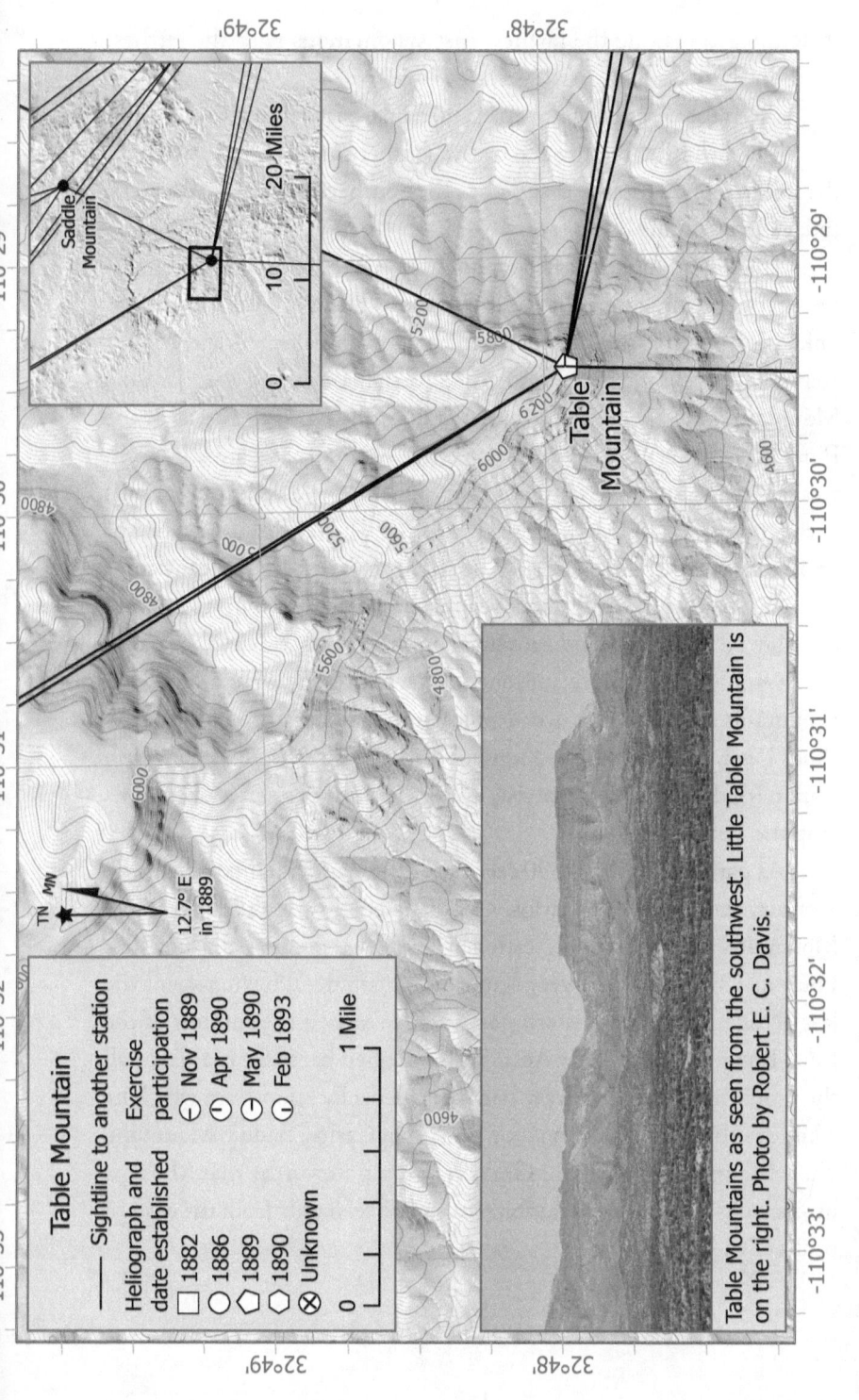

Map 60. Table Mountain heliograph station.

to southeast, making the farthest east synonymous with the farthest south in this case.

Given the descriptions by the lieutenants, particularly Reichmann's, as well as the visibility provided by the southernmost of these two hills, the heliograph station is mapped at the southern (or eastern) end of today's Little Table Mountain.[27]

The Pinal Mountains heliograph station was an important connection between the Gila Valley and the Tonto Basin.[28] This station was linked to six other stations: three to the north (Baker's Butte, Mount Reno, and Lookout Peak) and three to the south and east (Saddle Mountain, Table Mountain, and Mount Graham).[29] The areas of the Pinal Mountains visible from these other stations are generally (from west to east) a small area atop today's Signal Peak, a collection of hills at the top of Six Shooter Canyon (including Pinal Peak), and a line of small hills along the eastern ridge adjacent to Pinal Peak. Mean center analysis of the azimuth-distance vectors, based on data reported during the May 1890 exercise and contained on the Volkmar map, shows that the mean center of those stations connected to the May 1890 Pinal Mountains station is just over 1 mile south and slightly east of today's Pinal Peak.[30] As will be seen, while Pinal Peak was the heliograph station used for the May exercise, it was the third in a series of stations atop the Pinal Mountains.

As part of the April 1890 exercise, Lieutenant Cunliffe H. Murray, a cavalryman from San Carlos, was dispatched to evaluate the Pinal Mountains for a heliograph station. Murray had graduated from West Point in 1877 and was a very senior first lieutenant, having spent the last decade at various western posts.[31] He reached the summit of the Pinal Range on April 4. By April 5, after he had "carefully examined all the prominent peaks," Murray had identified a location from which he could see the stations at Lookout Peak, San Carlos, Saddle Mountain, Table Mountain, and Mount Graham.[32] It is noteworthy that Murray's list includes San Carlos, a station that can be seen only from the eastern peaks of the range.

Visibility analysis from those five connecting stations reveals common visible areas near today's Pinal Peak, some spots about 0.2 miles east of Pinal Peak, and two places along the north-south-running ridge (the eastern ridge) about a half mile northeast of Pinal Peak. Murray also reported the station was "nearly a mile and a half southeast of the Bremen saw-mill."[33]

In the late 1800s, Bremen's sawmill was a large structure measuring perhaps 50 feet wide and 30 feet deep. There was an iron chimney and a track extending from the mill downstream. Behind the structure was the head of a box canyon with very steep slopes in all directions.[34] The 1902 US Geological Survey topographic map of this area shows buildings at the head of both Icehouse and Six Shooter Canyons.[35] According to a docent at the Gila County Historical Museum, Bremen's sawmill was near the head of Icehouse Canyon (See Map 61).[36]

All three of these common visible areas are southeast of and about 1.3 miles from the head of Icehouse Canyon. Murray also stated that his station was a quarter mile southwest of the potato ranch of a Mr. James. Unfortunately, the locations of the various farms along this ridge are lost to history. Nevertheless, the visibility analysis alone confines the possible locations to Pinal Peak and some nearby hills to the east and northeast.

After the April exercise, Lieutenant Robert D. Read of the Tenth cavalry was assigned by Captain Murray (Lieutenant Murray was promoted to captain on April 15, 1890) to post himself at the heliograph station on Pinal Peak for the May exercise.[37] Read loaded his personal buggy with camping gear and left San Carlos for Globe, anticipating that Lieutenant Henry C. Keene Jr., an infantryman who later served in Cuba and the Philippines and medically retired in 1906, already stationed on Pinal Peak, would send mules to Globe to collect him. However, Keene sent the detachment of mules and men by a route that bypassed Globe, completely missing Read.

Read spent the night and the following morning waiting in Globe. Realizing something had changed, he drove his buggy up the northern slope of the Pinal Mountains to Bremen's sawmill. Leaving his buggy and camping gear there, he hiked to the ridgeline above, which he recorded as being 750 feet higher in elevation and approximately three-quarters of a mile away. From there, he followed a trail for another mile and a quarter until he reached the camp of the signal party, where he found Keene.[38]

Keene's camp was situated on a 4-acre plot that Read identified as Pasco's ranch, possibly the same location that Murray had referred to as James's ranch. The ranch, in either case, appeared to have been abandoned for quite some time. A cabin stood on the property. A short distance from it, Read discovered a spring, likely today's Ferndell Spring.[39] Read reported that Keene established a heliograph station atop the same location used by Murray in April, on "a highpoint 600 yards, direct, southwest from camp."[40]

On the night of April 29, Read and Keene decided to move the station, likely in an effort to improve visibility toward the more eastern stations. The next day, Keene relocated the station to a high point on the eastern ridge of the mountains, approximately 800 yards (just under 0.5 miles) due east of the camp.[41] This new position provided excellent visibility to the east and enabled communication with all required stations. However, to connect with stations to the northwest, Keene had to move the heliograph several yards, and the views to the southwest were obstructed.[42]

Based on the descriptions by Murray, Keene, and Read, it is reasonable to conclude that the soldiers' camp was located somewhere along the top of the mountain, southeast of Icehouse Canyon, where there was at least enough level ground for a cabin and a small farm. The alternatives are areas of significant slope. Since Keene's station was due east of the camp, we can also reasonably conclude that he did not place the heliograph station too far north along the ridge. If Keene had positioned the heliograph farther north, the camp, to the west, would then have been located in the canyons or steep northern slopes of the range, which does not match the descriptions.

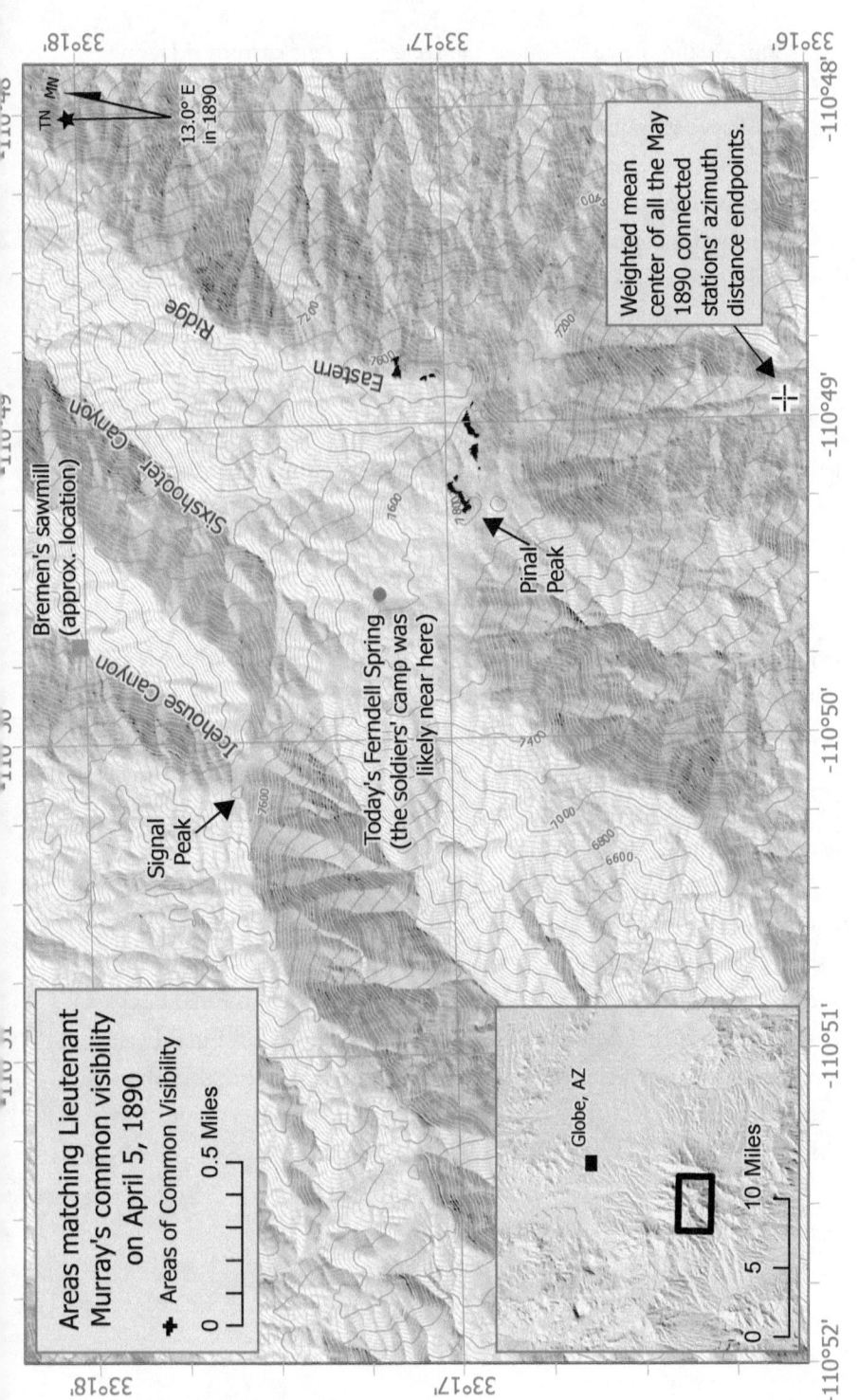

Map 61. Areas matching Lieutenant Murray's common visibility on April 5, 1890.

CHAPTER 9

Considering the common visible areas along the eastern ridgeline and the camp's relative position, we have mapped Keene's station on the first of several high points along the eastern ridge (see Map 62).

Because the heliograph on the eastern ridge needed to be moved to connect with certain stations to the northwest, Read surveyed other prominent points along the top of the range.[43] On May 2, 1890, he identified a high hilltop with unobstructed views in all directions, aside from a few trees, and established a heliograph station there. Using a barometer, he determined the hill's altitude to be 7,675 feet, noting that it was 100 feet higher than any other peak in the range.[44] The highest peak in the range, today's Pinal Peak, stands at 7,846 feet.[45] Read also described this peak as being 500 yards (about 0.3 miles) southeast of the camp. He spent the next several days building a wooden platform for the station.[46]

Based on Read's description, the height of Pinal Peak, the visibility analysis, and the mean center analysis, we have mapped this May 2, 1890, heliograph station atop today's Pinal Peak. These mapped locations represent the final heliograph stations used in the Pinal Mountains during the May exercise, including Read's station from May 2 and Keene's station along the eastern ridge. The only remaining location is the initial station used by Murray in April, which was likely situated just east of Pinal Peak.

Recall that three general areas met Murray's initial visibility requirements: Pinal Peak, sections along the eastern ridge, and an area east of Pinal Peak. Since Read initially used the same location that Murray did before moving to the eastern ridge and then to Pinal Peak, we can conclude that Murray's first station was neither on Pinal Peak nor along the eastern ridge. This leaves the area east of Pinal Peak as the likely possibility, and we have mapped this heliograph station atop a hill just 300 yards from Pinal Peak.

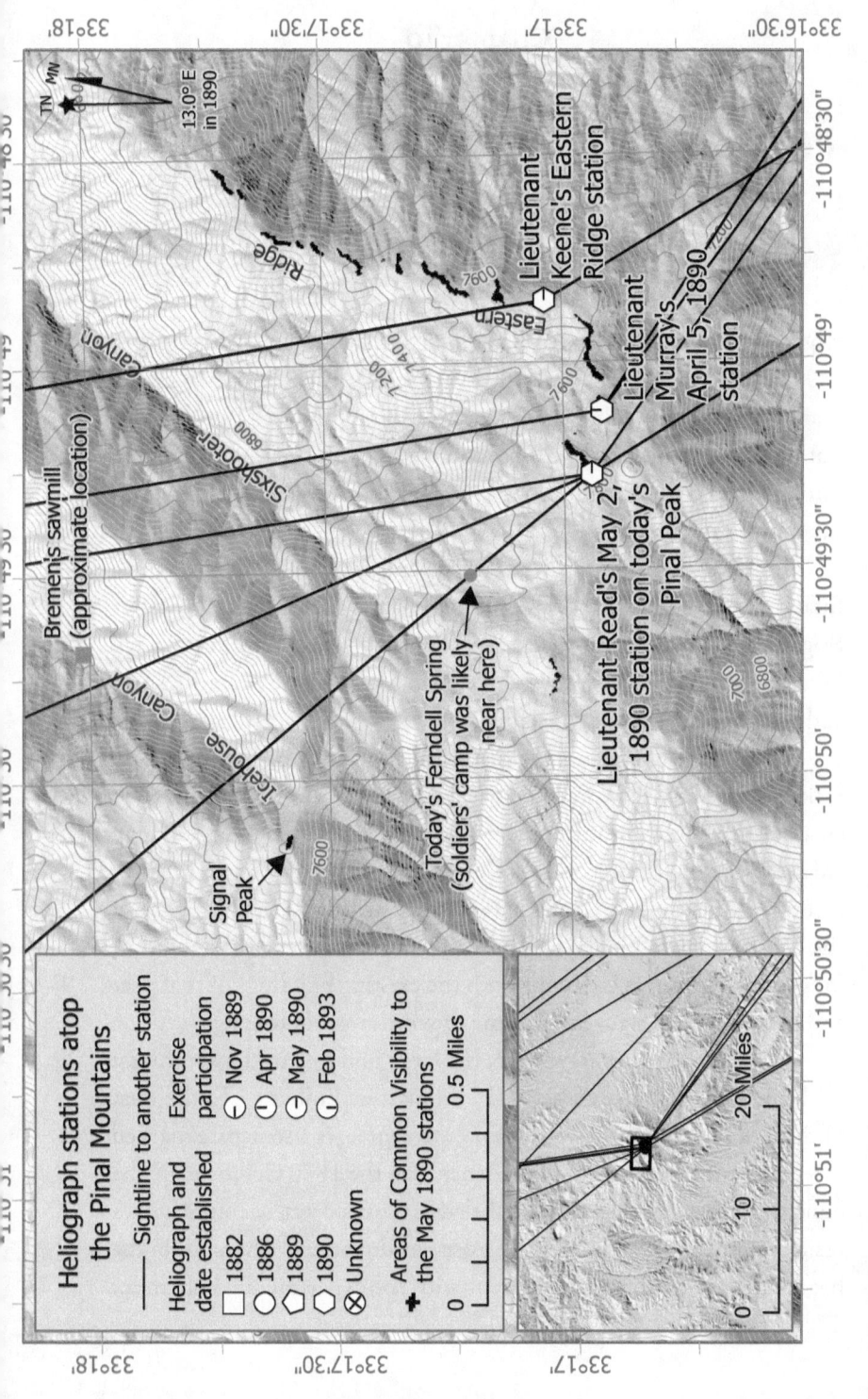

Map 62. Heliograph stations atop the Pinal Mountains.

Chapter 10

Fort Verde and Whipple Barracks

THE NORTHWESTERN STATIONS REPRESENTED THE TERMINUS OF THE 1890 heliograph line. They included Whipple Barracks, Bald Mountain, Squaw Peak, Little Squaw Peak, Fort Verde, and Baker's Butte. These stations were connected to the southern part of the heliograph system through the Tonto Basin stations, which included Mazatzal, Reno, and Lookout Peak.

In 1890 the headquarters and several companies of the Ninth Infantry were situated at Whipple Barracks, which was named after Brigadier General Amiel Weeks Whipple, who was killed in the Civil War. Whipple Barracks was established in May 1864 to protect miners and settlers near Prescott, Arizona.[1]

To better understand the historical geography of Whipple Barracks, I georeferenced three historical maps of the post: an 1883 map created by Lieutenant of Engineers Gustav J. Fieberger,[2] an 1890 map created by Lieutenant Charles L. Collins of the Twenty-Fourth Infantry (acting engineer officer),[3] and a 1902 map by the US Army Office of the Quartermaster General.[4] The 1902 map is particularly significant for georeferencing, as it depicts both the existing buildings of that year and proposed new structures, some of which remain today.

On the 1883 Fieberger map, the large building to the southwest of the parade grounds is labeled "Headquarters Building." To the west of this building (about 820 feet center to center) is a structure marked "Department Commander's Residence." On the 1890 Collins map, the building farther west, previously labeled as the department commander's residence in 1883, is now labeled "Post Headquarters." This shift in the headquarters buildings is consistent with the department commander

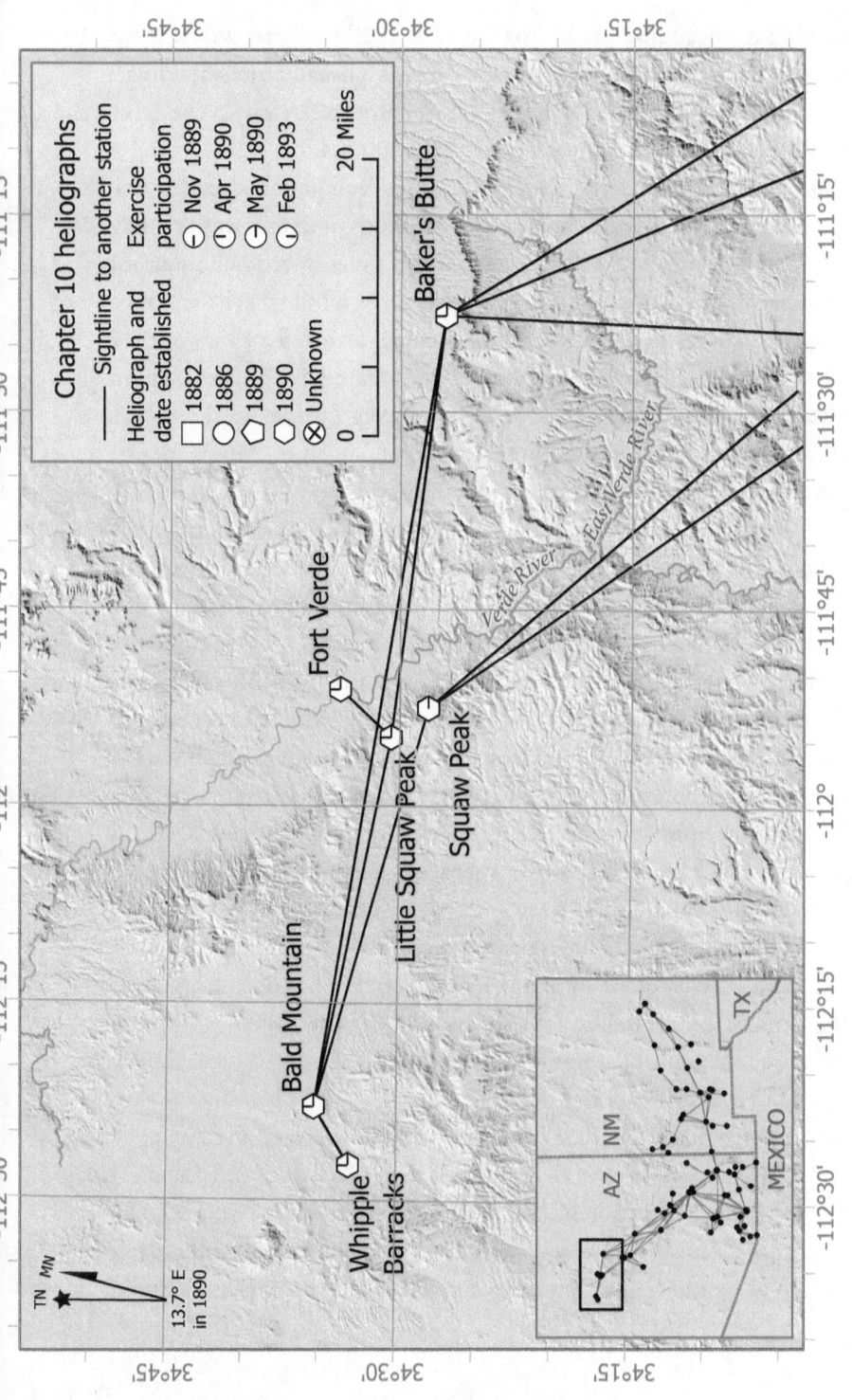

Map 63. Chapter 10 heliograph stations and connections.

moving his headquarters to Los Angeles in 1887, thereby freeing up his residence at Whipple Barracks to serve as a headquarters building.[5]

Lieutenant Tyson, a future US senator, referring to the May 1890 exercise, reported that the heliograph station was "located a short distance in front of the headquarters building" and had "a fine view of Bald Mountain."[6] He noted that "Bald Mountain was the only connected station and was seven and a half miles by road or about six miles by flash."[7] Very near the headquarters is a hill where the senior officers' quarters were located; it also provides an excellent view of Bald Mountain. The heliograph station has been mapped on this hill. Tyson likely used this same location while preparing for the April exercise, when he connected with Lieutenant Fenton, a future commander of Fort Myer, Virginia, on March 14, 1890.[8]

Whipple Barracks and Fort Verde are situated in mountainous terrain. Though they were only 34 miles apart, connecting them required the establishment of two additional heliograph stations: Squaw Peak (renamed Porcupine Mountain in September 2022) near Camp Verde and Bald Mountain (now known as Glassford Hill) near Whipple Barracks. In 1887 Glassford suggested Granite Mountain, located 8 miles northwest of Whipple Barracks, as a potential heliograph station to establish a connection to the east if necessary.[9] However, Lieutenants Tyson and Fenton recognized the practicality of using the closer Bald Mountain as the heliograph link out of the valley where Whipple Barracks was situated.[10]

Fenton set up a heliograph station on Bald Mountain on March 14, 1890. The Bald Mountain heliograph station has been mapped at the topographic crest of today's Glassford Hill, where there is visibility among all the connected stations: Whipple Barracks, both Little Squaw Peak and Squaw Peak, and Baker's Butte.

Baker's Butte, found on the edge of the Mogollon Rim, commands an excellent view in every direction. Both Glassford and Leonard Wood recognized its potential as a heliograph site.[11] On April 5, 1890, Lieutenant Ramsey ascended Baker's Butte and established a heliograph station as part of that month's exercise. From the top of Baker's Butte,

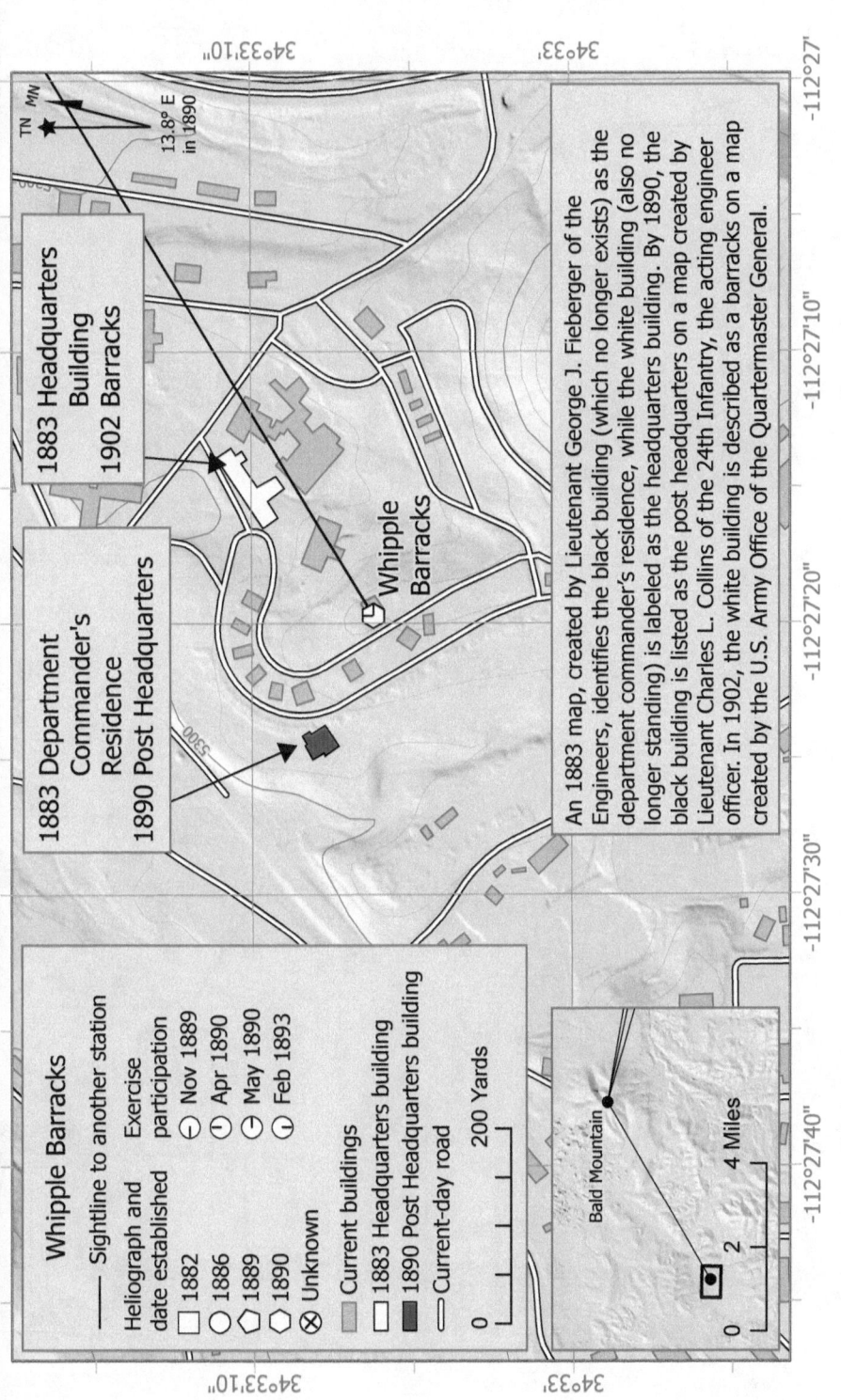

Map 64. Whipple Barracks heliograph station.

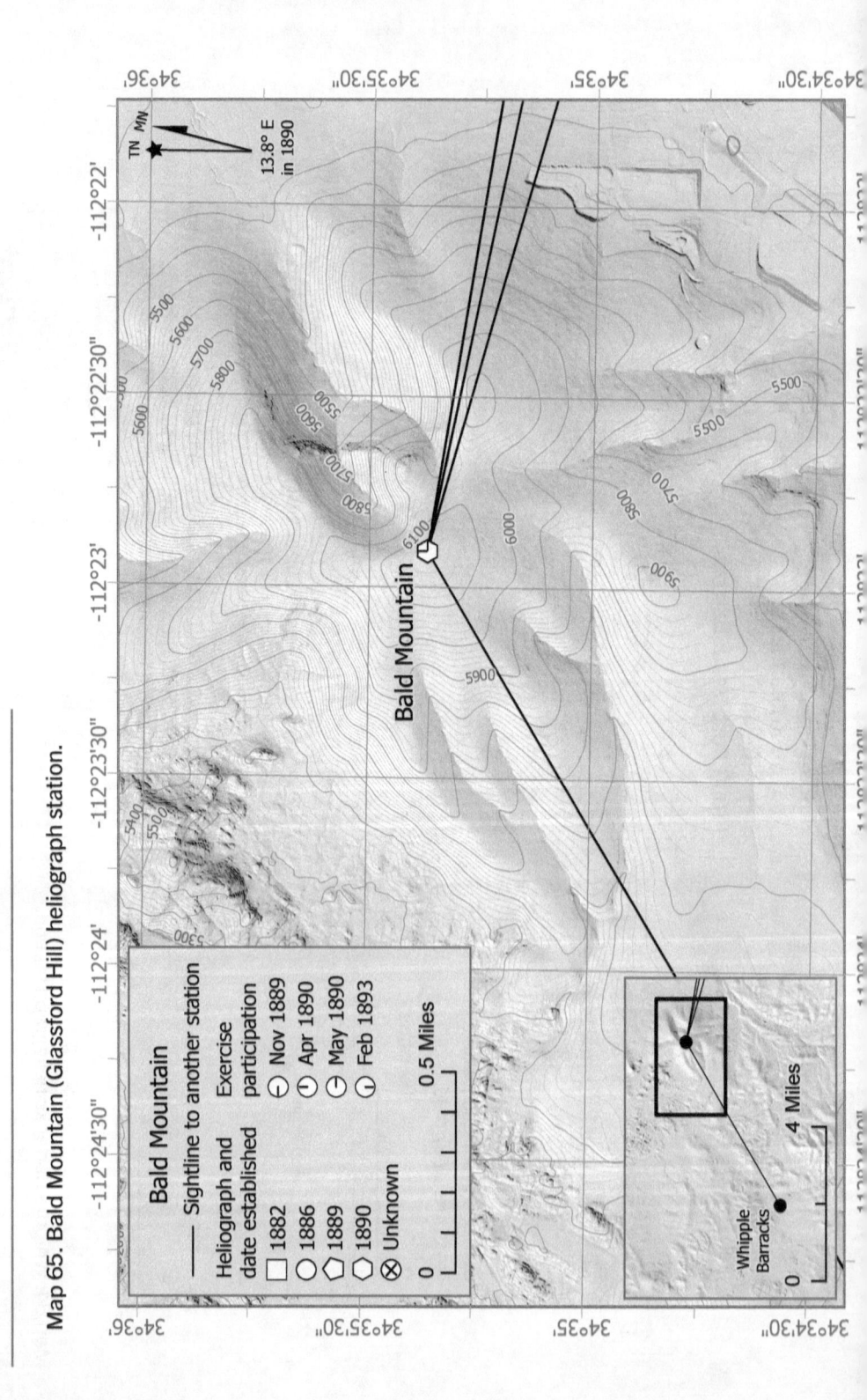

Map 65. Bald Mountain (Glassford Hill) heliograph station.

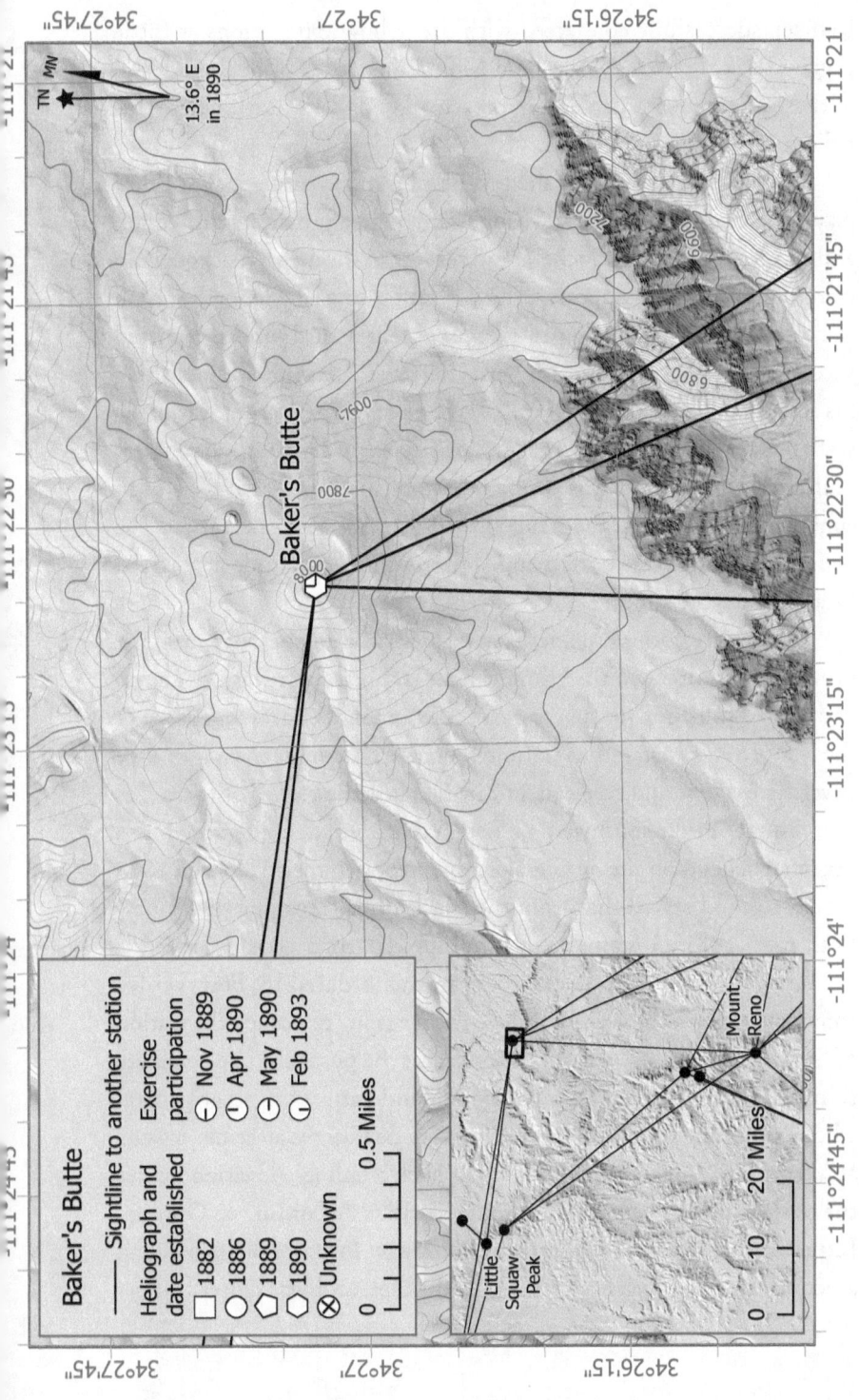

Map 66. Baker's Butte heliograph station.

Ramsey successfully connected with the heliograph stations at Little Squaw Peak, Lookout Peak, Bald Mountain, and Mount Reno.[12] All of these stations are visible from present-day Baker's Butte, where this heliograph station is mapped.

Fort Verde was established at its present location in 1871 and played a key role in General Crook's campaigns, providing protection and security to miners, ranchers, and settlers in the area.[13] Although no commissioned officer appears to have been staffing a heliograph station at Fort Verde during the April exercise, Lieutenant Duncan reported successful communication with Verde for several days after establishing his station on Little Squaw Peak.[14] The Fort Verde station was likely staffed by a corporal or sergeant positioned near the headquarters building. This station was probably established on April 5 or 6, in conjunction with Duncan's setup of the Little Squaw Peak station. The heliograph station has been mapped to the parade grounds next to the headquarters building.

Although Fort Verde actively participated in the May exercise, providing resources to the effort, a heliograph station was not set up and staffed full-time for that exercise. Lieutenant Ramsey, in charge of the May 1890 effort at that station, reported that some messages were nevertheless sent and received, but he had lost the record of them.[15]

Squaw Peak, southwest of Fort Verde, was also identified as a potential heliograph site by Glassford and Wood in their 1887 and 1888 reports. Wood stated that from the Black Hills (meaning the Squaw Peak area), the mountains around Whipple Barracks could be seen.[16] On April 6, 1890, Lieutenant Duncan was ordered by Fort Verde's commanding officer, Lieutenant J. R. Richards Jr., "to occupy the station of Squaw Peak, but if a suitable place nearer the post could be obtained, to use that."[17] Duncan, an infantryman and future brigadier general who was presented with the French Croix de Guerre after his actions at Verdun in World War I, found just such a suitable location nearer the post, which he called Little Squaw Peak.[18] According to Duncan, Little Squaw Peak was just as good as Squaw Peak for signaling, had a better camp with plenty of wood and water, and was closer to Fort

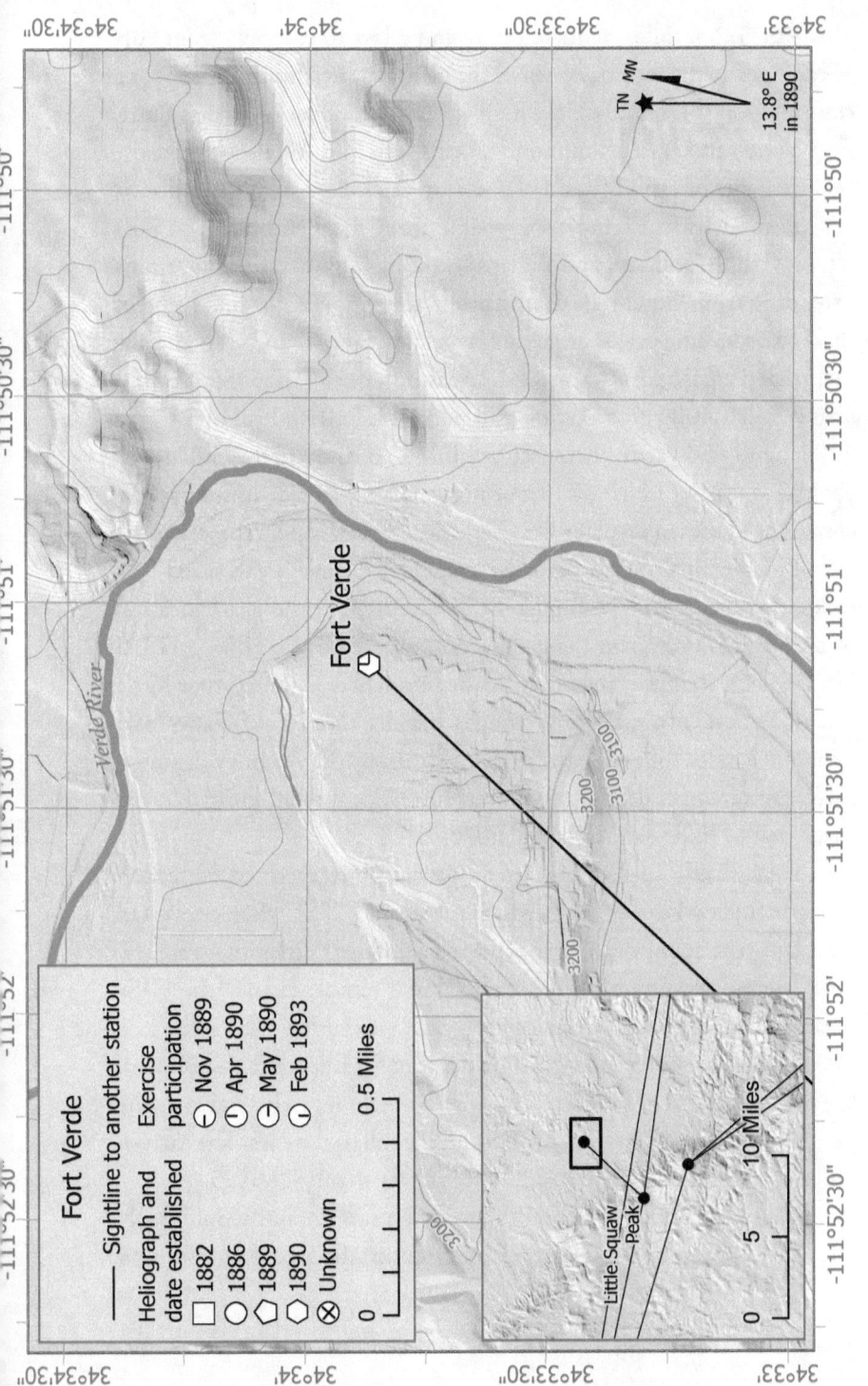

Map 67. Fort Verde heliograph station.

CHAPTER 10

Verde by 3 to 4 miles, reducing time and effort to resupply from Fort Verde—likely the rationale behind the commanding officer's desire for a closer station.[19] Duncan's Little Squaw Peak was visible to Baker's Butte, Fort Verde, and Bald Mountain and closer to Fort Verde—it met the requirements of visibility and of the Fort Verde commanding officer.[20]

A model identifying areas visible from the Black Hills to Fort Verde, Bald Mountain, and Baker's Butte highlights several locations extending from Squaw Peak northwest to Copper Canyon. Some of these areas include peaks; others are ridgelines. Since Duncan specifically mentioned a peak, this narrows the possibilities to Hill 5485, another nearby hill 1,400 yards east of Hill 5485, Hill 6076, Hill 6050, Hill 6276, a hill 450 yards northeast of Hill 6276, and another hill 1,020 yards southeast of Hill 6276. These unnamed hills are identified by their elevations as shown on older US Geological Survey topographic maps.[21]

Lieutenant Duncan reported that Little Squaw Peak saved 3–4 miles of difficult trail to Squaw Peak.[22] Although the exact location of this trail is unknown (as it does not appear on the 1887 or 1892 US Geological Survey maps), it can be reasonably assumed that Little Squaw Peak is within a 4-mile straight-line distance from Squaw Peak, or about 3 miles following the ridgeline southward. Within this range, three prominent hilltops stand out: Hill 5485, the hill located 1,400 yards east of Hill 5485, and Hill 6076.

Lieutenant Ramsey, stationed on Baker's Butte, reported a direction to Little Squaw Peak of 264°, which results in 277.6° when converted to today's true azimuth (he also reported a different direction to Squaw Peak, acknowledging both peaks).[23] This is almost exactly the GIS-measured direction between Baker's Butte and Hill 5485, which is 277.8°. If we consider hills within 1 degree of Ramsey's measurement, the hill to the east of Hill 5485 can also be included. This narrows the possible locations to two: Hill 5485 and the hill to the east. The eastern hill is at the end of a ridge and 300 feet lower than Hill 5485.

Hill 5485 appears to be in more open terrain and would likely offer better camping (as described by Duncan). So based on common

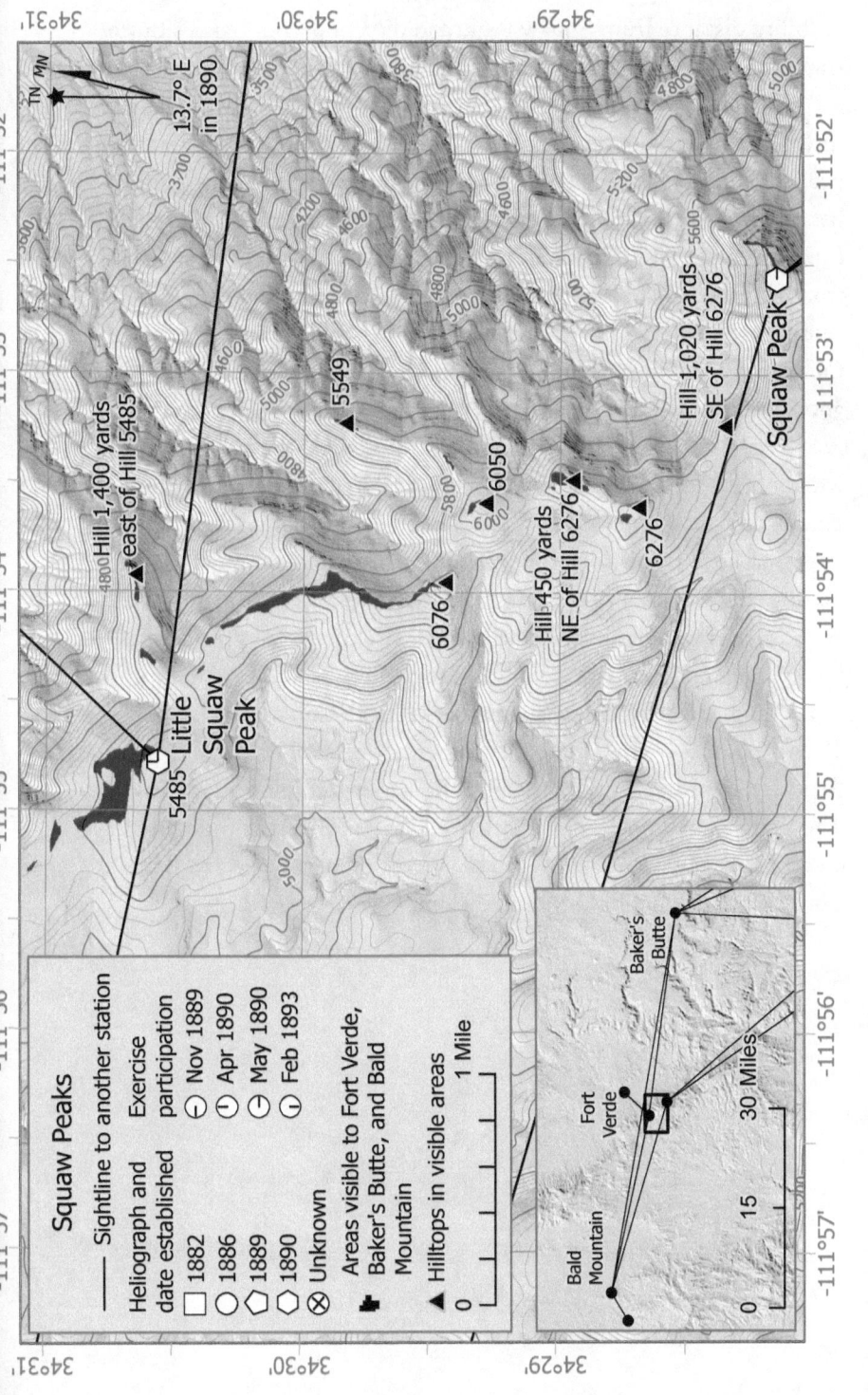

Map 68. Squaw Peaks heliograph stations.

visibility, distance from Squaw Peak, and direction from Baker's Butte, the Little Squaw Peak heliograph station is mapped atop Hill 5485.

While the Volkmar map and table do show "Little Squaw Peak," the directions given from other stations tend to point toward Squaw Peak proper. Additionally, the written reports concerning the May exercise list Squaw Peak (staffed by Sergeant A. J. Robinson) rather than Little Squaw Peak.[24] Visibility analysis confirms that Squaw Peak is visible from all four listed stations on the Volkmar map (Mazatzal, Baker's Butte, Fort Verde, and Bald Mountain). In his 1887 report, Lieutenant Glassford concluded that Squaw Peak was "easily ascended over a good, well-known trail, has water at the top, and commands a view of Fort Verde at its base . . . as well as Baker Butte and the San Francisco Mountains."[25] To that end, on March 14, 1890, Lieutenant Fenton on Bald Mountain connected with Lieutenant Ramsey on Squaw Peak.[26]

Sergeant Robinson, the noncommissioned officer detailed to occupy "Big Squaw Peak" during the May exercise, failed to find the abundance of water Glassford mentioned. The chore of supplying water for the men and animals would compel Robinson to move the camp to a spring nearer to Little Squaw Peak. As a result, except for April 30 and May 15—when he connected with Mazatzal Peak and Mount Reno—Robinson primarily occupied the station at Little Squaw Peak.[27] The Squaw Peak heliograph station is mapped near the topographic crest of today's Porcupine Mountain, where there are views to Baker's Butte, Mount Reno, Bald Mountain, and Fort Verde.

Chapter 11

The Pinaleño Mountains

A HELIOGRAPH LINK "CONNECTING ALL THE NORTHERN AND CENTRAL heliograph divisions with those of the south, south-east and south-west" was needed as the 1886 system was incorporated into the growing 1890 system.[1] At nearly 11,000 feet, the Pinaleño Mountains provided the base for that link. By 1890, four heliograph stations were used for exercises and testing atop the Pinaleños. Unfortunately, with a prominent peak on this range named Heliograph, little attention has been given to the three other, more important stations.

The Pinaleños extend about 40 miles from Underwood Canyon in the northwest to Railroad Pass, where today's Interstate 10 crosses, in the southeast. To the northwest of this range are the Santa Teresa Mountains, and to the southeast lie the Dos Cabezas and Chiricahua Mountains. Approximately 6.5 miles southwest of the Pinaleños' central ridge is Fort Grant, with an impressive elevation gain of more than 5,000 feet from the fort to the ridgeline.

On the southwestern slope of the Pinaleños, several streams have carved large canyons near Fort Grant. These canyons, listed from west to east, include Taylor Canyon, West and East Babcock Canyons, Jesus Canyon, Goudy Canyon, and the canyon formed by Grant Creek. The canyon formed by Grant Creek and its tributaries intersects the Pinaleño Mountain ridgeline between Webb Peak to the west and Hawk Peak to the east. This canyon is a significant and prominent terrain feature. Grant Creek served as the water source for Fort Grant, and Major Anson Mills, the fort's commanding officer, oversaw an elaborate water system modernization project at the fort. This project included an upper reservoir in Grant Creek canyon, a sewer system, fountains, and a lower reservoir at the post (named Lake Constance after Mills's daughter).[2]

Near the top of many of the canyons on the western side of the range, at around 9,000 feet, are flat areas in otherwise dramatic terrain. These "flats," pretty mountain meadows, have names like Riggs Flat, Peters Flat, and Hospital Flat. Hospital Flat owes its name to a "sanitary camp" built by the ever-industrious Major Mills near the top of Big Creek, east of the Grant Creek canyon.[3] He began construction of a log building there in the summer of 1888.[4] This camp was known to Lieutenant Neall as Alpina.[5]

Connecting Fort Bowie with the Gila Valley using heliographs was first achieved by Lieutenant Maus in 1882. He made the connection by skirting the Pinaleños to the south via Dos Cabezas and Ash Hill.[6] In 1887 Lieutenant Glassford proposed connecting Fort Thomas (along the Gila River) with Fort Grant, Willcox, and Fort Bowie by establishing a series of stations through Taylor Pass (in the northwestern part of the range) to Fort Grant, then using Dos Cabezas to reach Fort Bowie.[7] Because the central ridge of the Pinaleño Range is aligned with Fort Bowie to the southeast and Lookout Peak to the northwest, crossing this range with heliograph signals to a variety of other stations proved a challenging geometry problem, compounded by very high, rugged terrain.

From 1889 to 1893, there were four heliograph stations on the Pinaleño Mountains, two southeast and two northwest of Grant Creek canyon. The two on the southeast were present-day Heliograph Peak and Alpina. Established in November or early December 1889 by Lieutenant Keene, the Alpina heliograph station was situated just south of present-day Grant Hill. The Heliograph Peak station was established on December 15, 1889, by a sergeant. While Alpina and Heliograph Peak generally had good visibility and connected Fort Grant to stations farther east and southeast, they were blind to the northern stations in the Gila Valley. Another station was required to cross the ridge to the north.

On December 21, 1889, Lieutenant Eggleston established a station on present-day Merrill Peak, which connected across the northern ridge.

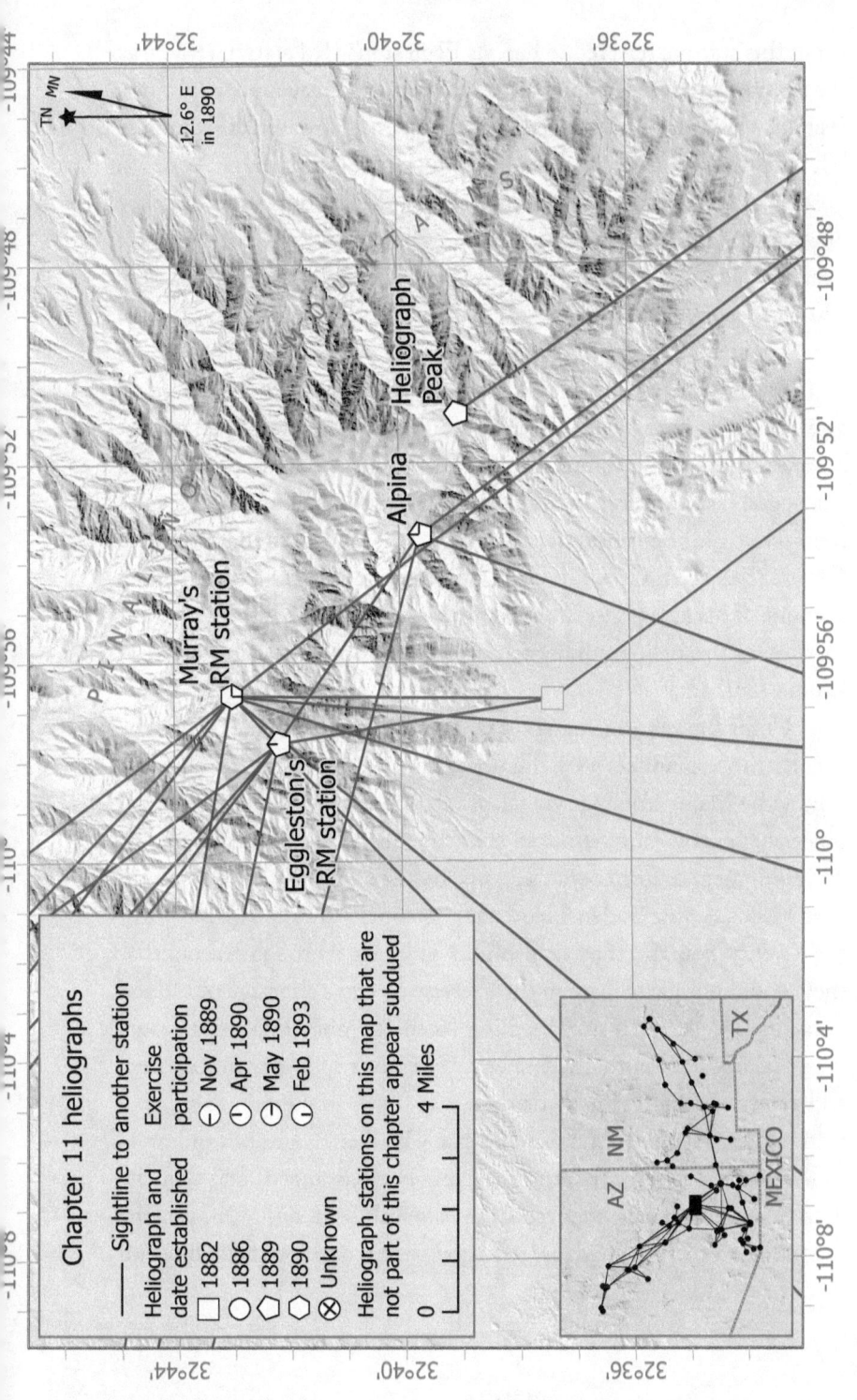

Map 69. Chapter 11 heliograph stations and connections.

CHAPTER 11

From this station, referred to here as Eggleston's RM station (RM was the army's generic call sign of a station atop the Pinaleño Mountains), visibility was nearly unrestricted in all directions. However, the station at Alpina was still needed to reach Fort Bowie. Months later, on April 22, 1890, Captain Murray moved Eggleston's RM station a short distance northeast to the southern of the two Grand View Peaks, where Bowie Peak was also visible, thereby eliminating the need for the Alpina station. This station, referred to here as Murray's RM station, was used for the 1890 and 1893 exercises.

The difficulty crossing this range with heliograph signals became apparent during the November 1889 concerted practice. As part of that exercise, Fort Grant's commanding officer was tasked with establishing a heliograph station atop the Pinaleño Range. This station was to be the central hub of the entire system. Its importance was such that Brigadier General Benjamin Grierson, the Department of Arizona commander, sent one of his aides-de-camp, Lieutenant John A. Perry, to Fort Grant to witness the heliograph system's performance from the station at Mount Graham.[8]

Unfortunately, during the exercise, heliograph signals from the Pinaleños were not seen by the other participants. After some questioning by Major Volkmar via telegraph, Lieutenant Keene, the signal officer for Fort Grant, admitted that the fort had not sent the people and equipment to form a station atop Mount Graham for the exercise.[9] Fort Grant's commander, Lieutenant Colonel Edward P. Pearson of the Twenty-Fourth Infantry, also told Volkmar (through Keene) that they would not participate in the exercise due to a shortage of trained personnel. Volkmar insisted, telling Keene to "obey implicitly [your] instructions."[10] Eventually, General Grierson's aide-de-camp, Perry, and a party led by Keene made their way up the mountain to set up a station. Perry reported that Keene's party had "no ... understanding as to the exact locations of stations having been arranged" and that the detachment was unable to successfully establish a station.[11] One gets the sense that Keene, who had just returned from ten days of field duty and now was caught between a lieutenant colonel and a major, was eagerly anticipating the end of the exercise.[12]

In response to Fort Grant's lackluster November heliograph performance, in December 1889 Lieutenant Eggleston, the ASO at San Carlos, was sent to Fort Grant to find a "location . . . connecting the Bowie and San Carlos Divisions, across the Graham mountains."[13] Eggleston arrived at Fort Grant on December 10. The next day he led a detachment of men, mules, horses, and heliographs up the western slopes of the mountain. He camped at a log cabin "built for sanitary purposes," marked "Alpina" on the map he was using. (This camp is present-day Hospital Flat.)[14]

After a delay due to weather,[15] on December 15 Eggleston divided his party. He and Lieutenant Neall from Fort Bowie searched for and found a location near where Lieutenant Keene had previously signaled Table Mountain to the northwest. Meanwhile, a sergeant led a detachment to a high peak near their camp. The sergeant's peak turned out to be an excellent location for connecting with Fort Bowie. Eggleston calculated its elevation to be 10,500 feet. This peak had a "fine view . . . to the east, south-east, south, south-west and west."[16] Neall mentioned that this peak was almost at the highest point of the range, with excellent views in all directions except to the north and northwest. Neall also gave compass bearings from this peak to various other peaks. His bearings to Table Mountain and Bowie Peak suggest that this was present-day Heliograph Peak.[17]

Referring to the station previously established by Keene, in his December 21, 1889, report, Neall noted that several peaks and stations were visible from it: Fort Bowie, Table Mountain, Cochise Stronghold (Fourr's Ranch North), Mount Animus, and Granite Mountain (near Stein's Pass). Neall also noted that Fort Grant can "easily be flagged," suggesting that Fort Grant was visible as well.[18] Eggleston described Keene's station as 3.25 miles by flash from Fort Grant.[19]

Areas atop the Pinaleño Mountains where there is common visibility to all the points mentioned by Neall are few. They are Heliograph Peak, some hilltops to the southeast of Heliograph Peak, some areas to the west of today's Hawk Peak, and a small peak and associated ridge just to the south of present-day Grant Hill (see Map 69).

Eggleston recorded the elevation of the sanitary camp as 9,375 feet and that of Keene's station as 8,400 feet.[20] This elevation difference, with Keene's station lower than the camp, eliminates the higher peaks (Hawk and Heliograph) as the station's location, as they are higher than the camp. Additionally, there is only one area with common visibility reasonably close to Eggleston's 3.25-mile distance from Fort Grant: the small peak and ridge just south of Grant Hill, 4.1 miles distant.[21] Moreover, Eggleston gave the distance from Keene's station to the sanitary camp as about 1 mile.[22] The distance from today's Hospital Flat to the small peak hill south of Grant Hill is almost exactly 1 mile. Additionally, Neall's directions to Cochise Stronghold and Table Mountain suggest that the small hill south of Grant Hill was Keene's station as well.[23] Given all this, the location of the station used by Keene is mapped at the crest of the small peak south of Grant Hill.[24] Eggleston named this peak Alpina after the nearby stream and camp.[25]

Eggleston, while happy enough with Alpina's performance connecting to Bowie Peak, pointed out that connecting with the stations to the north, in the Gila Basin and farther, was impossible from this or any other peak in this portion of the range, because "just to the north and north-west, and rising about 500 feet above it, is a long ridge densely timbered and cutting off all view to the Gila Valley, or any point north. This ridge ... comprise[s] the highest points in the Graham Range."[26] If one stands on Alpina today and looks to the northwest, the ridgeline seen extends from Mount Graham (on the viewer's right) and leftward to present-day Hawk Peak, Webb Peak, and Merrill Peak.

Eggleston, based on his reconnaissance and information from Neall, noted that "further examination of the country made during the next two or three days [December 16 and 17] showed that there was no point of the range west of the Cañon immediately back of the post [Fort Grant] and in which the stream has its head which supplies the reservoir, from which Bowie Peak can be seen."[27] Here, Eggleston

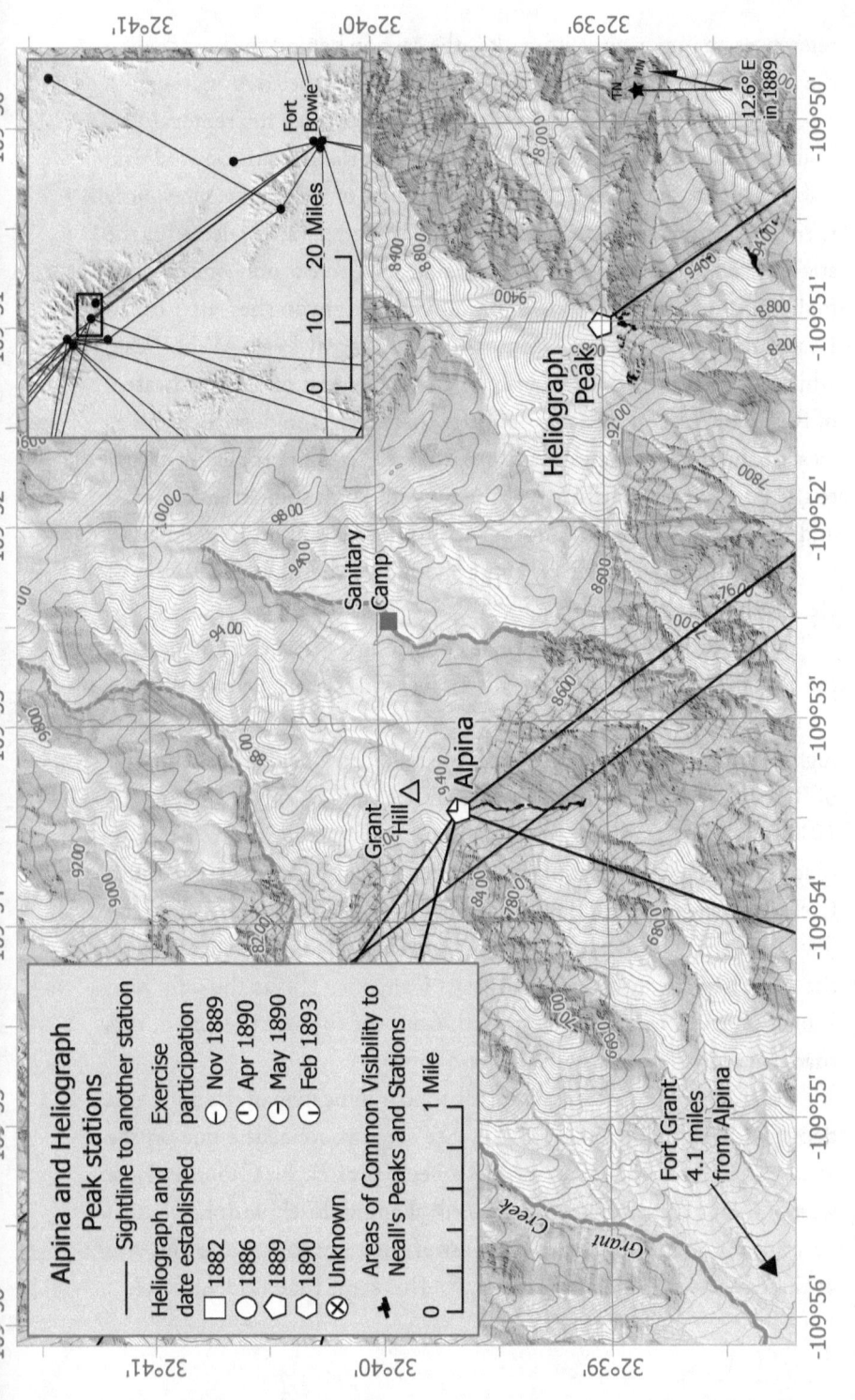

Map 70. Alpina and Heliograph Peak heliograph stations.

references an important landmark—the canyon behind the post feeding the reservoir—current-day Grant Creek Canyon (labeled A on Map 67).

Before leaving Alpina, Eggleston marked trees for removal (for visibility) and "marked with stone and stakes the lines to Bowie Peak, Table Mountain, *and to the high point west of the Cañon previously referred to* [emphasis added] and to which was afterwards selected as the station with which to get across the range."[28] There is only one obvious high point on the range west of Grant Creek Canyon (hereafter called Reservoir Cañon, as Eggleston called it): Merrill Peak. Webb Peak, while conspicuous, is located north and a bit west of the headwaters of Reservoir Cañon, and it is doubtful that Eggleston described this as west of the canyon. Hawk Peak and Mount Graham proper are clearly to the north, if not northeast, of the canyon. With some trees removed and the cairns set, Eggleston returned to Fort Grant on December 18 to replenish supplies and prepare to climb to the high point west of Reservoir Cañon.

On December 20, 1889, Eggleston, along with a detachment from the Tenth Cavalry (provided by Fort Grant) and carrying enough supplies for five days, began climbing a trail near Fort Grant.[29] His goal was the high point he had seen from Alpina. He followed an old trail to the west of Reservoir Cañon and "two or three miles" east of Taylor Cañon.[30] A 1911 US Army map of Fort Grant and the Pinaleños shows a trail or road leaving the fort, heading west for a short distance, crossing Grant Creek, and then turning almost due north along Goudy Creek.[31] After a few miles, the trail veers sharply west and then heads north again along the spine separating Goudy and Jesus Canyons. This is the only road shown on the 1911 map going north from the fort. At the end of this road, there is a feature marked "wood camp."

The 1942 US Geological Survey topographic map of the same area depicts a road following essentially the same route as the one on the 1911 map.[32] The road starts about 3 miles east of Taylor Cañon and just west of Grant Creek (Reservoir Cañon), aligning closely with Eggleston's description of an "old trail to the west of the 'reservoir' cañon and two or three miles east of Taylor's cañon."[33] This same road is identified in a historical narrative of the Pinaleño Range as an old military road, built in 1872 to satisfy Fort Grant's demand for lumber.[34]

While this road seems to be the one Eggleston took, the scale of the 1911 map appears to be off. The cartographer, J. M. Hilton, was uncertain about some of the features distant from the fort, noting that Mount Graham itself was an approximate location. Furthermore, although Hilton's road generally aligns with the 1942 road in relation to key features, notably Goudy Creek, the bend to the west is more pronounced on the 1911 map. This then casts some doubt on the road's position. However, the same map does show a feature labeled "plot of ground in upper left hand fork of Grant Creek," which is clearly east of the depicted road and wood camp.[35] This suggests that the road is not following Grant Creek or any of its tributaries.[36]

Given the correlation of features between the 1911 Hilton map and the 1942 map, along with the distance from the trail to Taylor Wash and Eggleston's description, it is likely that this was the road Eggleston followed when he re-ascended into the range on December 20, 1889. While Eggleston does not mention the wood camp, scaling the distance from the fort places it about 5.7 miles away, line-of-sight. Measuring the same distance in the GIS places the wood camp about half a mile northeast of Merrill Peak.

Once Eggleston ascended the canyon, he headed directly to the high point and established a heliograph station, where visibility was "superb."[37] From this station, he listed more than a dozen visible peaks,[38] commenting that this "station would answer all requirements and necessarily would become the main one." While this location may have been superb, Eggleston still needed the station at Alpina to reach Bowie Peak.[39] Visibility analysis from Bowie Peak shows that the southern Grand View Peak, Webb Peak, and Hawk Peak are clearly visible. Because of this, we can eliminate them as the location for Eggleston's station.

Places where one can see the peaks Eggleston listed as visible and where Bowie Peak is not visible are two: Merrill Peak and the area just west of Hawk Peak. With just these two areas, it is tempting to conclude that instead of being physically unable to see Bowie Peak, Eggleston just didn't see it, perhaps because the heliograph operators were not at the Bowie Peak station in December or it was lost to sight to trees or a complex of other hills and peaks of the Dos Cabezas Range. However, in April 1890, when Captain Murray was at Eggleston's RM station and after ensuring that the heliograph station at Bowie Peak was in operation, Murray was also unable to reach Bowie Peak directly.[40]

The visible areas just to the west of Hawk peak are almost due north of Alpina and would not have been considered west of Reservoir Cañon. Furthermore, with Hawk Peak only a mile distant, the obvious choice would have been just to occupy the peak itself. Given the visibility to Eggleston's listed peaks and stations (plus Alpina and Fort Grant), lack of visibility to Bowie Peak, and Merrill Peak's location to the west of the Reservoir Cañon, it is reasonable to conclude that Eggleston's RM station was atop Merrill Peak, with its excellent views in almost every direction and the advantage of being very near the top of Goudy Canyon, thereby keeping logistics close with Fort Grant.

Merrill Peak's disadvantage as a heliograph station is that it could not directly connect with Bowie Peak, an important element of the network. Furthermore, because of its alignment with Mount Turnbull and Lookout Peak, heliograph operators on Lookout Peak, another important station, had difficulty locating Eggleston's RM station. In a report of Eggleston's (January 6, 1890) written in response to the Lookout Peak station's difficulty in finding RM station, he noted that RM, Mount Turnbull, and Lookout Peak are "almost on a direct line."[41] Indeed, the angle created by Merrill Peak, Lookout Peak, and Mount Turnbull is only 0.56°, though all the peaks in this area have roughly the same alignment with Mount Turnbull and Lookout Peak.

In late April, Captain Murray left San Carlos "for Fort Grant . . . to investigate and report upon the failure of the signal detail recently at Graham (R M) station to communicate with Lookout [Peak]."[42] On April 19, Murray and Lieutenants Dade and Stockle climbed to Eggleston's RM station. There Murray uncovered the same reason for not seeing Lookout Peak that Eggleston had described in his (apparently unread) January 6 report: Mount Turnbull was almost directly in line

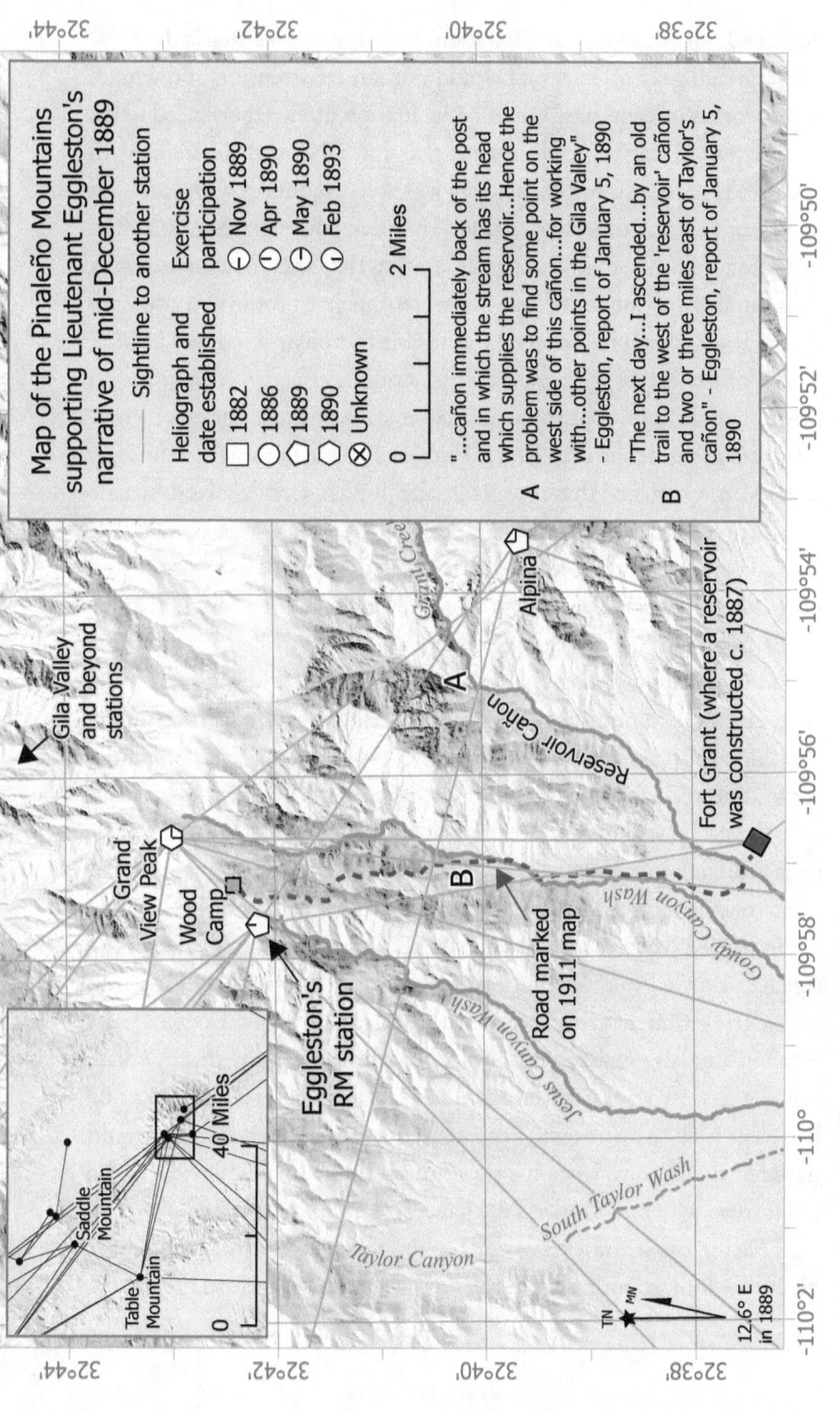

Map 71. Eggleston's RM station on Merrill Peak.

between Lookout Peak and the station. Murray observed that "a flash just over the dull gray of Turnbull would not attract attention and would, if seen for an instant, pass for an hallucination often experienced while intently examining a profile against the sky."[43] Nevertheless, on April 20, after repeated attempts, Murray was able to connect with Lookout Peak from Eggleston's RM station. That day, Murray sent a telegram to Major Volkmar in Los Angeles, stating that they had successfully connected with Lookout Peak but were unable to communicate with Bowie Peak, which he believed was visible but obstructed by haze.[44]

Unfortunately, Murray's reporting on his efforts to connect with Bowie Peak is somewhat confusing. After sending the telegram on April 20, stating that he thought Bowie Peak was visible but was obscured by haze, Murray learned that it was actually "hidden from view by Apache Peak" (present-day Government Peak in the Dos Cabezas Range is aligned between Eggleston's RM station and the Bowie Peak station) and not masked by haze as initially thought.[45] Further complicating matters, in the middle of the paragraph where Murray described his attempts to communicate with Bowie Peak on April 20, he wrote, "Late on the 20th instant, I communicated with the detail on that peak."[46] Based on the sentence's placement in the narrative, it suggests that he successfully communicated with Bowie Peak from Eggleston's RM station as well.

Several possibilities arise to explain this. First, Murray may have been referring to Lookout Peak and mistakenly placed the sentence in the wrong context. Second, he could have sent Lieutenant Dade (who accompanied Murray and Stockle on this trip) to Alpina to relay the signal to Bowie Peak. Third, he could have seen Bowie Peak. However, considering that Murray expressed uncertainty about Bowie Peak's location and later realized it was blocked from view by an intermediate hill, the first two possibilities seem most plausible. Furthermore, if he had clear visibility of Bowie Peak, it seems very unlikely that he would have felt the need to leave his position on the following day to "find a point from which it [Bowie Peak] and Lookout could be seen."[47]

Facing those difficulties connecting to Lookout Peak and recognizing the limitations of Eggleston's RM station's reliance on the Alpina station to connect with Bowie Peak, Captain Murray sought a better location. He contacted Lieutenant Neall at Fort Bowie and ensured that the Bowie Peak station was operational. Then, On April 21, he focused his efforts on finding a vantage point from which both Bowie and Lookout Peaks could be seen.[48] The following day, April 22, "such a station was found, thus removing the necessity for Alpina station."[49]

This new station, Murray's RM station, is located atop the more southern of the two nearby Grand View Peaks. Although the US Geological Survey currently labels the northern of these two peaks as Grand View Peak, decades ago the organization labeled the southern, slightly higher peak as Grand View Peak.[50] It is this southern peak where Murray placed his RM station.

Eleven other stations were required to connect to the Mount Graham station.[51] The overlapping visibility diagrams from the eleven stations show several areas visible, including the shoulder of Hawk Peak and some adjacent areas to the west, areas to the west of Webb Peak, and the southern Grand View Peak.[52]

However, while visibility modeling suggests that Bowie Peak is visible from these locations, practical experience shows that visibility to Bowie Peak is often obstructed and in many cases completely obscured by trees on intervening crests. These conditions were likely the same in 1890. Since Captain Murray's primary reason for moving the RM station from Merrill Peak was to establish a direct connection with Bowie Peak, ensuring clear visibility to Bowie Peak must have been a key factor in selecting the new RM station.[53]

After identifying the new RM station on April 22, Murray traveled to the Alpina station to evaluate other potential sites. The following day he returned to his new station from Alpina via Mount Graham proper.[54] While he thought Mount Graham would be an ideal peak for a heliograph station, he nevertheless continued back to the station he chose on April 22. This observation effectively rules out Mount Graham as the location of the new RM station.

All the officers posted at the Pinaleño station during the May 1890 exercise reported that Murray's RM station was about 1.5 miles northeast of the previous station (Eggleston's RM station on present-day Merrill Peak) and about a mile from the camp used by the soldiers of both Eggleston's and Murray's RM stations, the "old log camp."[55] This geometry fits well with directions and distances among Eggleston's RM station, the wood camp, and the southern of the two Grand View Peaks.[56] The distance from Merrill Peak to the southern Grand View Peak is 1.4 miles at 45° true (32.4° magnetic). The distance from the wood camp to the Grand View Peak is approximately 0.9 miles in nearly the same direction (23° magnetic).

Most telling, however, is the azimuth and distance from Fort Grant to Murray's RM station. The listed azimuth is 346.5° magnetic, which becomes 359.0° true after applying the magnetic declination in 1890 of 12.5° east (for Fort Grant). The listed distance is 6 miles. When a line segment is drawn from Fort Grant at 359.0° true for 6 miles, the end plots almost directly south and approximately 0.4 miles from the southern Grand View Peak. Indeed, if the line were extended it would come within 200 yards of the visible areas atop the southern Grand View Peak. Given the relatively short distance of 6 miles—resulting in only a small margin of error due to the limitations of the prismatic compasses used at the time (see chapter 1)—the azimuth and distance from Fort Grant strongly indicate that the southern Grand View Peak was the location of Murray's RM station.

On the other hand, mean center analysis of the eleven azimuth-distances from the listed connections places the center nearer to Webb Peak, plotting less than half a mile southwest of Webb Peak and about 1.1 miles southeast of the Grand View Peaks. However, due to the dispersion of the azimuth-distance terminal points, the standard distance circle is quite large, encompassing Eggleston's RM station, Grand View Peak, and Hawk Peak. This broad circle reduces the relative importance of the mean center location, as it doesn't strongly favor one specific peak over the others.

Also supporting a Webb Peak location is Lieutenant Dade's remarks that the views from Murray's RM station are excellent with the "entire horizon ... unobstructed" except for a quadrant of about 60° north of

Bowie Peak.[57] This description fits very well with Webb Peak's view obstructed by Hawk Peak to the east of 62°, whereas the views from the southern Grand View Peak are less in concert with Dade's observation. Depending on where the observer is located on Grand View Peak, the view's obstruction to the east ranges from about 75° to an intermittently obstructed view of about 127°.[58]

The taller Webb Peak has excellent views of Fort Bowie—the reason Captain Murray moved the station in the first place. These factors—the mean center analysis, Dade's blocked field of view, and the clear view of Bowie Peak—suggest a Webb Peak location for Murray's RM station.

However, despite its advantages, Webb Peak is not visible from Fort Grant, meaning a relay would be needed to establish a heliograph connection. Given that Murray was trying to avoid relaying through Alpina to reach Fort Bowie, this added relay seems an improbable solution. Furthermore, Volkmar stated in both his map and text that Fort Grant was directly connected to Murray's RM station.[59] Additionally, the geometry described by the lieutenants does not align well with Webb Peak. Webb Peak is 1.8 miles from the wood camp and 2.3 miles in a straight line from Eggleston's RM station on Merrill Peak, farther if one considers routing around the considerable obstacle created by Gaudy Canyon. In any case, it is farther than "about a thirty minutes' walk north from camp," as reported by Lieutenant Perry.[60] Lastly, the direction from Webb Peak to Merrill Peak is more west than the reported southwest.

Therefore, considering the close alignment with the azimuth and distance from Fort Grant, the visibility to all the connected stations, and the geometric relationships reported by the lieutenants, Murray's RM station is most likely located on the southern of the two Grand View Peaks.[61] The southern peak features several rock outcrops separated by small, marginally level areas. Visibility modeling shows that the area around two of these rock outcrops offers visibility to all eleven stations. However, a closer examination reveals that trees on the shoulder of Webb Peak obscure the view of Bowie Peak from one of these areas. So the most suitable location for the station is the southern outcrop of the southern of the Grand View Peaks.[62]

CHAPTER 11

The RM stations presented here were in use for only a short period. Eggleston's RM station was used in December 1889 to confirm viability and possibly again in early spring, then once more in April during that month's exercise and to troubleshoot the connection to Lookout Peak. Murray's RM station was used in May 1890 as well as during the February 1893 exercises. Heliograph Peak was used by the US Army for just one day, on December 15, 1889. The name Heliograph Peak likely originated after World War I, when the Forest Service used the peak for heliograph purposes.[63] Although Murray and Eggleston do not mention Alpina being used after the RM stations were established, during the May exercise it may have been initially occupied and then abandoned. Late in the exercise, it may have been used to connect with Fort Huachuca.

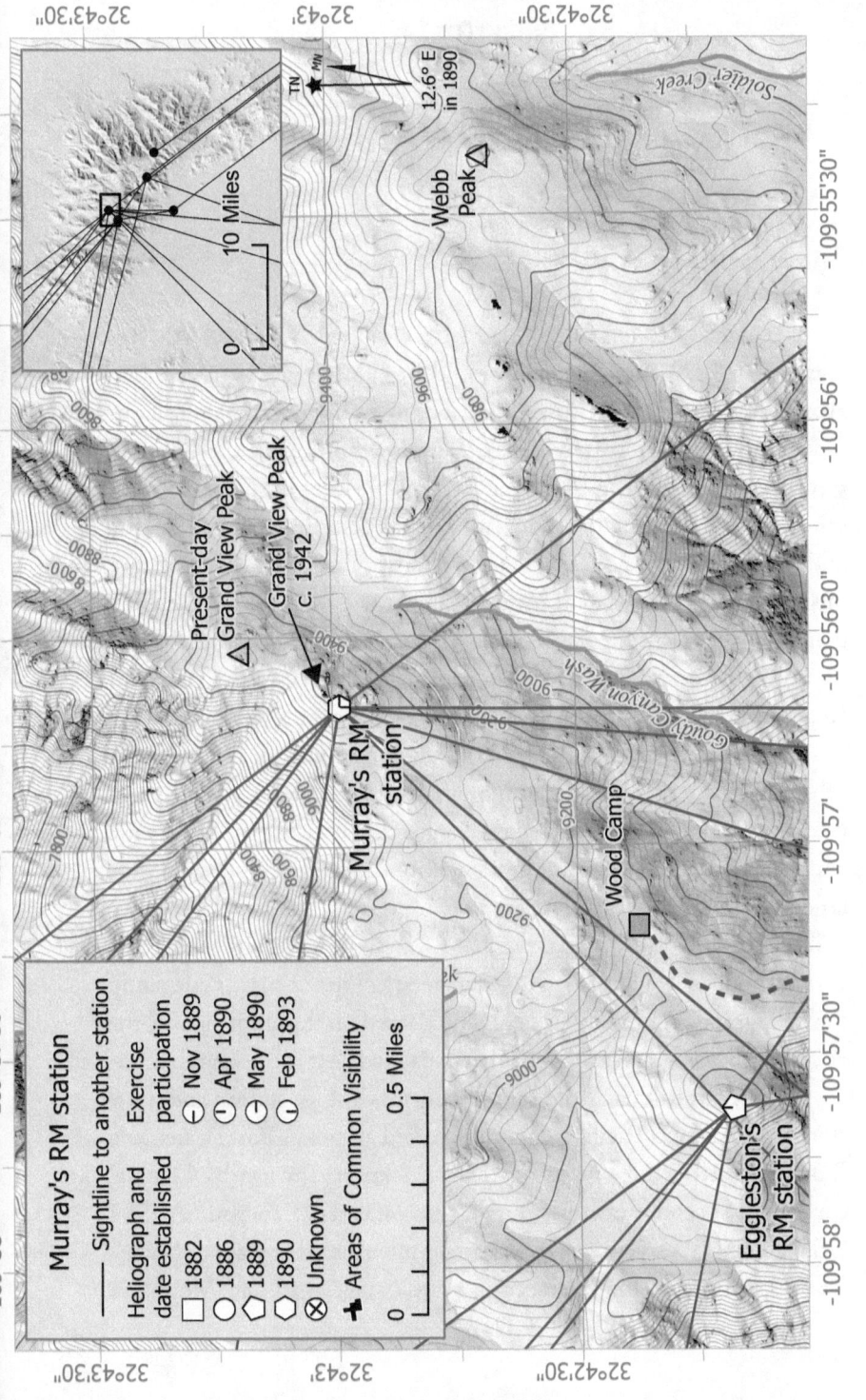

Map 72. Murray's RM station on Grand View Peak.

Chapter 12

Fort Lowell

FORT LOWELL WAS ESTABLISHED AT ITS CURRENT, EASTERN TUCSON location in 1873, after being relocated from a site closer to Tucson.[1] According to the Fort Lowell Museum, the fort was named after Brigadier General Charles Russell Lowell, who died in the Battle of Cedar Creek during the Civil War. Parts of the fort's hospital and some officer housing still stand today. Two reports commissioned by Tucson and Pima County provide extensive descriptions of the historic fort, including maps of its layout. Unfortunately, these maps do not include a heliograph station.[2] However, a telegraph station was identified on the north side of today's Fort Lowell Road, about 300 feet west of Craycroft Road, near the northwest corner of the former parade ground.[3]

It makes sense that the heliograph station was located near the other signals personnel and equipment at the telegraph office, as this proximity would have facilitated the connection between the heliograph and telegraph systems. From the vicinity of the telegraph station, there is good visibility to the east, where the connecting stations were located, and the row of cottonwood trees that flanked the southern edge of the parade grounds would not have obstructed the lines of sight. Gale set up the heliograph on November 17, 1889.[4] Based on this information, the Fort Lowell heliograph station has been mapped near the telegraph office.

In the late 1800s, Mountain Spring was a stage station and hotel along the Southern Pacific Mail Line, located at today's Posta Quemada Ranch.[5] By 1879 there was a hotel at the station. The nearby Colossal Cave would become an attraction for the residents of Tucson.[6] A 1904 US Geological Survey topographic map identifies this place as Shaw's Ranch, situated at the intersection of the stage road and Mountain

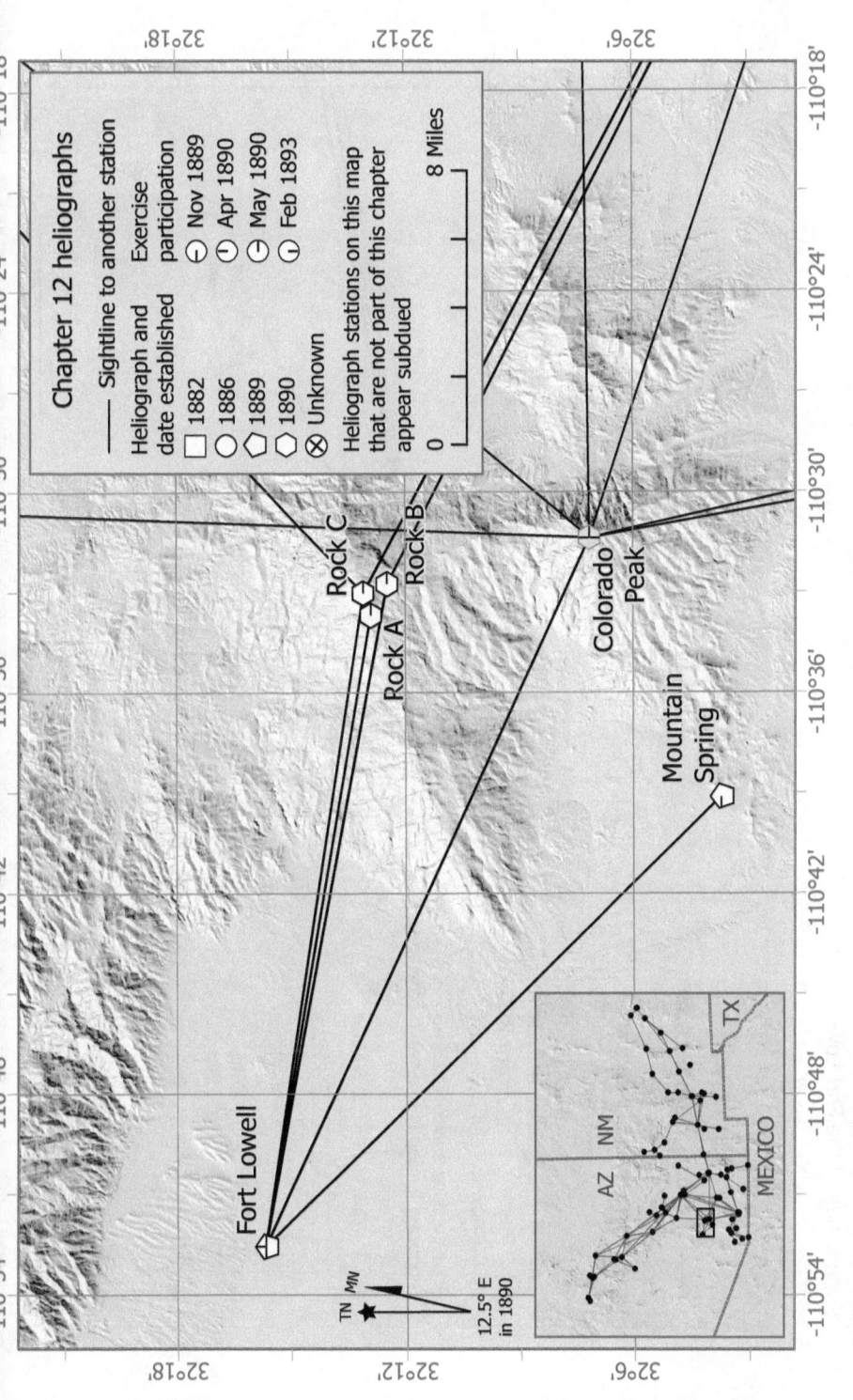

Map 73. Chapter 12 heliograph stations and connections.

Map 74. Fort Lowell heliograph station. The subdued background is the current US Geological Survey map of the Fort Lowell area.

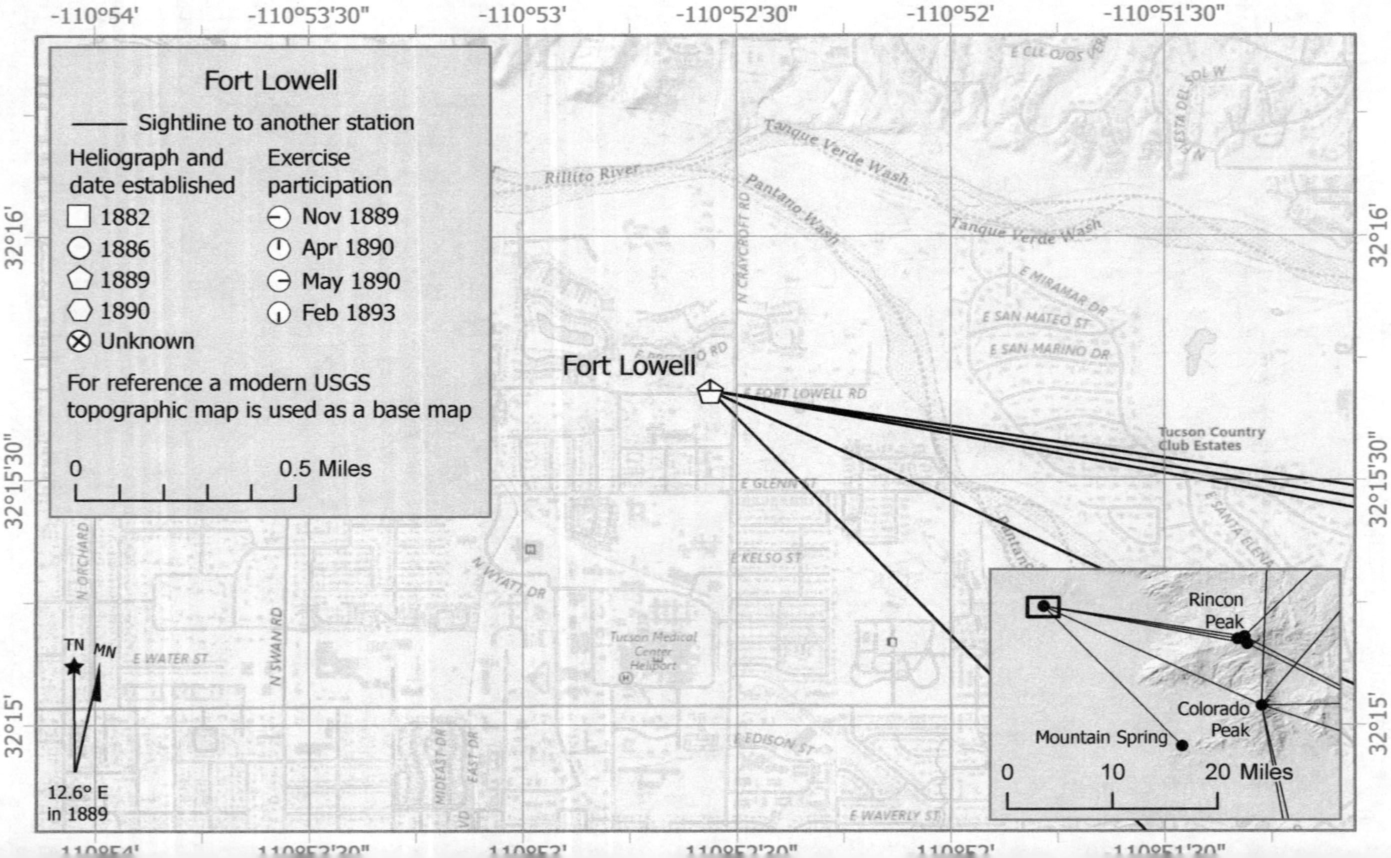

Spring Canyon (now Posta Quemada Canyon).[7] On the 1893 *Official Map of Pima County*, the place is simply labeled "Mountain Spring."[8]

In November 1889, Lieutenant Gale determined that "a hill about 1 1/2 mile N. N. W. [337.5° magnetic] from Mountain Spring" would satisfy the connection requirements between Fort Lowell and stations farther east, principally Fourr's Ranch North.[9] Gale sent Corporal Smart, Troop E, Fourth Cavalry, to establish a heliograph station atop that hill.[10] While Smart successfully connected with Fort Lowell from the station, he was unable to establish communication with Fourr's Ranch North despite his best efforts.[11]

In his January 20, 1890, report of the November exercise, Gale provided directions to several nearby visible peaks from the hill where Smart had set up the heliograph. These included Fort Lowell, Cochise Stronghold, the highest point of the Whetstone Mountains, the hill west of Tucson (today's Wasson Peak), Picacho Peak, and the high point in the Santa Rita Mountains (today's Mount Wrightson).[12] An observer can see all these peaks from three different locations near Mountain Spring: the top of today's Pistol Hill (to the northwest, about a mile from Mountain Spring), a small area on the southern slope of Pistol Hill, and a hilltop on a ridge 1.5 miles northeast of Mountain Spring. Since Gale specified that the station was on a hill northwest of Mountain Spring, the Mountain Spring heliograph station has been mapped atop today's Pistol Hill. The difficulty in connecting to eastern stations from Pistol Hill likely prompted Gale to establish a station atop the much higher Colorado Peak in March (see the Fort Bowie chapter).

Near the end of the May 1890 exercise, the spring supplying water to the heliograph station on Colorado Peak ran dry. Corporal L. P. Gouldman, then in charge of the Colorado Peak heliograph station, contacted Lieutenant Gale at Fort Lowell for instructions.[13] Gale, after consultation with Major Volkmar at Fort Bayard, instructed the corporal to close that station and, with Gale's assistance, establish a new station in the Rincon Mountains to the north of Colorado Peak.

On May 11, Gale sent mules to Colorado Peak to extract Gouldman and bring him and his team back to Mountain Spring, where they stayed

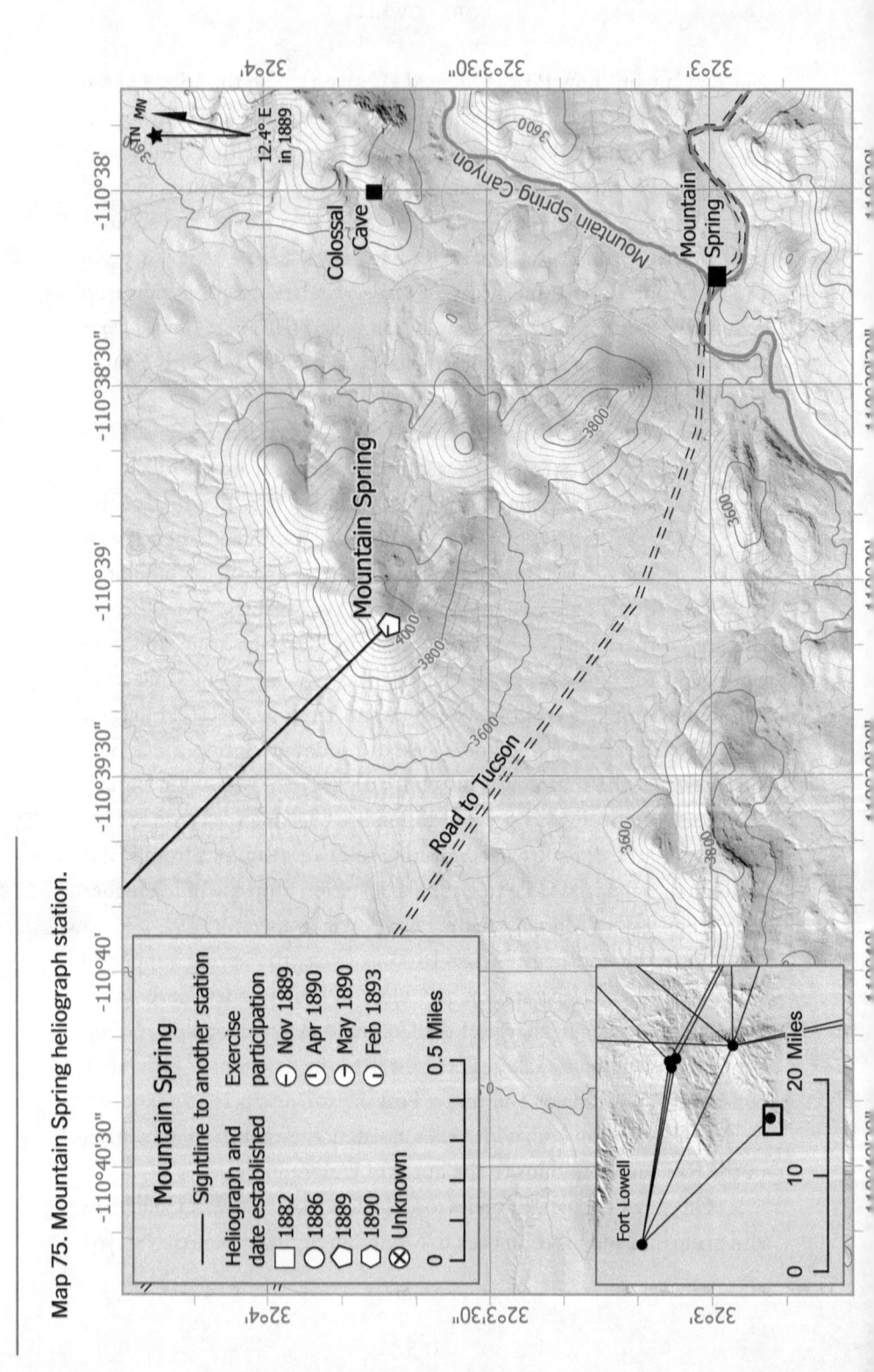

Map 75. Mountain Spring heliograph station.

overnight. The next morning, Gale and Gouldman ascended into the Rincon Mountains. Near the summit, several large, prominent rock outcrops are found. Some of these outcrops are large enough to have names: Saucer Rock (from 1890), Spud Rock (from 1904), and the more modern Helen's Dome, Duckbill, and Man Head. (For reference, Spud Rock is approximately 100 yards across.) Gale described these outcrops as "prominences," indicating their conspicuous nature.[14] He and Gouldman evaluated four of these rocks for potential heliograph stations, labeling them A, B, C, and D. They successfully connected with other stations from A, B, and C, but it appears that D was never used for communication.

At about noon on May 12, Gale and Gouldman established a station on Rock A. Gouldman remarked that from A they attempted to contact only Fort Lowell and Murray's RM station and then abandoned it to find a "more favorable place for a station."[15] The next day, May 13, they occupied Rock B, a mile to the southeast of Rock A. On May 14, Gale found another potential station, called Rock C, about a half mile east of Rock A. As the weather was bad, Gale instructed Gouldman to occupy Rock C on May 15. Then Gale returned to Fort Lowell.

If their intention was to mirror Colorado Peak's capability, there were six stations they needed to connect with (Fort Lowell, Fourr's Ranch North, Fort Huachuca, Bowie Peak, Table Mountain, and Murray's RM station). Since Fort Huachuca was to the south and Table Mountain to the north of Rincon's east–west ridge, any station selected that could connect to all six would be on or very near that east–west ridgeline in between. Indeed, any station visible to both Table Mountain and either Fort Huachuca or Fourr's Ranch North needed to be on or very near that east–west ridgeline.

On May 15, the last day of the exercise, Corporal Gouldman took position on Rock C and successfully connected with Fort Lowell, Murray's RM station, and Fourr's Ranch North. Gouldman reported that all six stations "worked from Colorado Peak can be worked at this point except Huachuca. This [the Huachuca station] however could be worked if the station there were moved half a mile west of the present site."[16] Visibility analysis reveals several small areas atop the Rincon Mountains where all the stations, except Fort Huachuca, are visible. These areas include a small spur located 0.3 miles northeast of today's Mica Mountain (the range's highest peak), some scattered rocks farther northeast, and today's Spud Rock, situated approximately 0.4 miles west of Mica Mountain. Among these options, only Spud Rock can be considered a "prominence." This heliograph station was called Rincon by the 1890 exercise participants.[17]

Visibility analysis also confirms Gouldman's prediction: If the Huachuca station were moved west, then Rock C would be visible from that point as well. If the Huachuca station were relocated across Huachuca Canyon to the adjacent ridge (the station was on the ridgeline east of Huachuca Canyon at the time), Spud Rock would be visible while the other locations (the spur and scattered rock outcrops northeast of Mica Mountain) would remain out of sight.[18] Based on Gouldman's description and the visibility analyses, Rock C has been mapped at present-day Spud Rock.

Rock A was about a half mile west of Rock C.[19] According to Gale, Rock A was visible from all of the six required stations "but loses Bowie Peak and Fourr's Ranch."[20] While most of the rock prominences east of Spud Rock (as well as Spud Rock) are visible to Fourr's Ranch North as well as Bowie Peak, none of the rock prominences west of Spud Rock are visible from Fourr's Ranch North. Indeed, no place on the range west of Spud Rock is visible from Fourr's Ranch North.

Today's Helen's Dome (not the one near Fort Bowie), a prominent feature located 0.68 miles west of Spud Rock, is visible from Fort Lowell, Murray's RM station, and Table Mountain but not from the heliograph station at Fort Huachuca. However, it is visible from the slightly higher peaks in the Huachuca Range. Interestingly, Corporal Gouldman, who had previously connected to Fort Huachuca from Colorado Peak and was familiar with the station's location, made no mention of Fort Huachuca's visibility from Rock A. This suggests that Rock A may align with Gale's description only if "Fort Huachuca" refers to the hills and peaks above the fort, rather than the heliograph station itself.

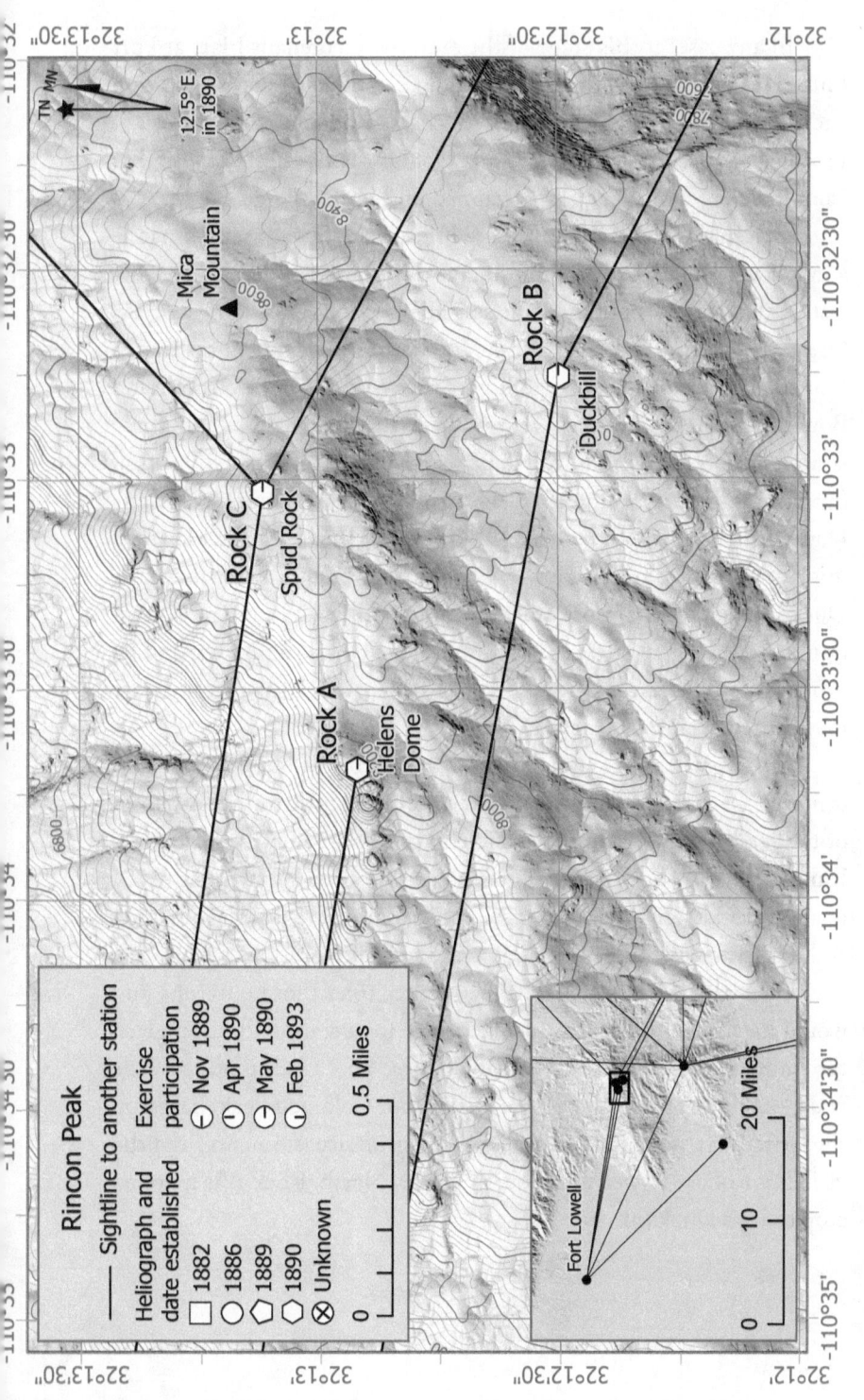

Map 76. Rincon Peak heliograph stations.

In any case, at this stage of the exercise, Lieutenant Hart at Fort Huachuca was only attempting to connect with Murray's RM station and did not attempt to connect with the stations on Rincon Peak.[21] Based on the descriptions of Gale and Gouldman as well as the visibility from Helen's Dome, Rock A is mapped on Helen's Dome.

Gale and Gouldman both noted that Rock B was approximately a mile away from Rock A. Gouldman's description added a direction, stating that B was "about one mile southeast of A."[22] According to Gale, his visibility from Rock B was limited to Fort Lowell and Fourr's Ranch North. He mentioned that visibility to Huachuca might be possible and Bowie might be seen if trees were removed to the east.[23] Notably, there is no mention of visibility to Murray's RM station. Today's Duckbill (another large rock prominence), located southeast of Rock A (Helen's Dome), is visible to both Fort Lowell and Fourr's Ranch North. However, it is not visible from Fort Bowie, Fort Huachuca, or Table Mountain. Duckbill is situated almost exactly a mile from Rock A and at a direction of 105° magnetic (in 1890). Given Duckbill's prominence and its proximity to the stated distance from Rock A, it is a strong candidate for being identified as Rock B.

On the other hand, at a direction from Rock A of 105°, Duckbill is more east than southeast and rock outcrops a little to the southeast of today's Manning Camp are visible to both Fort Lowell and Fourr's Ranch North. These small outcrops are very near southeast (129° magnetic) and 1 mile from Rock A. The visibility as well as the direction and distance make these outcrops a good fit for Rock B as well.

That said, Duckbill is 400 feet higher than those outcrops and would have been the more obvious choice for a station location given the limited amount of time available (three and a half days) to Gale and Gouldman to find alternative stations for Colorado Peak. Based on the spatial relationship with Rock A, the greater prominence, and the visibility to Fort Lowell and Fourr's Ranch North, Rock B is mapped atop today's Duckbill.

Chapter 13

Epilogue

UNDOUBTEDLY, THERE WERE HELIOGRAPH STATIONS USED BY THE ARMY in the Southwest between 1882 and 1893 that are not documented here. In 1882 Lieutenant Maus had a system in place to connect Fort Bowie, Fort Grant, and soldiers stationed upstream along the Gila River. It is therefore reasonable to conclude that the camps along the Gila had heliographs as well. In a 1926 correspondence between Colonel Dravo and Charles Gatewood (mentioned in the introduction), Dravo confirmed that during the 1886 campaign, "Every organization in the field had a heliograph with it."[1] It is likely that most of those heliographs were actively used.

Using the Spencer map, the list of those camps (not covered previously) includes Horse Spring, New Mexico; Fairview, New Mexico (near Winston); a station east and a bit south of Fort Selden; Cloverdale, New Mexico; and Bisbee, Arizona (though this may be the same as the Henry Forrest's Ranch).

Fort Apache and Fort Mojave reported using heliographs for practice in 1890, although they were not part of that year's two heliograph exercises. Fort Bayard also reported practicing with flags at a 10-mile range. Fort Marcy in New Mexico experimented with distances of up to 20 miles using a 2-inch heliograph.[2] In 1892 there were reports of a heliograph survey of the important military districts and hopes that Clough Overton, Frank Greene, and William Volkmar would collaborate on a "perfect heliograph map."[3]

While the army's expenditures for military signals more than doubled from 1886 to 1891 (rising from $2,511 to $6,392), the majority of the Signal Corps expenditures during that time went to monitoring and reporting the weather, averaging $150,721 yearly between 1886 and 1891.[4] On July 1, 1891, the US Weather Service moved from the Signal Corps to the Department of Agriculture.[5] This shift allowed Greely to redirect more funds toward heliograph manufacturing and deployment to the field. In 1893 the Signal Corps spent $18,790 on military signaling. In conjunction with this increase, 120 heliographs were contracted, built, and delivered to the Signal Corps.[6] These instruments, combined with the procurement of other signal equipment, prompted Greely to report, "For the first time in its history, the army is equipped with sufficient [signal] instruments for ordinary practice and instruction."[7]

Heliograph use would continue for several years, including limited use in the Spanish-American War, the Philippines, China, and the Veracruz Expedition of 1914, and heliographs are mentioned in cables during World War I. For a time, wherever the army went, the heliograph went as well.[8] According the General Arthur MacArthur's signals officer, the heliograph was particularly useful for island-to-island communications in the Philippines in cases where the expense of a cable was too much to connect "smaller islands, not sufficiently important to warrant a cable."[9] In November 1916, members of the First North Carolina Infantry Brigade and the Michigan Signal Corps deployed to the US–Mexican border as part of the Mexican Punitive Expedition, set up a heliograph on Mount Franklin, Texas, and connected with their camp below as well as Fort Bliss.[10] This may have been the last operational use of the heliograph within the United States. In 1913, in a successful test conducted by the Forest Service, heliographs returned to the Pinaleños at Webb Peak and connected to a ranger station below as well as to Eureka Peak in the Galiuro Mountains to the west.[11] A 1918 cable from the adjutant general to General Pershing in France mentions the use of heliographs to report personnel matters.[12]

Advances in electrical systems and wireless and wired telegraphy, combined with the army's operational move into areas where the terrain and climate were different from the trans-Mississippi west, especially the Southwest, led to a diminished reliance on the heliograph and more reliance on newer technologies. From the turn of the century on, the

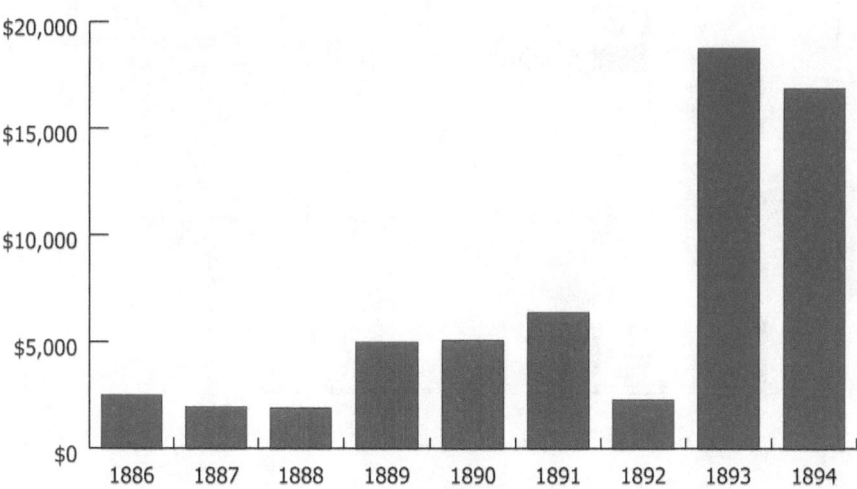

Figure 15. US Army Signal Corps expenditures on "military signaling" between 1886 and 1894. (Source: Annual reports of the chief signal officer.)

Figure 16. Members of the First North Carolina Infantry and Michigan Signal Corps, deployed as part of the Mexican Punitive Expedition, at a heliograph station along the US–Mexican border in November 1916. Courtesy of the State Archives of North Carolina.

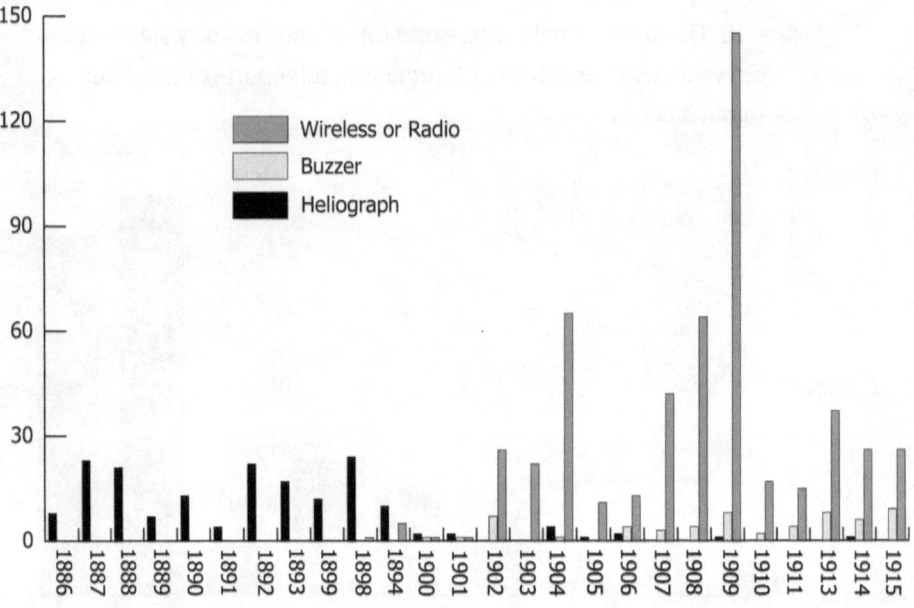

Figure 17. Number of times the words heliograph, buzzer, and wireless or radio are mentioned in the various chief signal officer annual reports.

heliograph received fewer and fewer mentions in the chief signal officer's annual reports, where more and more ink was devoted to telegraphy and wireless telegraphy.

In 1898 Greely, in his report to the secretary of war, stated that Colonel James Allen (a future chief signal officer) had "devoted much attention to the system of wireless telegraphy with a view to the adoption by the Signal Corps."[13] In 1899 Lieutenant George Squire spearheaded the effort to develop a wireless system. On September 30, 1899, as a result of that effort, wireless telegraphy connected the lighthouse at Fire Island, Long Island, and a lightship (an anchored ship used in lieu of a lighthouse) 10 miles distant.[14] In 1905 the Signal Corps was "perfecting" a portable wireless communication system, and by 1908 wireless telegraphy was being employed in the United States, the Philippines, and Cuba.[15] The 1908 kit of the army's communications companies contained two wireless telegraphs sporting a nominal range of 20 miles.[16] Signal Corps Communication Company I, deployed to Cuba, created a wireless network between the shore near Havana and the fleet in Havana's harbor. Company I also connected wirelessly, using a long antenna, to Key West, 90 miles distant.[17] By the end of the first decade of the 1900s, wireless systems, radio, had moved out of the laboratory and into the hands of deployed commanders.

Wired systems advanced as well. Smaller and lighter telegraph and telephone systems were designed and deployed. By 1908 "buzzer lines," which were reliable and able to operate over poorly shielded wires,[18] had "developed to a point where they are so mobile and reliable in operation as to inspire the confidence of the line of the army which they are designed to serve."[19] While these buzzer lines were primarily for larger units, as early as 1899 efforts were undertaken to make wired technology light and mobile enough to satisfy the needs of smaller, lighter forces. One such system, the D-type kit, an impedance buzzer, weighed less than 10 pounds, with a mile of aluminum-bronze wire weighing 8 pounds.[20] The lightweight buzzer had developed into a cavalry wire set by 1907.[21]

Both wired and wireless telegraphy were methods of communication "against which darkness, fog, and weather conditions [the chief limitations of the heliograph] avail nothing."[22] Moreover, wired communication had the added advantage of not requiring a line of sight, which flags, heliographs, and to varying degrees wireless communication needed. Heliographs required line of sight and sunlight—both are functions of terrain, time, and weather. While the desert Southwest provided advantageous terrain and weather for heliographs, that advantage diminished as the army began operating in wetter, greener environments: Cuba and the Philippines—operations coincident with technological advances in wired and wireless communications. These advances combined with different operational environments effectively ended the army's use of the heliograph.

Heliographs are still present in the desert Southwest. At any of the still-standing forts with a visitor center or museum, you might find one set up, usually in a corner, with information placards nearby. If there isn't an actual heliograph, a map, description, and photographs display their history and use. In any case, they are there, still communicating their part of history to anyone interested.

Appendix 1

Profiles

Below are short profiles of the soldiers (and one civilian), mostly lieutenants, who located or otherwise assisted in establishing a heliograph station or who had an impact on the system as a whole.[1] For every one of these officers, there were several, if not dozens, of enlisted men who worked tirelessly to turn the officer's plans and schemes into something that worked. Unfortunately, with few exceptions, the names of these soldiers were not recorded by history.

Frank Herman Albright graduated from West Point in 1887 as an infantry officer. He served at Fort Huachuca from November 1888 to October 1891 and was promoted to first lieutenant in 1895. In 1899 he rose to the rank of captain and continued his service on the frontier. Between 1907 and 1909 he was deployed to the Philippines. By 1911 Albright had been promoted to major and commanded battalions in Washington state and Hawaii.[2] He graduated from the Army War College in 1915 and was promoted to brigadier general by 1917.[3] During World War I, he served in France as a brigade commander.[4] Albright retired in 1919 and died in 1940. He is buried in San Diego.[5]

William Black enlisted in the US Army Signal Corps' weather service on October 7, 1871, and served until August 30, 1879. On September 1, 1879, he accepted a commission as a second lieutenant in the Twenty-Fourth Infantry.[6] Assigned to Fort Bayard on June 5, 1888, he was placed in charge of the heliograph during the May 1890 exercise.[7] Black served throughout the West and in the Philippines until his retirement as a major in the Thirteenth Infantry in 1905. He was later called back to active duty, serving in Cuba in 1906 and in Oklahoma in 1908. He died from heart disease in Oklahoma City on November 25, 1911, at the age of sixty-one.[8]

Andre Walker Brewster (in some records, his middle initial is connected to his first name, rendering "Andrew") was born in Hoboken, New Jersey, on December 9, 1862.[9] He was commissioned as a lieutenant in the Tenth Infantry in January 1885 and was posted at Fort Union, New Mexico, that same year.[10] Brewster served on the frontier, in Cuba, during the China Relief Expedition (Boxer Rebellion), and in World War I as inspector general of the American Expeditionary Force. He retired as a major general in 1925. He died on March 27, 1942, and is buried in Arlington, Virginia. As a captain of the Ninth Infantry in Tientsin, China, he was awarded the Medal of Honor on September 15, 1903.[11]

William Alexander Campbell enlisted in 1885 and was commissioned a second lieutenant in 1889.[12] He served at Fort Huachuca and Fort McDowell before attending the Infantry and Cavalry School from 1891 to 1893.[13] Following this, he was assigned to Frankford Arsenal for a course of instruction in ordnance duties.[14] From September 1895 to November 1898, he served as a professor of military tactics in Illinois, after which he was posted at Fort Crook, Nebraska.[15] Campbell served in the Philippines, where he was stationed at Arayat, Pampanga, Luzon, as a captain and the post's quartermaster.[16] In 1906 he was mentioned in a *Detroit Free Press* article as a civilian commander of the Spanish War Veterans organization.[17]

Adna Romanza Chaffee enlisted in the Sixth Cavalry in the summer of 1861. In 1862 he was promoted to first sergeant and then to second lieutenant. After the Civil War, he transferred to the western frontier, and by 1882 he had been brevetted as a lieutenant colonel. His actions at the Battle of Little Dry Wash in Arizona were noteworthy, as this was one of the last significant force-on-force engagements with the Apaches.[18] Chaffee commanded during the Santiago campaign and later served as chief of staff to the military governor in Cuba. During the Boxer Rebellion, he led the US contingent in China. He also commanded the Department of the Philippines and served as its governor. In 1904 Chaffee became the chief of staff of the US Army, a position he held until his retirement in 1906. He died on November 1, 1914.[19]

Alexander Lucian Dade, born on July 18, 1863, graduated from West Point in 1887. He served at Fort Bayard, Fort Grant, and San Carlos before transferring to the Infantry and Cavalry School at Fort Leavenworth, Kansas, in August 1891.[20] Dade later served in Cuba and completed multiple tours in the Philippines. After returning to the United States in 1908, he was assigned to the Army War College in August 1909.[21] In 1911 he was stationed in Wyoming and Texas, and he served in Mexico during the Mexican Punitive Expedition, from March 1916 to February 1917, earning two Silver Stars.[22] From November 1917 to February 1918, he commanded the Air Division of the Signal Corps.[23] Dade retired as a colonel on June 23, 1920, and died on January 8, 1927, in Hopkinsville, Kentucky. He was posthumously promoted to brigadier general on June 21, 1930.[24]

Edward Everett Dravo was born on November 23, 1853, in Pittsburgh, Pennsylvania. He graduated from West Point with the class of 1876 and headed west to join Troop I, Sixth Cavalry, at Fort McDowell, Arizona.[25] When the Sixth Cavalry was reassigned in 1884, Dravo moved to Fort Bayard, New Mexico. He served in the Department of Colorado and Texas from 1892 to 1898 and later deployed to the Philippines and the China Relief Expedition. By 1905 Dravo had risen to the rank of colonel. He returned to the Philippines in 1907. After coming back to the United States in 1909, he served for another four years before retiring in 1913. Recalled to active duty in 1916, Dravo organized a logistics center in Kansas City and was relieved from active service at the end of World War I. He died in Paris, where his daughter lived, on July 30, 1932.[26]

George Brand Duncan graduated from West Point in 1886 and served in the West at Fort Wingate, Fort Union, Whipple Barracks, and Fort Mojave in the Ninth Infantry.[27] Duncan served in the Spanish-American War in Cuba and the Philippines, as well as along the US–Mexican border from 1914 to 1917. During World War I, he was promoted to brigadier general and commanded the Twenty-Sixth Infantry. For his actions at the Battle of Verdun, he was awarded the French Croix de Guerre. He went on to command the First Brigade of the First Division before his promotion to major general. As a major general he commanded the Seventy-Seventh and Eighty-Second Divisions during World War I.[28] Duncan retired in 1924 and died in 1950.[29]

Millard Fillmore Eggleston graduated from West Point in 1877 and was assigned to the Tenth Cavalry. He served at several forts in Texas before being assigned as a first lieutenant to Fort Verde in 1887. Selected for the Infantry and Cavalry School of Application at Fort Leavenworth, he attended from September 1887 to July 1889. After returning to Arizona, where he served at Fort Apache and San Carlos, he resigned his commission on June 9, 1890.[30] He then moved to Ashland, Oregon, where he became a journalist, mining engineer, municipal judge, secretary of the board of trade, and president of the Ashland Commercial Club. In 1911 Eggleston was elected to the Oregon State Legislature. He died later that year in Salem, at the age of fifty-six, on February 3.[31]

Charles Wendell Fenton was born in Michigan in 1865. He graduated from West Point in 1888 and initially served at Fort Whipple, Arizona, along with several other western posts before being deployed to Puerto Rico in 1899. Returning to the United States, he completed tours in New Mexico and South Dakota. He deployed to the Philippines in 1903 and again from 1910 to 1912, stationed at Camp Overton. Fenton attended the Army War College in 1915 and subsequently served as the commander of Fort Myer, Virginia, until his death on January 15, 1918, from meningitis.[32]

Alvarado Mortimer Fuller enlisted in the US Army in 1870 and retired as a colonel in 1918. He began his military career as a private in the Second Cavalry at Omaha Barracks on January 1, 1870. Fuller was commissioned as a second lieutenant in 1879, promoted to first lieutenant in 1888, and became a captain in 1898.[33] He retired on April 1, 1918, setting a record for length of service. His career was noted in the *Army and Navy Journal of 1918* and reported in the *Catholic Advance* article "Maj. Alvarado M. Fuller Promoted to Be Colonel" on September 28,

1918.³⁴ Fuller saw service in Cuba and the Philippines. He was also a futurist writer; his novel *A.D. 2000* was published in 1890. Fuller died in Topeka, Kansas, in January 1924.³⁵

George Henry Goodwin Gale was born in 1858 in Maine. He graduated from West Point in 1879 and was assigned to the Fourth Cavalry. After being stationed at various posts in the western United States, he arrived at Fort Huachuca in 1884 and was promoted to first lieutenant the same year. From 1884 to 1888, Gale served as a mathematics professor at West Point.³⁶ He subsequently held a variety of posts until his deployment to the Philippines in July 1898 as a captain. After returning to the United States in 1899 on sick leave, he was transferred to the Fifth Cavalry and later the Ninth Cavalry. Gale returned to the Philippines twice between 1903 and 1907. He was promoted to lieutenant colonel in 1907 and then to colonel in 1912. He retired in 1914 but was recalled to active duty during World War I. Gale died in Brussels on May 1, 1920.³⁷

Charles Baehr Gatewood graduated from West Point in 1877 and was assigned to Camp Apache, Arizona, as a lieutenant in the Sixth Cavalry. In 1879, as commander of Company A, composed of Native American scouts, he participated in operations against Victorio.³⁸ In July 1886, when General Miles received word that hostile Apaches were open to surrender, he sent Gatewood south into Sonora with two Native American scouts on a "mission to enter the hostile camp and demand their surrender."³⁹ After coordinating with Henry Lawton, Gatewood entered the camp on August 24 and delivered the demand for surrender. The following day, Geronimo arrived at Lawton's camp to discuss peace.⁴⁰ For the next four years, Gatewood served as an aide to General Miles. In 1892 he was wounded in a dynamite incident at Fort McKinley, Wyoming. Gatewood died of cancer at Fort Monroe, Virginia, on May 20, 1896.⁴¹

Quincy O'Maher Gillmore graduated from the US Military Academy in June 1873 and, after a period of leave, was posted at Fort Sill, Oklahoma. From 1875 to 1880, he served on frontier duty in Texas.⁴² From 1880 to 1882 he was an instructor of tactics at West Point and also served as the post's commissary officer and artillery commander.[43] Gillmore was stationed in New Mexico from November 30, 1885, to July 19, 1886. In January 1886, he served as the commanding officer of Troop G and Camp Boyd near Hillsboro, New Mexico.[44] In 1895 Gillmore was placed in a government hospital, and he was medically retired from active service in 1896.[45] He later continued his military service with the New Jersey Volunteer Infantry, reaching the rank of brigadier general. Gillmore died on July 14, 1923.[46]

After three years at the US Naval Academy (1871–1873), William Alexander Glassford enlisted as a private and later advanced to sergeant before being commissioned as a lieutenant in the Signal Corps in 1879.[47] While serving in Arizona as the department's signal officer, he helped lay the groundwork for heliograph communications by evaluating potential sites, work that partly informed the locations used during the 1890 exercise.[48] An innovator, Glassford became involved with signaling balloons, which eventually led him into the emerging field of aviation.[49] He retired in 1916 as the commanding officer of Rockwell Field, North Island, California. Glassford retired to Phoenix and died in San Francisco on August 7, 1931.[50]

L. P. Gouldman remains largely a mystery. He is briefly mentioned in the 1890 report of the chief signal officer as a participant in the heliograph exercise, but beyond this little is known about him as a soldier.[51] Interestingly, an L. P. Gouldman appears in Washington, DC, newspapers around the turn of the century as a police officer, though it is unclear if this is the same individual.[52]

Frank Greene, born on March 16, 1849, was the architect of the 1893 heliograph system and a member of the US Army Signal Corps.[53] Greene enlisted in 1876 and received a commission as a second lieutenant in 1882.[54] As a lieutenant colonel, he deployed to Santiago, Cuba, and in 1912 to the Philippines.[55] Greene retired prior to World War I but was recalled to active duty to serve as the Signal Corps superintendent of the Army Transport Service. In December 1917, he was relieved from the Army Transport Service and assigned as the signal officer at Western

Department headquarters in San Francisco.[56] Greene died from a stroke on August 9, 1924, while on a road trip north of Pasadena, California.[57]

Frank Carter Grugan enlisted in the US Army during the Civil War and remained in service afterward, stationed in the West as a cavalryman.[58] In 1873 he was assigned to detached service with the Signal Corps in Washington, DC, as an instructor.[59] Grugan served with the Signal Corps for several years before switching to the artillery in 1879.[60] In 1881 he was awarded a patent for the Grugan heliotrope, a fixed-mirror type that used a screen instead of shifting the mirror. This design strongly influenced the army's "service heliograph."[61] Along with Lieutenants Maus and Greene, Grugan also led the Signal Corps board that in 1888 recommended the service heliograph.[62] In 1898 he left the army as a major after his health began to fail during service in Cuba. Grugan died in New York City in 1917.[63]

William Horace Hart was born in Minnesota on March 20, 1864. He graduated from West Point in 1888 and was initially assigned to the Twentieth Infantry, serving at Fort Assinniboine, Montana.[64] In early 1889 he was reassigned to the Fourth Cavalry in the Department of Arizona, likely at Fort Huachuca, where he became involved in the heliograph systems in 1889 and 1890. Hart left Arizona with the Fourth Cavalry in mid-1890 and continued to serve on the frontier until January 1899, when he deployed to Cuba as part of the Army of Cuban Occupation. In 1898 he transferred to the Quartermaster Corps, with which he held various positions and served in France during World War I.[65] In 1922 Hart was appointed quartermaster general with the rank of major general. He died at Walter Reed Hospital on January 2, 1926.[66]

Henry Walter Hovey started his military career as a private in Company B, Seventh New York Regiment.[67] On November 23, 1880, Hovey was commissioned as a second lieutenant in the Twenty-Fourth Infantry.[68] He served at Fort Bayard, New Mexico, from 1888 to 1895, followed by a posting at Fort Huachuca in 1896. Hovey then spent four years in Skagway, Alaska, before being stationed at Fort Missoula. In 1904 he deployed to the Philippines, returning to the United States in

1907.⁶⁹ After settling in Vermont, Hovey became the inspecting officer of the National Guard and commander of Norwich University. He died unexpectedly from a heart condition on November 14, 1908.⁷⁰

Evan Malbone Johnson Jr. began his military career as a private and, forty years later, retired as a general. He enlisted in the army in 1882, quickly rose to the rank of sergeant, and received a commission in 1885, which came with a transfer to the Tenth Infantry on the western frontier. Johnson served in Cuba and in the Philippines, where he became the provincial governor of Romblon as a major. He was rapidly promoted, becoming a lieutenant colonel in 1914, a colonel in 1916, and a brigadier general in 1917. On December 1, 1917, he assumed command of the Seventy-Seventh US Infantry Division, a position he held until May 10, 1918. Johnson died in Paris on October 13, 1923, a year after retiring.⁷¹

Henry Clay Keene Jr. was born at sea off the coast of Peru on the ship his father commanded. He graduated from West Point in 1881 as a second lieutenant in the Twenty-Fourth Infantry and began his frontier duty at Fort Reno in 1886.⁷² Keene served at various posts, including Fort Grant, Fort Bayard in New Mexico, and San Carlos in Arizona. Later he saw service in Santiago, Cuba, and the Philippines.⁷³ He retired from active duty in 1906 due to a disability but took a commission as a major in the United States Guard in 1918.⁷⁴ Keene died on March 24, 1940, in Brookline, Massachusetts.⁷⁵

Joseph Dugald Leitch graduated from West Point in 1889 as an infantry lieutenant and was immediately sent to frontier duty at Fort Bayard, New Mexico, where he served from October 1889 through December 1891. He was stationed at various posts across the western United States before deploying to Cuba from Fort Douglas, Utah, in 1898. After contracting yellow fever, Leitch was transferred back to the United States and returned to Fort Douglas in October 1898.⁷⁶ In 1901 he deployed to the Philippines as a captain, returning to the United States in 1902. Leitch subsequently served in Nebraska, Kansas, Iowa, and Virginia before commanding a company in the Twenty-Fifth Infantry from November 1906 to June 1907. He deployed again to the

Philippines in September 1908 and returned in 1909.[77] After attending the War College from 1913 to 1914, he again served in the Philippines in 1916 and 1917. Upon his return from the Philippines, Leitch served with the Fortieth Infantry in Minnesota and joined the general staff, both in 1917. In 1918 he was promoted to brigadier general and major general, briefly holding commands before reverting to colonel in 1919.[78] He served as inspector and chief of staff of the American Expeditionary Force in Siberia from 1919 to 1920.[79] Following Siberia, he held commands in the Philippines and the United States until his retirement. He retired as a major general at Fort Lewis, Washington, on March 5, 1928.[80] Leitch died in San Francisco in 1938.[81]

William T. Littlebrant graduated from West Point in 1888 and was assigned to the Nineteenth Infantry. He initially served in Colorado before transferring to the Tenth Cavalry.[82] From 1891 to 1898, he held various posts across the western United States, including in Arizona, Montana, North Dakota, and Utah, before being transferred to Havana, Cuba, in 1898.[83] Littlebrant served in Cuba until his transfer back to the United States in 1902. In 1903 he was deployed to the Philippines, where he commanded the First Squadron, Twelfth Cavalry, from 1904 to 1905. He later deployed to Cuba with the Eleventh Cavalry from 1907 to 1909. After serving at various posts in the United States and the Philippines, he became a student at the Army Service Schools in 1911.[84] Promoted to major in 1912, Littlebrant became the superintendent of Yosemite National Park in 1913. In 1917 he was promoted to colonel, and he commanded the Eighty-First Field Artillery from 1917 to 1918. Promoted to brigadier general in 1918, he assumed command of Camp McClellan, Alabama, on March 20, 1919. He held the position until his death on July 2, 1919.[85]

Marian Perry Maus was born in Maryland in 1850 and graduated from West Point in 1874. He served with the First Infantry during the Nez Perce Expedition, where he received the Silver Star for his actions at Bear Paw Mountain. In 1882 Maus was transferred to Arizona.[86] In January 1886, General George Crook sent an expedition into Mexico to capture Geronimo, with Captain Emmet Crawford in command and

Lieutenant Maus leading a fifty-man scout company under Crawford. Near the headwaters of the Agos River, deep in Mexico, Crawford's forces found and attacked Geronimo's camp, only to discover it had been abandoned. During occupation of the empty camp, Mexican forces attacked and killed Captain Crawford. Lieutenant Maus took command, convinced the Mexican forces to abandon their hostilities, and initiated negotiations with Geronimo.[87] For these actions, Maus was later awarded the Medal of Honor.[88] Throughout his career, Maus served in a variety of roles, including staff positions and infantry commands. He deployed to the Philippines and was promoted to colonel in 1904. In 1907 Maus was assigned as the commander of the Department of California. Two years later he was promoted to brigadier general and commanded the Department of Columbia. He retired in 1913 and died on February 9, 1930.[89]

Nelson Appleton Miles was born near Westminster, Massachusetts, on August 8, 1839. He joined the Twenty-Second Massachusetts Infantry as a first lieutenant in 1861 and quickly rose through the ranks during the Civil War, mustering out as a brigadier general in 1866.[90] In 1868 he was appointed colonel of the Fifth Infantry and sent to the western frontier, where he fought in campaigns against various Native American tribes from Texas to Montana.[91] Promoted to brigadier general in 1880, Miles replaced General Crook as commander of the Department of Arizona on April 12, 1886.[92] Miles accepted Geronimo's surrender on September 4, 1886.[93] After Geronimo's surrender, Miles continued to rise swiftly through the ranks, becoming chief of staff of the US Army on September 29, 1895.[94] He held this position until his retirement on August 8, 1903.[95] Historian Robert Utley, in his *Encyclopedia of the American West*, describes Miles as "perhaps the most effective and successful of all the military leaders of the Indian Wars."[96] Miles died in Washington, DC, on May 15, 1925, and is buried in Arlington, Virginia.[97]

Cunliffe Hall Murray graduated from West Point in 1877 and spent thirty-nine years in the US Army Cavalry.[98] He served in a variety of locations, including Arizona, the Presidio in San Francisco, and the Philippines, where he was the aide-de-camp to General Otis.[99] Throughout his career, he was also stationed in Washington, DC, Chicago, Kansas, Iowa, Vermont, Ohio, and Nebraska. In 1914 he returned to the Philippines, where he commanded the Seventh Cavalry.[100] He retired in August 1916 as a colonel. He died in March 1936 in Tenafly, New Jersey.[101]

John Mitchell Neall was commissioned as a second lieutenant of cavalry from West Point on June 13, 1883, and was subsequently assigned to Fort Bayard, Fort Cummings, Fort Grant, Fort Thomas, and Fort Bowie.[102] In 1890 he participated in the May heliograph exercise at Fort Bowie before transferring to Fort Lowell and later to Fort Bidwell, California.[103] From November 1890 to September 1894, Neall taught military science and tactics at the University of Nevada, Reno. After his teaching assignment, he served at the Presidio in San Francisco and then at Yosemite National Park before returning to the Presidio in November 1896.[104] On February 16, 1899, Neall "disappeared, leaving a deficiency in his accounts with the Post Canteen."[105] He was court-martialed later that year and subsequently dismissed from the army for failing to properly account for funds, conduct unbecoming an officer, and unauthorized absence.[106]

William W. Neifert was born in Pennsylvania on January 27, 1865. He left home at an early age and worked various railroad jobs before enlisting in the US Army Signal Corps on November 20, 1885.[107] He was transferred to the West, where he operated the heliograph station atop Little Baldy Peak through September 1886.[108] Neifert also served at Fort Reno in the Oklahoma Territory, as well as in San Antonio and Brownsville, Texas; Norfolk, Virginia; and Vineyard Haven, Massachusetts, before leaving the Signal Corps. He later became a major in the Signal Reserve and served with the Pennsylvania National Guard. In his civilian career, Neifert worked as a meteorologist for the Department of Agriculture. He died at the age of eighty-six on March 23, 1951.[109]

Clough Overton graduated from West Point in 1888 and was initially assigned to Fort McDowell, Arizona.[110] After participating in the May 1890 heliograph exercise, he was transferred to Fort Walla

Walla, Washington, arriving in August 1890 and serving there until September 1895. He was then appointed as an instructor of tactics at Maryland Agricultural College. He served in Cuba in 1898 and fought at the Battle of San Juan and the siege of Santiago.[111] In October 1898, he was temporarily assigned from Fort Washakie, Wyoming, to Galveston, Texas, as the assistant mustering-out officer of volunteers.[112] Overton spent the next few years on the frontier, including commanding Fort Washakie in 1900, before being transferred to the Philippines as the commander of Company D, Fifteenth Cavalry, in November 1902.[113] He was killed in action on May 15, 1903.[114] A new US Army post in Mindanao was christened Camp Overton that same year.[115]

Richard Bolles Paddock passed the examination to become a second lieutenant in the US Army in October 1883 and was subsequently assigned to the Thirteenth Infantry.[116] Reassigned to the cavalry, in 1889 he was stationed at Fort Stanton, New Mexico, where Second Lieutenant John J. Pershing was also posted.[117] That same year, Paddock married Pershing's sister. The couple had two children: a son, Richard B., and a daughter, May. While serving under General Chaffee in Tientsin, China, during the Boxer Rebellion, Paddock died of pneumonia on March 9, 1901.[118]

John Joseph Pershing graduated from West Point in the summer of 1886 as a second lieutenant in the Sixth Cavalry. He served at Forts Bayard, Stanton, and Wingate until September 1890, followed by assignments in South Dakota and Nebraska through September 1891. Pershing then became a professor of military science at the University of Nebraska.[119] From October 1895 to June 1897, he served with the Tenth Cavalry before transferring to West Point as an assistant instructor of tactics, a role he held until May 1898. From June to August 1898, Pershing participated in the campaign against Santiago de Cuba.[120] He next worked in the Office of the Assistant Secretary of War in Washington, DC, until September 1899.[121] Following service in the Philippines through 1903, Pershing quickly ascended the ranks to general. He commanded the Mexican Punitive Expedition from March 15, 1916, to February 6, 1917.[122] In 1917 President Woodrow

Wilson selected him to command the American Expeditionary Forces in World War I, where he held the rank "general of the armies."[123] He died on July 15, 1948, and is buried in Arlington Cemetery.[124]

Frank DeWitt Ramsey graduated from West Point in 1885 and served at various posts, including Fort Russell, Wyoming; Fort Wingate and Fort Union, New Mexico; and Whipple Barracks, Arizona, before being assigned to Fort Verde in July 1888.[125] In 1891 he was a professor of military science at West Virginia University.[126] He served in the Santiago campaign from April to June 1898 and deployed to the Philippines in 1899. In 1900 he later served in the China Relief Expedition. Returning to the Philippines, he became General Chaffee's aide-de-camp in late 1901, a position he held until 1903. Ramsey was then assigned to the General Staff Corps in Washington, DC, where he served until his untimely death at the age of forty-four in 1906.[127]

Robert Doddridge Read was born in Tennessee in 1854.[128] Read graduated from West Point in 1877 as a lieutenant in the Tenth Cavalry. Before serving in Arizona, he was stationed in Texas at Fort Clark, Fort Davis, and Camp Rice. In 1883 he attended the Cavalry and Infantry School at Fort Leavenworth, Kansas.[129] After completing his studies, Read was assigned to frontier duty at Fort Thomas, San Carlos, and Fort Grant, Arizona. As a captain, he served in Cuba and later deployed twice to the Philippines, returning to the United States in 1909.[130] In 1910 Read, now a colonel, took command of the Tenth Cavalry before assignment to the Third Cavalry at Fort Sam Houston in Texas in the latter half of 1911.[131] He held various duties until his retirement in August 1914 due to a disability contracted in the line of duty.[132] After retiring, he lived in Los Angeles until his death (in San Francisco, where he failed to recover from a surgery) in 1919. He is buried at the San Francisco National Cemetery.[133]

Carl Reichmann, born Karl Friedrich Wilhelm Reichmann in Germany in 1857, struggled in college due to "his fondness for strong beer, women, and dueling."[134] In 1881 he emigrated to the United States and enlisted in the US Army the same year.[135] In 1884 Reichmann was commissioned as a second lieutenant in the Twenty-Fourth

Infantry.¹³⁶ His initial assignments took him to Fort Reno, Fort Sill, and San Carlos, followed by Fort Leavenworth and then several years at Madison Barracks in New York. Reichmann deployed to the Philippines in 1900 and then served at Fort McPherson, Georgia, before returning to the Philippines in 1904 for two more years. From 1906 to 1907, he deployed to Cuba. In 1908 he returned to Madison Barracks, followed by assignments in Washington, DC, and Schofield Barracks from 1913 to 1916.¹³⁷ In 1917 he was selected for promotion to brigadier general, but the promotion did not go through due to unfounded suspicions of German sympathies.¹³⁸ On May 1, 1925, after leading the army's recruiting effort in the Northwest and purchasing land in the lake region of northern Minnesota, Reichmann retired.¹³⁹ He died on October 27, 1936 in Minneapolis.¹⁴⁰

A. J. Robinson is mentioned briefly in the *History of the Ninth US Infantry* as the one who established the Squaw Peak heliograph station.¹⁴¹ He was discharged in April 1891 and subsequently lost to history.

Robert Sherwood was a member of the US Army Signal Corps. He established several New Mexico heliograph stations when Lieutenant Dravo was absent due to illness. Indeed, Sherwood set up five of the New Mexico stations (Alma, Camp Henely, Lyda (Mule) Spring, Separ, and Siggins's Ranch). Little more is known about him.

Thomas Horace Slavens was born in Indiana in 1863 and graduated from West Point in 1887. He served on the frontier at Fort Lowell, Fort Wingate, Fort Huachuca, and San Carlos, where he commanded a troop. In 1901 Slavens was assigned to Fort Leavenworth, where he became an honor graduate of the Infantry and Cavalry School, later serving as an instructor.¹⁴² He served as the aide-de-camp to Major General Greene in Cuba and participated in several actions in the Philippines and served as the aide-de-camp to General MacArthur.¹⁴³ In 1904 Slavens joined the Quartermaster Corps and later deployed to the Philippines and the Mexican Punitive Expedition. In 1918, as a colonel, he deployed to France, where he was reassigned to the infantry, commanding the Fifty-First Infantry during the Meuse-Argonne Offensive, Verdun, and Recey-sur-Ource.¹⁴⁴ He then commanded the Eleventh Brigade, Sixth

Division at Eller, Germany, with the Army of Occupation.[145] Promoted to brigadier general in 1923, Slavens commanded the Twenty-First Brigade of the Hawaiian Division before retiring in 1927.[146] He died on Christmas Eve 1954 and is buried in Fort Sam Houston National Cemetery in Texas.[147]

Alexander Smart served on the frontier at Fort Lowell in 1889. Very little can be found about his life before or after that service. However, in 1893, he was cited for "skill, courage, fortitude and energy" in a search for lost citizens in the Bitterroot Mountains of Idaho.[148] *The San Francisco Call and Post* of March 2, 1913, reports a Sergeant Alexander Smart retiring from US Army active service.[149]

William Harvey Smith graduated from West Point in 1883 and was commissioned in the Tenth Cavalry. He initially served on frontier duty in Texas before being transferred to Fort Verde, Arizona, in 1885, followed by assignments at Fort Apache in 1888 and 1889.[150] Smith held various posts in the Southwest, including at Fort Grant, Fort Apache, Fort Wingate, and San Carlos. He was later transferred to Fort Custer, Montana, in April 1892 and attended the Infantry and Cavalry School at Fort Leavenworth from August 1895 to June 1897, graduating with honors.[151] Smith was next assigned to Fort Assinniboine, Montana, until April 1898. He served in Georgia and Florida before deploying to Cuba, where he was killed in action "commanding and leading his troop in the assault on San Juan Hill" on July 1, 1898.[152]

George Edward Stockle graduated from West Point in 1888 as a second lieutenant of infantry and was assigned to the Twenty-Fifth Infantry at Fort Missoula, Montana. In February 1889 he transferred to the Tenth Cavalry, where he commanded a troop of Apache scouts at Fort Apache, Arizona.[153] Stockle served at various posts in Arizona and Montana and later, as a first lieutenant, he became a professor of military science and tactics at Washington Agricultural College in Pullman. He then served in South Dakota, Nebraska, Wyoming, Colorado, and Montana before being sent to Huntsville, Alabama, in 1898, followed by a deployment to Cuba, where he served until February 1901.[154] In 1905, as commander of Troop H, Eighth Cavalry, Stockle deployed to the Philippines, returning in 1907.[155] As a major with the Ninth Cavalry, he was again deployed to the Philippines in 1911, serving as a quartermaster. In 1916 Stockle commanded the Second Squadron, Twelfth Cavalry, at Fort Russell, Wyoming, and took part in the Mexican Punitive Expedition in Columbus and Hachita, New Mexico, and Boca Grande, Mexico. As a colonel, he commanded the Third Squadron and the subdistrict of Columbus, New Mexico, later commanding the Twelfth Cavalry in 1917 and 1918.[156] After thirty years of service, Stockle retired on December 6, 1918.[157] In retirement, he organized the Department of Military Science at Washington State College. He died in San Francisco on March 15, 1933.[158]

Lawrence Davis Tyson was born in 1861 and graduated from West Point in 1883. In 1886 he married into a wealthy East Coast family. Tyson resigned his commission in 1896 to pursue a career in law, though he remained active in the volunteers, eventually rising to the rank of brigadier general. During World War I, he commanded the Fifty-Ninth Brigade, Thirtieth Infantry Division, leading his troops in action at Ypres and in the Somme sector. In 1925 Tyson was elected as a US senator for Tennessee, a position he held until his death in 1929.[159]

William Jefferson Volkmar was born in Philadelphia on June 29, 1847. He served in the Civil War as a sergeant in the Thirty-Third Pennsylvania Infantry from June 19 to August 4, 1863.[160] Volkmar graduated from West Point on June 15, 1868, as a second lieutenant of cavalry and went on to serve with the Fifth Cavalry in Kansas, Colorado, and Nebraska. He was promoted to first lieutenant in March 1870 and later transferred to Fort McDowell, Arizona. In 1875 Volkmar served as the chief signal officer for the Department of Missouri and then as aide-de-camp to General Pope. From December 1876 to June 1879, he returned to frontier duty at Fort D. A. Russell, Wyoming.[161] In 1881 he was the aide-de-camp to General Sheridan at Fort Leavenworth, Kansas. In 1888 he became the adjutant general of the Department of Arizona as well as the chief signal officer from 1889 to 1890.[162] He then served as the adjutant general and chief signal officer for the Department of Missouri from October 1890 to July 1891. Promoted to lieutenant colonel in 1893 and colonel in 1898, Volkmar did a final tour as adjutant general of the Department of Colorado from 1896 to April 28, 1900.[163] Volkmar died in Passaic, New Jersey, on March 4, 1901.[164]

Daniel Williams was on the initial rolls of Company H, Tenth Cavalry, which was organized on July 21, 1867, at Fort Leavenworth, Kansas.[165] By the time of the 1890 heliograph exercise, he had served with the Tenth Cavalry for twenty-three years, likely seeing action throughout the West. Rosters list a Private Daniel Williams who served with the Tenth Cavalry during the Spanish-American War.[166]

Edmund Wittenmyer graduated from West Point in 1887 and was commissioned as a second lieutenant in the Ninth Infantry.[167] He served at San Diego, on the frontier at Fort Wingate, and later in New York and Chicago. In 1893 Wittenmyer was assigned as a student to the Infantry and Cavalry School at Fort Leavenworth, Kansas. After graduating he was stationed at Sheridan, Illinois, Fort Bayard, and Fort Huachuca.[168] In 1899 he served in Cuba before deploying to the Philippines with the Fifteenth Infantry in 1900, returning in 1903.[169] He was assigned to the Fifth Infantry in 1905 and later to the Instruction Brigade, deploying to Cuba again in 1906 and returning in 1909.[170] After several years in the United States, Wittenmyer returned to Cuba as the military attaché in Havana, serving through 1914.[171] He was promoted to brigadier general by 1917 and deployed to France in command of a brigade. Promoted to major general in 1918, he commanded the Seventh Division.[172] Wittenmyer retired as a major general after forty years of service and died on July 3, 1937.[173]

Leonard Wood was born in 1860 in New Hampshire. He earned a doctor of medicine degree from Harvard Medical School in 1884. He received an appointment as a surgeon in the US Army on January 5, 1886. While serving in the Arizona Territory, Wood was awarded the Medal of Honor for his actions during the campaign against Geronimo.[174] In 1898 he became colonel of the First Volunteer Cavalry (the Rough Riders) and led the regiment to Cuba, commanding them until his promotion to brigadier general later that year.[175] In 1899 he became the military governor of Cuba. Wood was later transferred to the

Philippines, where he commanded the Department of Mindanao from 1903 to 1905 and then commanding the Division of the Philippines through 1908.[176] From 1910 to 1914 he held the position of chief of staff of the US Army. Between 1914 and 1917, he commanded the Department of the East, followed by the Central Division from 1919 to 1921. After retiring from active service in 1921, Wood became the governor general of the Philippines, a role he held from 1921 to 1927. He died in Boston on August 7, 1927.[177]

Appendix 2

Fort Selden

A 1963 article in *New Mexico Magazine* by Burt Jenness describes a heliograph station atop Mount Robledo that flashed messages between Fort Selden and Fort Bliss.[1] In 1969 *Reclamation Era* included a photograph of a "sturdy iron heliograph stand" cemented into a rock and mortar base atop today's Lookout Peak.[2] Supporting documentation for Fort Selden's National Historic Landmark designation in 1970 mentioned that "in the early 1880s, a heliograph was set up on nearby Monte Robledo, enabling communication with Cook's Peak, Mount Franklin, Fort Stanton, and Fort Bliss."[3] A 1975 archaeologist's report also asserted the presence of a heliograph station at Lookout Mountain, based on the discovery of "two upright heavy gage steel tubes solidly set in rock and mortar bases. Both have small, hinged gates near the top and on one the opening faces Fort Selden."[4] Allan J. Holmes's 2009 *Fort Selden 1865–1891* identifies a heliograph station "on a peak in the Robledo mountains (now referred to as Signal Peak by the locals, but Lookout Peak by the map makers)."[5] Holmes's description of the station's connections mirror the 1970 National Historic Landmark documents. A US presidential proclamation in May 2014 affirmed that "an 1880s US military heliograph station, the remains of which still stand at Lookout Peak in the Robledo Mountains, transmitted Morse code messages during the army's western campaigns."[6]

Despite these numerous accounts, there is no mention of the station at Lookout Peak by period heliographic practitioners of the US Army. Lieutenant Marian P. Maus, who in 1882 connected Fort Bowie with

Fort Grant and other stations along the Gila River, does not reference Fort Selden in connection with a heliograph.[7] General Nelson A. Miles, credited with establishing the 1886 system, also did not include Fort Selden, Lookout Peak, or Mount Robledo in his heliographic scheme;[8] nor did Lieutenant Edward E. Dravo, whom Miles tasked with setting up the New Mexico portion of the 1886 system.[9] In late 1889 Lieutenants Henry W. Hovey and Richard B. Paddock, who established the southern line connecting Fort Cummings to Fort Stanton, made no mention of a heliograph station atop Lookout Peak or Mount Robledo, although Paddock did attempt, unsuccessfully, to connect with Fort Selden via heliograph from the San Andreas Mountains (now called the San Andres Mountains).[10] Fort Selden was not included in the 1890 or 1893 heliograph exercises.[11]

None of these practitioners mentioned the use of large steel or iron pipes as mounts. In fact, officers who established heliograph stations throughout Arizona and New Mexico between 1882 and 1893 never reported using steel pipes or mortar. Heliographs, designed to be lightweight and portable, typically sat atop small, light tripods, although some of the stations in the 1886 system were mounted on locally sourced timbers

Period mapmakers Eugene J. Spencer (1886) and William J. Volkmar (1890) included Fort Selden on their maps but did not mark it as a heliograph station, as they did for other stations.[12] However, Spencer's map does include a "camp for scouting & observation" near what is today's San Augustin Pass. According to Colonel Dravo in 1926, soldiers stationed there had a heliograph with them.[13]

An unnamed and undated reference in the records of the Bureau of Land Management suggests that the Lookout Peak station was referred to as the Rincon station, named after the farming community located 19 miles to the north. This explanation attempts to account for the lack of US Army references to a Lookout Peak station, as there are numerous mentions of the Rincon NM heliograph station in army records. Lieutenant Paddock, who established the Rincon NM station in late December 1889, described its location as "on a hill N.12 1/2°

W. from R. R. depot at Rincon, N.M. and about 1 mile distant."[14] This description does not match the geography or geometry of Lookout Peak near Fort Selden.

Citing a February 1889 report from Lieutenant James E. Brett, Holmes asserts that there was but one known heliograph message sent from Fort Selden. In this message, Brett reports connecting by heliograph with his detachment patrolling in the nearby San Andreas Mountains.[15] Unfortunately, the document Holmes references cannot be found in the National Archives. Furthermore, between January 1888 and December 1889, there is no mention of heliograph work in Fort Selden's post returns or record of events.[16] Nevertheless, accepting Holmes's assertion as true, the soldiers likely made their connection from the nearest high spot with a clear view of the San Andreas Mountains, which would have been Lookout Peak, as visibility analysis shows that the Doña Ana Mountains obscure the view from lower peaks, preventing a direct connection from Fort Selden.

From Lookout Peak, visibility is poor to the south and west. The taller Mount Robledo blocks the southern view while the Sierra de las Uvas Mountains obscure the west. A heliograph atop Lookout Peak would have been blind to many other heliograph stations in either the 1886 or 1890 systems. Both Fort Stanton and Fort Bliss required an intermediate station to maintain a connection, and the Fort Cummings station was not visible from Lookout Peak. Additionally, Fort Bayard and the Pinos Altos stations, as well as Deming, were out of sight. However, the Little Florida and Hillsboro heliograph stations were visible, as was Cook's Peak. Despite Cook's Peak being visible, there was no heliograph station there except for a failed attempt at long-range signaling toward the end of the 1890 exercise.[17] The heliograph station used by Fort Cummings was a short distance to the south of the fort.[18]

Two questions arise. First, why is there so much reporting about Lookout Peak and the iron pipes? It is quite possible that a heliograph was indeed placed at Lookout Peak at some point. Perhaps earlier in the 1880s, when there were more soldiers stationed at Fort Selden, a heliograph on Lookout Peak maintained communication with patrolling detachments in the Rio Grande Valley and the San Andreas Range. The presence of the iron pipes provides a tangible reference point, something concrete to observe and write about, which could explain the continued interest in and reporting on the site. The second question is: What are these mysterious steel and iron pipes cemented to the earth?

In 1882 an astronomical team used a heliostat to take 216 "splendid" photographs of the transit of Venus from a location in the Cerro Robledo Mountains.[19] The heliostat, which had two sections—a plate holder and the heliostat itself—was mounted on heavy metal piers that were "sunk into the ground" and separated by 12 meters.[20] Comparing photographs provided by the Bureau of Land Management with diagrams and drawings from reports on the 1882 Venus transit effort, it seems likely that the iron pipes found at Lookout Peak were the piers used to hold the two ends of the heliostat, not parts of a heliograph.[21]

Heliostats aside, given Holmes's assertion of a heliograph-to-heliograph connection from a station near Fort Selden, the poor visibility from the fort itself, and the excellent visibility to the San Andreas Mountains from Lookout Peak, combined with the "camp for scouting & observation" near San Augustin Pass shown on the Spencer map, the Fort Selden heliograph station is mapped on Lookout Peak.

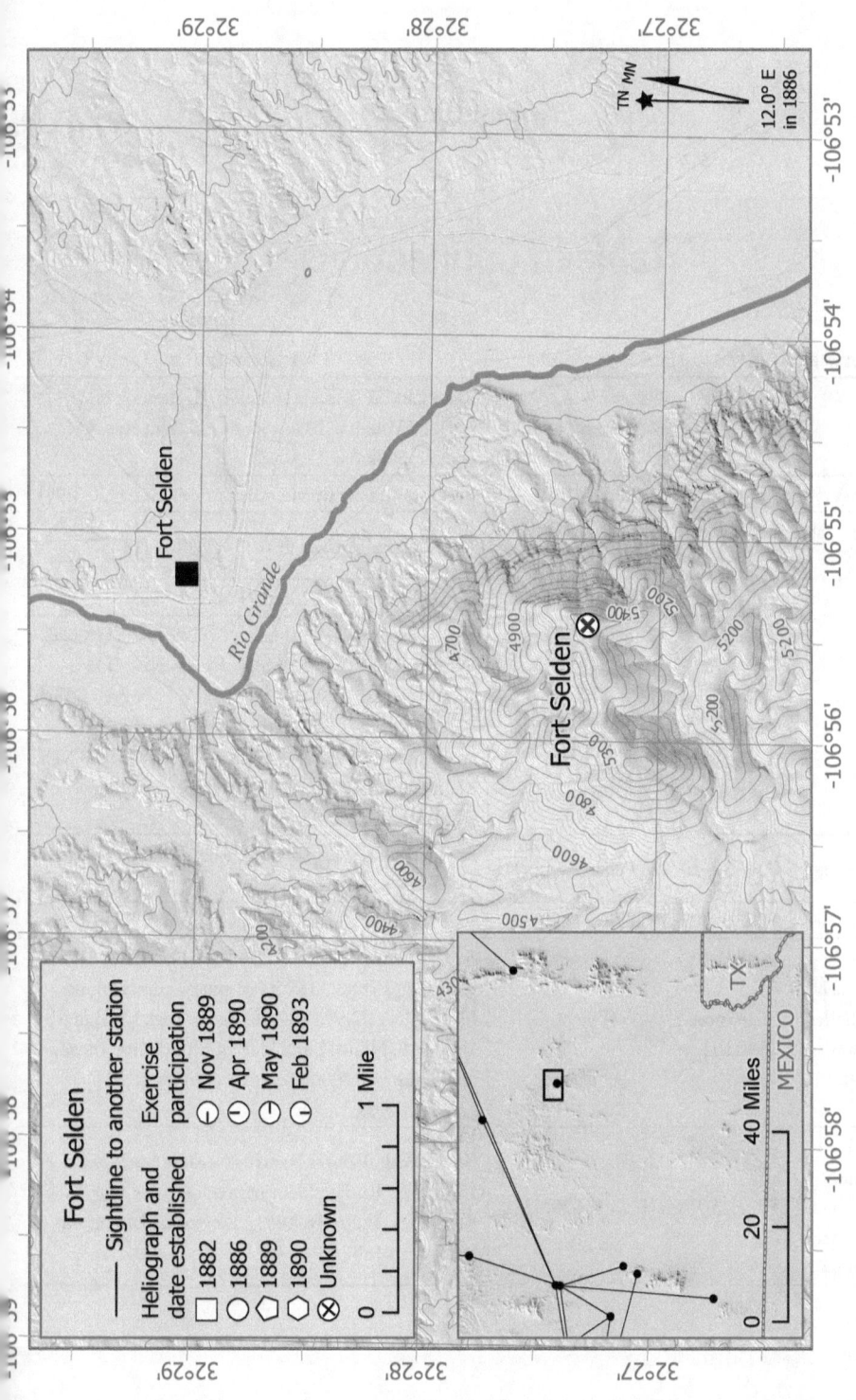

Map 77. Fort Selden heliograph station.

Appendix 3

Table of Connections

Station (also known as)	Who First Established and Date	Distant Stations
Fort Grant	Maus, 7/10/82 (Hazen 1883:94)	Dos Cabazas (Hazen 1883:94), Eggleston's RM station (Volkmar 1890c:36, Dade), Murray's RM station (Volkmar 1890b:8)
Dos Cabezas	Maus, 7/20/82 (Hazen 1883:94)	Fort Grant, Fort Bowie, Ash Butte (Hazen 1883:94)
Ash Butte (Ash Peak)	Maus, likely early August 1882 (Hazen 1883:94; Willcox 1883:152)	Dos Cabazas (Hazen 1883:94; Willcox 1883:152)
White's Ranch	Fuller, 4/29/86 (Fuller 1890b:2)	Bowie Peak, Emma Monk (Fuller 1890a)
Bowie Peak	Fuller, 4/29/86 (Fuller 1890a)	Dos Cabazas (likely) (Hazen 1883:94), Fort Bowie, White's Ranch, Emma Monk, Stein's Pass, Bowie station (Fuller 1890a), Colorado Peak, Fourr's Ranch North, Murray's RM station (Volkmar 1890b:49, Neall), Camp Henely (Volkmar 1890b:61, Rhodes), Heliograph Peak (Eggleston 1890b:7), Alpina (Eggleston 1890b:8)
Fort Bowie	Fuller, 4/29/86 (Fuller 1890b:1)	Bowie Peak (Fuller 1890a), Helen's Dome (Fuller 1890b:2)
Helen's Dome	Fuller, 4/29/86 (Fuller 1890b:1)	Fort Bowie (Fuller 1890b:2)
Camp Henely (Bessie Rhodes Mountain, Henely Peak)	Sherwood, 5/9/86 (Dravo 1890d:1)	Separ (Dravo 1890c), Stein's Pass, Fort Bayard, Fort Cummings 1886, Hachita Mining Camp, Deming, Pinos Altos (Dravo 1890b), Bowie Peak (Volkmar 1890a:18, Neall), Little Florida Mountains (Hovey and Paddock 1890:2), Fort Cummings 1890 (Volkmar 1890b:63, Leitch)
Emma Monk (extreme northern point of the Swisshelm Mountains)	Fuller, 5/9/86 (Fuller 1890b:2)	Bowie Peak, White's Ranch, Antelope Springs, Old Camp Rucker, Henry Forrest (Fuller 1890a), Huachuca (Fuller 1890b:3), Limestone Mountain (Glassford 1887:7)

TABLE OF CONNECTIONS

Station (also known as)	Who First Established and Date	Distant Stations
Separ	Sherwood, 5/9/86 (Dravo 1890d:1)	Camp Henely (Dravo 1890c)
Old Camp Rucker (Camp Rucker, Rucker, Fort Rucker)	Fuller, 5/13/86 (Fuller 1890b:2)	Emma Monk (Fuller 1890a)
Fort Cummings 1886	Dravo, 5/16/86 (Dravo 1890b)	Camp Henely, Lockhart's Well, Lake Valley, Deming (Dravo 1890b), Heatley's Well (Dravo 1890d:2)
Antelope Springs	Fuller, 5/22/86 (Fuller 1890b:3)	Emma Monk, Fort Huachuca 1886 (Fuller 1890a)
Fort Huachuca 1886	Fuller, 5/21/86 (Fuller 1890b:3)	Antelope Springs, Fourr's Ranch South, Little Baldy Peak (Fuller 1890a), Emma Monk, Elgin (Fuller 1890b:3), Gardner's Ranch (Fuller 1890b:4)
Hachita Mining Camp (Eureka)	Dravo, 5/29/86 (Dravo 1890d:2)	Camp Henely (Dravo 1890b)
Elgin	Fuller, 5/29/86 (Fuller 1890b:3)	Fort Huachuca 1886 (Fuller 1890b:3)
Heatley's Well (Greasewood Hills)	Chaffee, 5/30/86 (Dravo 1890d:2)	Fort Cummings 1886 (Dravo 1890d:2)
Gardner's Ranch	Fuller, 6/2/86 (Fuller 1890b:4)	Fort Huachuca 1886 (Fuller 1890b:4)
Tubac	Fuller, 6/8/86 (Fuller 1890b:4)	Little Baldy Peak (Fuller 1890a)
Lockhart's Well	Dravo, 6/11/86 (Dravo 1890d:2)	Fort Cummings 1886 (Dravo 1890b)
Little Baldy Peak (Josephine Peak)	Fuller, 6/12/86 (Fuller 1890b:4)	Fort Huachaca 1886, Tubac, Crittenden (Fuller 1890a), Nogales, Calabasas (Neifert 2001:560)
Nogales	unknown, 6/13/86 (approx.) (Neifert 2001:560)	Little Baldy Peak (Neifert 2001:560)
Calabasas	unknown, 6/13/86 (approx.) (Neifert 2001:560)	Little Baldy Peak (Neifert 2001:560)
Lake Valley	Johnson, 6/15/86 (Dravo 1890d:2)	Fort Cummings 1886, Hillsboro (Dravo 1890b)
Hillsboro (Camp Boyd)	Gillmore, 6/20/86 (Dravo 1890d:2)	Lake Valley (Dravo 1890b), Fort McRae (Dravo 1890a:2)
Henry Forrest	Fuller, 6/20/86 (Fuller 1890b:5)	Emma Monk (Fuller 1890a)
Stein's Pass	Fuller 6/27/86 (Fuller 1890b:5)	Fort Bowie (Fuller 1890a), Camp Henely (Dravo 1890b)
Fourr's Ranch South (Cochise Stronghold)	unknown, 6/27/86 (Fuller 1890b:5–6)	Fort Huachuca 1886 (Fuller 1890a)

APPENDIX 3

Station (also known as)	Who First Established and Date	Distant Stations
Deming	Fuller, 6/28/86 (Dravo 1890d:3)	Fort Cummings 1886, Camp Henely, Pinos Altos (Dravo 1890b)
Fort Bayard	Dravo, 6/28/86 (Dravo 1890d:3)	Pinos Altos, Camp Henely (Dravo 1890b)
Crittenden	Fuller, 7/9/86 (Fuller 1890a)	Little Baldy Peak (Fuller 1890a)
Bowie station	Fuller, 7/14/86 (Fuller 1890a)	Bowie Peak (Fuller 1890a)
Pinos Altos	Sherwood, 8/14/86 (Dravo 1890d:3)	White House, Fort Bayard, Deming, Henely (Dravo 1890b)
Limestone Mountain	unknown, 8/15/86 (approx.) (Glassford 1890:7)	Emma Monk, San Bernardino Ranch (Wood and Lane 2009:108; Utley 2012:216)
White House	Sherwood, 8/20/86 (Dravo 1890d:3)	Pinos Altos, Siggins's Ranch (Dravo 1890b)
Alma (Camp Maddox)	Sherwood, 8/23/86 (Dravo 1890d:3)	Siggins's Ranch (Dravo 1890b)
Siggins's Ranch (Sundial Mountain, Siggen's Ranch, Siggin's Ranch)	Sherwood, 8/26/86 (Dravo 1890d:4)	Alma, White House (Dravo 1890b), Lyda Spring (Dravo 1890d:4)
San Bernardino Ranch	Lawton, 8/30/86 (Wood and Lane 2009:108)	Limestone Mountain (Wood and Lane 2009:108; Utley 2012:216)
Lyda Spring (Mule Spring)	Sherwood, 9/15/86 (Dravo 1890d:4)	Siggins's Ranch (Dravo 1890d:4)
Fort McRae	Pershing, 11/16/86 (Dravo 1890a:2)	Hillsboro, Dripping Springs (Dravo 1890a:2)
Dripping Spring (Silver Top Mountain)	Pershing, 11/19/86 (Dravo 1890a:3)	Fort McRae, Nogal (Dravo 1890a:2–3)
Nogal (Vera Cruz Mountain)	Pershing, 11/21/86 (Dravo 1890a:3)	Dripping Spring, Fort Stanton 1886 (Dravo 1890a:3)
Fort Stanton 1886	Pershing, 11/23/86 (Dravo 1890a:3)	Nogal (Dravo 1890a:3)
Lookout Peak (Aztec Peak)	Wood, 8/28/88 (Wood 1890:1)	Mount Reno (Volkmar 1890c:12, Overton), Mazatzal Peak (Volkmar 1890b:28, Overton), Baker's Butte (Volkmar 1890c:4, Ramsey), Pinal Mountain, Pinal East (Volkmar 1890b:37, Read), Pinal April (Volkmar 1890b:36, Read), Murray's RM station (Volkmar 1890b:45, Dade), Eggleston's RM station (Volkmar 1890c:33, Murray), Saddle Mountain (Volkmar 1890c:25, Reichmann), Mount Turnbull (Wood 1890:1)

TABLE OF CONNECTIONS

Station (also known as)	Who First Established and Date	Distant Stations
Mount Turnbull (Eggleston station)	unknown, 8/29/88 (Wood 1890:1)	San Carlos, Fort Thomas (Eggleston 1890a:2), Lookout Peak (Wood 1890:1; Wood reported seeing flashes from Mount Turnbull on August 29 while he was atop Lookout Peak)
San Carlos	Eggleston, 11/7/89 (Eggleston 1890b:1)	Triplets, Mount Turnbull (Eggleston station), Turnbull station (Eggleston 1890a:1–3), Saddle Mountain (Volkmar 1890c:24, Reichmann)
Triplets (Mount Triplet)	Eggleston, 11/7/89 (Eggleston 1890a:1)	San Carlos (Eggleston 1890a:1)
Fort Thomas	unknown, 11/9/89 (Eggleston 1890a:2)	Mount Turnbull (Eggleston station) (Eggleston 1890a:2)
Alpina	Keene then Eggleston, 11/15/89 (approx.) and again in December (Eggleston 1890b:7)	Table Mountain (Eggleston 1890b:7), possible connection with Fort Huachuca 1890 (Volkmar 1890b:53, Hart), Bowie Peak (Eggleston 1890b:8; while Bowie Peak was clearly seen by both Eggleston and Neall, it does not appear that a connection was actually made, though both Eggleston and Murray imply that one was), Eggleston's RM station (Eggleston 1890b:7; again, this connection is implied)
Fourr's Ranch North (4F, Cochise, Stronghold, Fourr's)	Hart, 11/16/89 (Volkmar 1890a:24–25, Hart)	Fort Huachuca 1889 and early 1890, Bowie Peak, (Volkmar 1890a:25, Hart), Fort Huachuca 1890, Murray's RM station (Volkmar 1890b:51, Allbright), Rock B (Volkmar 1890b:58, Gale), Rock C (Volkmar 1890b:57, Gouldman), Colorado Peak (Volkmar 1890b:54, Bartsch)
Mountain Spring (Pistol Hill)	Smart, 11/16/89 (Volkmar 1890a:27, Gale)	Fort Lowell (Volkmar 1890a:27, Gale)
Fort Lowell	Gale, 11/17/89 (Volkmar 1890a:27, Gale)	Mountain Spring (Volkmar 1890a:27, Gale), Colorado Peak (Volkmar 1890b:54, Bartsch), Rock A, Rock B, Rock C (Volkmar 1890b:56–57, Gouldman)
Fort Huachuca 1889 and early 1890	Albright, 11/17/89 (approx.) (Volkmar 1890a:7, 25)	Fourr's Ranch North (Volkmar 1890a:25, Hart), Colorado Peak (Gale 1890:3)
Turnbull station	Glassford, 11/27/89 (Eggleston 1890a:3)	San Carlos (Eggleston 1890a:3)
Sierra Blanca	Paddock, 12/4/89 (Hovey and Paddock 1890: 7)	Fort Stanton 1889, San Andreas (Volkmar 1890b:66, Brewster), San Nicholas Peak (Hovey and Paddock 1890:7) (assumed), Tularosa (assumed) (Volkmar 1890b:65, Paddock)
Fort Stanton 1889	Paddock, 12/4/89 (Hovey and Paddock 1890:7)	Sierra Blanca (Volkmar 1890b:66, Pershing)

APPENDIX 3

Station (also known as)	Who First Established and Date	Distant Stations
Table Mountain (Little Table Mountain)	Dade, 12/9/89 (Dade 1890:1)	Pinal Mountain (Volkmar 1890b:37, Read), Pinal April (Volkmar 1890b:36, Read), Murray's RM station (Volkmar 1890b:44, Dade), Eggleston's RM station (Eggleston 1890b:10), Colorado Peak (Volkmar 1890b:54, Bartsch), Alpina (Eggleston 1890b:7), Saddle Mountain (Volkmar 1890c:24, Reichmann)
Heliograph Peak	unknown, 12/11/89 (Eggleston 1890b:7)	Bowie Peak (Eggleston 1890b:7)
San Nicholas Peak (Big Brushy, Black Brushy)	Paddock, 12/15/89 (approx.) (Hovey and Paddock 1890:7)	Sierra Blanca (Hovey and Paddock 1890:7). This is assumed—he had stations at both sites and he would have been very unlikely to leave this station without first checking its ability to connect back to Sierra Blanca.
Eggleston's RM station (Merrill Peak)	Eggleston, 12/20/89 (Eggleston 1890b:10)	Table Mountain (Eggleston 1890b:10), Saddle Mountain, Pinal April, Fort Grant (Volkmar 1890c:36–37, Dade), Lookout Peak (Volkmar 1890c:33–34, Murray; Volkmar 1890b:33, Reichmann), Alpina (Eggleston 1890b:7)
Little Florida	Paddock, 12/21/89 (Hovey and Paddock 1890:2)	Camp Henely (Hovey and Paddock 1890:2)
Rincon NM	Paddock, 12/25/89 (Hovey and Paddock 1890:3–4)	Fort Cummings, San Andreas (Hovey and Paddock 1890:4–5)
San Andreas	Paddock, 12/31/89 (Hovey and Paddock 1890:4)	Rincon NM (Hovey and Paddock 1890:5), Sierra Blanca (Volkmar 1890b:66, Brewster), Fort Cummings (Volkmar 1890b:63, Leitch), Tularosa (Volkmar 1890b:65, Paddock)
Whipple Barracks	Tyson, 3/14/90 (Fenton 1890:2)	Bald Mountain (Volkmar 1890c:1, Fenton)
Bald Mountain (Glassford Hill)	Fenton, 3/14/90 (Fenton 1890:2)	Little Squaw Peak, Baker's Butte (Volkmar 1890c:1 Fenton), Whipple Barracks, Squaw Peak (Fenton 1890:2)
Squaw Peak (Porcupine Peak)	Ramsey, 3/14/90 (Fenton 1890:2)	Mazatzal Peak, Mount Reno (Volkmar 1890b:26 Robinson), Bald Mountain (Fenton 1890:2)
Colorado Peak	Gale, 3/18/90 (Gale 1890:1)	Fort Lowell, Table Mountain, Bowie Peak, Fourr's Ranch North, Fort Huachuca 1890, Murray's RM station (Volkmar 1890b:54, Bartsch), Fort Huachuca 1890 and early 1890 (Gale 1890:3)

TABLE OF CONNECTIONS

Station (also known as)	Who First Established and Date	Distant Stations
Mount Reno (Mount Ord)	Overton, 4/3/90 (Volkmar 1890c:11, Overton)	Fort McDowell, Lookout Peak (Volkmar 1890c:11–12 Overton), Mazatzal Peak (Volkmar 1890c:13 Overton), Murray's RM station (Volkmar 1890b:45, Dade), Squaw Peak (Volkmar 1890b:26, Robinson), Baker's Butte (Volkmar 1890b:26, Duncan), Pinal Mountain (Volkmar 1890b:37, Read)
Fort McDowell	Overton 4/3/90 (Volkmar 1890c:11, Overton)	Mount Reno (Volkmar 1890c:12, Overton), Mazatzal Peak, Spur (Volkmar 1890c:13, Overton)
Pinal April	Murray, 4/5/90 (Volkmar 1890c:30, Murray)	Saddle Mountain (Volkmar 1890c:24, Reichmann), Lookout Peak, Eggleston's RM station (Volkmar 1890c:30, Murray)
Fort Verde	unknown 4/5/90 (Volkmar 1890c:9, Duncan)	Little Squaw Peak (Volkmar 1890c:10, Duncan)
Baker's Butte	Ramsey, 4/5/90 (Volkmar 1890c:4, Ramsey)	Lookout Peak (Volkmar 1890c:4, Ramsey), Pinal Mountain, Mount Reno, Little Squaw Peak (Volkmar 1890b:26–27, Duncan), Bald Mountain (Volkmar 1890c:1, Fenton)
Little Squaw Peak	Duncan, 4/6/90 (Volkmar 1890c:9–10, Duncan)	Fort Verde, Baker's Butte, Bald Mountain (Volkmar 1890c:10, Duncan; Volkmar 1890c:4, Ramsey)
Saddle Mountain (Stanley Butte)	Reichmann, 4/7/90 (Volkmar 1890c:24 Reichmann)	Pinal April, San Carlos, Table Mountain, Lookout Peak (Volkmar 1890c:24–25, Reichmann), Eggleston's RM station (Volkmar 1890c:36, Dade)
Spur	Overton, 4/9/90 (Volkmar 1890c:13, Overton)	Fort McDowell (Volkmar 1890c:13, Overton)
Mazatzal Peak (North Peak, North Peak proper)	Overton, 4/10/90 (Volkmar 1890c:13, Overton)	Mount Reno, Fort McDowell (Volkmar 1890c:13, Overton), Lookout Peak, Squaw Peak (Volkmar 1890b:28, Overton)
Murray's RM station (Grand View Peak)	Murray, 4/22/90 (Volkmar 1890c:34, Murray)	Rock C (Volkmar 1890b:57, Gouldman), Bowie Peak (Volkmar 1890b:49, Neall), Table Mountain, Lookout Peak, Pinal Mountain, Colorado Peak, Fourr's Ranch North, Fort Huachuca 1890, Mount Reno (Volkmar 1890b:44–46, Dade), Fort Grant (Volkmar 1890b:8)
Fort Huachuca 1890	Hart, 4/24/90 (Volkmar 1890b:52, Hart)	Colorado Peak (Volkmar 1890b:54, Bartsch), Alpina (possible), Fourr's Ranch North (Volkmar 1890b:52–53, Hart), Murray's RM station (Volkmar 1890b:46, Dade)

APPENDIX 3

Station (also known as)	Who First Established and Date	Distant Stations
Fort Cummings 1890	Lietch, 4/27/90 (Volkmar 1890b:63 Leitch)	San Andreas, Camp Henely, Rincon NM (Volkmar 1890b:63, Leitch)
Pinal East Ridge	Keene, 4/30/90 (Volkmar 1890b:36, Read)	Lookout Peak, Table Mountain (Volkmar 1890b:37, Read)
Pinal Mountain	Read, 5/2/90 (Volkmar 1890b:37, Read)	Lookout Peak, Table Mountain, Mount Reno, Murray's RM station (Volkmar 1890b:37–38, Read), Baker's Butte (Volkmar 1890b:26, Duncan)
Rock A (Helens Dome [not near Bowie])	Gale and Gouldman, 5/12/90 (Volkmar 1890b:56, Gouldman)	Fort Lowell (Volkmar 1890b:56, Gouldman)
Rock C (Spud Rock)	Gale and Gouldman, 5/13/90 (Volkmar 1890b:58, Gale)	Fort Lowell, Murray's RM station, Fourr's Ranch North (Volkmar 1890b:57, Gouldman)
Rock B (Duckbill)	Gale and Gouldman 5/13/90 (Volkmar 1890b:58, Gale)	Fourr's Ranch North, Fort Lowell (Volkmar 1890b:58, Gale)
Tularosa (Tularosa Hill, Tulerosa)	Paddock, 5/16/90 (Volkmar 1890b:65, Paddock)	Reconnaissance to determine suitability as a bridge between San Andreas and Sierra Blanca in case of dust or bad weather (Volkmar 1890b:65, Paddock)

Sources

Dade, Alexander L. 1890. "Report of a Reconnaissance of Table Mountain, Galiuro Range." In *Instructions for Signal Officers of Heliograph Divisions & Stations*, by William J. Volkmar. Los Angeles: Department of Arizona.

Dravo, Edward E. 1890a. "Heliograph Stations in New Mexico. Descriptive Notes by 1st Lieut. E. E. Dravo, 6th Cavalry, A. S. O. January 26, 1887." In *Instructions for Signal Officers of Heliograph Divisions & Stations*, by William J. Volkmar. Los Angeles: Department of Arizona.

Dravo, Edward E. 1890b. "Monthly Report of Heliograph Lines. District New Mexico July 31, to August 31, 1886." In *Instructions for Signal Officers of Heliograph Divisions & Stations*, by William J. Volkmar. Los Angeles: Department of Arizona.

Dravo, Edward E. 1890c. "Report of Heliograph Lines. District of New Mexico, to July 31, 1886." In *Instructions for Signal Officers of Heliograph Divisions & Stations*, by William J. Volkmar. Los Angeles: Department of Arizona.

TABLE OF CONNECTIONS

Dravo, Edward E. 1890d. "Report of the Heliograph System, District of New Mexico. E. E. Dravo, 1st Lieut. 6th Cavalry, A. S. O., Officer in Charge. Fort Bayard, New Mexico, September 20, 1886." In *Instructions for Signal Officers of Heliograph Divisions & Stations*, by William J. Volkmar. Los Angeles: Department of Arizona.

Eggleston, Millard F. 1890a. "Reconnaissance for Stations in San Carlos Heliograph Division, by 1st Lieut. M. F. Eggleston, 10th Cav., A. S. O. San Carlos, A. T., November 30, 1889." In *Instructions for Signal Officers of Heliograph Divisions & Stations*. Los Angeles: Department of Arizona.

Eggleston, Millard F. 1890b. "Report of Reconnaissance for Connecting Mt. Graham with the Bowie and San Carlos Heliograph Divisions." In *Instructions for Signal Officers of Heliograph Divisions & Stations*, by William J. Volkmar. Los Angeles: Department of Arizona.

Fenton, Charles W. 1890. "Reconnaissance for Intermediate Stations Connecting Whipple and Verde Heliograph Divisions, by 2d Lieut. C. W. Fenton, 9th Infantry, A. S. O. Office of Acting Signal Officer, Post of Whipple Barracks, A. T., March 15th, 1890." In *Instructions for Signal Officers of Heliograph Divisions & Stations*, by William J. Volkmar. Los Angeles: US Army.

Fuller, Alvarado M. 1890a. "Consolidated Report of Heliograph Line. Division of Arizona from May 1st to Sept 30th, 1886, during the Continuance of the Line. 2d Lieutenant Alvarado M. Fuller, 2d Cavalry, Acting Signal Officer in Charge of Heliograph Service." In *Instructions for Signal Officers of Heliograph Divisions & Stations*, by William J. Volkmar. Los Angeles: Department of Arizona.

Fuller, Alvarado M. 1890b. "Heliograph Report of A. M. Fuller, 2d Lieut. 2d Cavalry, A. S. O., in Charge of Line. Fort Huachuca, A. T., September 30, 1886." In *Instructions for Signal Officers of Heliograph Divisions & Stations*, by William J. Volkmar. Los Angeles: Department of Arizona.

Gale, George H. G. 1890. "Report of Reconnaissance for Stations in Lowell Heliograph Division." In *Instructions for Signal Officers of Heliograph Divisions & Stations*, by William J. Volkmar. Los Angeles: Department of Arizona.

Glassford, William A. 1890. "Heliograph and Signal Work. Extracts from Report of Lieut. Glassford, Signal Corps. July 17, 1887." In *Instructions for Signal Officers of Heliograph Divisions & Stations*, by William J. Volkmar. Los Angeles: Department of Arizona.

Hazen, William B. 1883. "Annual Report of the Chief Signal Officer, United States Army, to the Secretary of War for the Fiscal Year Ending June 30, 1882." Washington, DC: GPO.

Hovey, Henry W., and Richard B. Paddock. 1890. "Reports of Concerted Reconnaissances for Short Line Connecting Bayard and Stanton Heliograph Divisions, N.M." In *Instructions for Signal Officers of Heliograph Divisions & Stations*. Los Angeles: Department of Arizona.

Neifert, William W. 2001. "Trailing Geronimo by Heliograph." In *Eyewitnesses to the Indian Wars 1865–1890*. Vol. 1, *The Struggle for Apacheria*, edited by Peter Cozzens, 557–61. Mechanicsburg, PA: Stackpole Books.

Utley, Robert M. 2012. *Geronimo*. New Haven, CT: Yale University Press.

Volkmar, William J. 1890a. "Report of Concerted Practice Department Heliograph System, in November 1889." In *Instructions for Signal Officers of Heliograph Divisions & Stations*, by William J. Volkmar. Los Angeles: Department of Arizona.

Volkmar, William J. 1890b. "Report of General Practice of the Heliograph System, Dept. of Arizona, in May, 1890: Under Direction of Wm. J. Volkmar, Assistant Adjutant General, Chief Signal Officer." Washington, DC: Signal Office.

Volkmar, William J. 1890c. "Reports of Preliminary Reconnaissances and Concerted Practice Between Arizona Divisions, of the Department Heliograph System, during April, 1890." In *Instructions for Signal Officers of Heliograph Divisions & Stations*, by William J. Volkmar. Los Angeles: Department of Arizona.

Willcox, Orlando B. "Report of Bvt. Maj. Gen. O. B. Willcox." In *Annual Report of the Secretary of War. 1882/83*, Vol. 1, by Robert T. Lincoln. Washington, DC: GPO, 1882.

Wood, Leonard. 1890. "Reconnaissance for Heliograph Stations in Arizona and New Mexico. Extracts from Report of 1st Lieut. Leonard Wood, Med. Dept., Fort McDowell, A. T., October 20, 1888." In *Instructions for Signal Officers of Heliograph Divisions & Stations*, by William J. Volkmar. Los Angeles: Department of Arizona.

Wood, Leonard, and Jack C. Lane. 2009. *Chasing Geronimo: The Journal of Leonard Wood, May–September 1886*. Lincoln: University of Nebraska Press.

Notes

Chapter 1

1. Mills, *The Military Heliograph and Its Use in Arizona and New Mexico*, 2; Reade, "About Heliographs," 98. Reade provides a rather good historical description of the use of reflected light by militaries.
2. Reade, "About Heliographs," 107.
3. Myer, *Annual Report of the Chief Signal Officer to the Secretary of War for the Year 1877*, 148.
4. Myer, 136, 508; United States Patent Office, "Frank C. Grugan."
5. Miles, *Personal Recollections and Observations of General Nelson A. Miles* (hereafter cited as *Recollections*), 481; Raines, *Getting the Message Through*, 78n120; Myer, *A Manual of Signals*, 243. Raines cites Miles's assertion of using heliographs as merely claims, as these heliographic events are absent from other, more official reports. While the information is not in Myer's report to the secretary of war, his 1879 signals manual references the heliograph's recent use on the western plains.
6. Myer, *A Manual of Signals*, 242–60.
7. William B. Hazen, *Annual Report of the Chief Signal-Officer to the Secretary of War for the Fiscal Year Ending June 30, 1881*, 5.
8. In 1887 Lieutenant William A. Glassford proposed a series of heliograph links through Taylor Canyon (near the northwestern part of the Pinaleños). In 1889 and 1890 Lieutenant Millard F. Eggleston and Captain Cunliffe H. Murray established connections to the Gila Valley by traversing over the top of the Pinaleños.
9. Glass, *The History of the 10th Cavalry*, 132; Horn, *Life of Tom Horn*, 197. Between the 1882 and 1886 systems, there is mention of an 1885 system. In 1885, during General Crook's "water-hole" campaign, heliographs were reportedly used within a communication network extending from Fort

Bowie to Nácori, Mexico, through seven intermediate stations. This assertion comes from Tom Horn's autobiography, in which he frequently mentions their use. Additionally, in a 1920 letter, General Leonard Wood—who in 1885 was a lieutenant serving in the Arizona Territory and later became chief of staff of the army—mentioned that communication during this campaign "was principally by telegraph or courier, with a certain number of heliographs." However, official reports from Crook in 1885 and 1886, as well as from chief signal officer General Hazen, do not document such a system. The lack of corroborative evidence in formal military records suggests that while heliographs were used, their operational scale, as reported by Horn, might not have been as extensive as claimed. The absence of further details precludes a detailed mapping of such a system.

10. Miles, *Recollections*, 481–82.
11. Kelly, *Talking Mirrors at Fort Bowie*. Included are excerpts of correspondence between Charles Gatewood and Colonel Edward. E. Dravo from 1926.
12. Dravo, "Report of the Heliograph System, District of New Mexico. E. E. Dravo, 1st Lieut. 6th Cavalry, A. S. O., Officer in Charge. Fort Bayard, New Mexico, September 20, 1886," 4 (hereafter cited as "September 20, 1886, Report"); Dravo, "Heliograph Stations in New Mexico. Descriptive Notes," 2–3 (hereafter cited as "Descriptive Notes"); Fuller, "Heliograph Report of A. M. Fuller (hereafter cited as "September 30, 1886, Report"). Dravo got off to a late start due to his prior assignment to a court-martial.
13. Myer, *A Manual of Signals*, 87. Myer describes the reasonable daylight range for flags (4 x 4 foot flags on a 12-foot pole) as 15 miles, although distances of up to 25 miles were possible. By contrast, the median range for heliograph connections in these systems (1882–1893) was 31.8 miles, with 23 of the 126 connections exceeding 50 miles.
14. Fuller, "Consolidated Report of Heliograph Line"; Dravo, "Report of Heliograph Lines"; Dravo, "Monthly Report of Heliograph Lines." These data are extracted from tabular reports submitted by both officers; Fuller reported 2,240 messages, and Dravo reported 754.
15. Rolak, "General Miles' Mirrors," 151, 153. Kelly, *Talking Mirrors at Fort Bowie*, Appendix C, 3; Smith, *Fort Huachuca*, 66, 125. Rolak discusses heliograph messages concerning logistics, personnel, weather conditions, and command communication, stating that "the military value of such activity was certainly

minimal." I disagree with this assessment. Logistics, personnel status, weather conditions, and command communication are essential components of military operations and should not be dismissed simply because they did not directly result in the deployment of a strike force against hostile Natives. Kelly's account is based on Dravo's memory, as his records were destroyed in the San Francisco fire. Smith refers to a movement of forces in May and again on June 5, when Colonel Beaumont (Fort Bowie) sent a cavalry troop to Antelope Springs from the east.

16. This difference is smaller than expected, however. Typically, when the radius is doubled, a fourfold increase in area is anticipated. In a featureless landscape, visibility area increases by the square of the radius, meaning that doubling the radius should result in four times the area. However, as the terrain becomes more compartmentalized and the relief more dramatic, the advantage of a longer-range visual system like the heliograph diminishes.

17. This includes all the heliographs established in Arizona and New Mexico except those used temporarily. The stations used include Bowie Peak, White's Ranch, Fort Bowie, Camp Henely, Emma Monk, Old Camp Rucker, Fort Cummings, Antelope Springs, Fort Huachuca, Hachita Mining Camp, Tubac, Lockhart's Well, Little Baldy Peak, Nogales, Calabasas, Lake Valley, Hillsboro, Henry Forest, Stein's Pass, Fourr's Ranch, Deming, Fort Bayard, Crittenden, Bowie Station, Pinos Altos, Limestone Mountain, White House, Alma, and Siggins's Ranch. Temporary stations not in this analysis include Eglin, Gardner's Ranch, Heatley's Well, Helen's Dome near Bowie, Separ, and San Bernardino Ranch.

18. During this time, there were thirty-three links (connections between two heliograph stations) among the thirty stations mentioned. If flags were used, then links shorter than 15 miles could be bridged using signal flags at the stations themselves, without the need for additional flag stations. Links between 15 and 30 miles would require one additional signal flag station, while links between 30 and 45 miles would need two. The only link exceeding 45 miles—Pinos Altos to Deming at 48.6 miles—would require three additional signal flag stations.

19. Miles, *Recollections*, 483; Fuller, "Consolidated Report of Heliograph Line." Miles states that each station required two or three operators, one to five guards, and couriers as needed. Rations were supplied by the nearest post,

and water, if not readily available, was hauled by mule. Based on Fuller's reporting, the average number of soldiers at each of the Arizona sites was 4.6.

20. William C. Endicott, *Report of the Secretary of War*, 72–73 (hereafter cited as *Secretary of War's 1886 Report to Congress*). In the "Report of the Lieutenant General of the Army," Philip Sheridan includes a specific restraint on Miles, instructing him to use only regular troops.

21. Endicott, *Secretary of War's 1886 Report to Congress*, 166.

22. Dempsey, *Doctrine for the Armed Forces of the United States*, GL-5; US Air Force, *Air Force Doctrine Publication (AFDP)* 3-30, 69; US Army, *FM 6-0*, 1–2. The US Joint Chiefs of Staff define command and control (C2) as "the exercise of authority and direction by a properly designated commander over assigned and attached forces in the accomplishment of the mission." The US Air Force elaborates, stating that "fluid horizontal and vertical information flow enables effective C2 throughout the chain of command. This information flow, and its timely fusion, enables optimum decision-making." Similarly, the US Army explains that "control is the regulation of forces and warfighting functions to accomplish the mission in accordance with the commander's intent," with key components including "direction, feedback, information, and communication."

23. Endicott, *Secretary of War's 1886 Report to Congress*, 167 (report of General Miles).

24. Utley, *Geronimo*, 193.

25. Utley, 195.

26. Utley, 195.

27. Utley, 196.

28. Altshuler, *Cavalry Yellow & Infantry Blue*, 351.

29. "Post Return for Fort Huachuca," May 1886, NAID: 561324, RG94, Microfilm Serial: M617, Microfilm Roll: 490, National Archives; Altshuler, *Cavalry Yellow & Infantry Blue*, 198–99. "Post Return for Fort Huachuca" shows Lawton "Comd'g Battalion Troops in the field, since 5th [of May 1886]." Lawton enlisted as an infantryman during the Civil War, rising to the rank of lieutenant colonel by the end of the conflict. In 1866 he was commissioned as a second lieutenant in the regular army. He later served as an engineer officer for the District of New Mexico before transferring to the cavalry

at Fort Stanton in 1883. By the following year, he was stationed at Fort Huachuca. Lawton's cavalry battalion, which included a cavalry troop, twenty infantrymen, twenty Native scouts, and two pack trains, pursued Geronimo until his surrender. Lawton was promoted to major general and later served in Cuba, but he was killed in action in the Philippines in December 1899.

30. Utley, *Geronimo*, 198.
31. Endicott, *Secretary of War's 1886 Report to Congress*, 169, 166. The quote is from Miles's General Field Order No. 7.
32. Smith, *Fort Huachuca*, 66: "On May 20th Major Eugene B. Beaumont, 4th Cavalry sent a heliograph message to Thompson at Bowie from heliograph station number three. In it he stated that Colonel Shafter's troops were the only ones near enough a band reported in the Dragoon Mountains to pursue and intercept them"; Kelly, *Talking Mirrors at Fort Bowie*, Appendix C, 3.
33. Rolak, "General Miles' Mirrors," 151; Endicott, *Secretary of War's 1886 Report to Congress*, 177; Bigelow, *On the Bloody Trail of Geronimo*, 207; Spencer, *Outline Map of the Field of Operations Against Hostile Chiricahua Indians* (hereafter cited as *Spencer Map of 1886*). I question whether the station at Antelope Springs initiated this report. According to Lawton, the Apaches were sighted emerging from the Whetstone Mountains and heading south. The distance from Antelope Springs to the eastern side of the Whetstones is 30 miles, an extremely long distance for observation. Furthermore, Lieutenant John Bigelow, who was pursuing the Apaches from the north, indicated that he proceeded west around the point of the mountain, suggesting that both he and the Apaches were on the western side of the Whetstones. This suggestion is further supported by the Spencer map, which also shows the pursuit on the western side. If they were indeed on the western side, observation from Antelope Springs would have been impossible. Given this, it is far more likely that, if the report was initiated by heliograph, it came from the station located about a mile west of Gardner's Ranch. This station was temporarily placed there by Fuller on June 2 and then moved to Little Baldy Peak on June 11.
34. Utley, *Geronimo*, 196.
35. Rolak, "General Miles' Mirrors," 156.
36. I see the continued expansion of the system as an acknowledgment that Miles was uncertain whether the fight or pursuit would cross back over the

border. This uncertainty is entirely reasonable and prudent from a military perspective, as nothing is assured in war.

37. Utley, *Frontier Regulars*, 387–89.
38. Miles, *Recollections*, 538.
39. Glassford, "Heliograph and Signal Work," 1 (hereafter cited as "1887 Reconnaissance").
40. Wood, "Reconnaissance for Heliograph Stations in Arizona and New Mexico" (hereafter cited as "1888 Reconnaissance"); Utley, *Frontier Regulars*, 387–88. Wood, a medical doctor, was part of Lawton's battalion that pursued Geronimo deep into Mexico between May and September 1886.
41. Miller, *The Heliograph, a Bibliography*, 2. Miller's bibliography includes "War Department, United States Signal Service. Instructions for Using the Heliograph of the United States Signal Service, by R.E. Thompson, Washington, D.C: 1888."
42. Greely, *Annual Report 1888*, 4-5.
43. Greely, 5.
44. Funds expended for the US Army Signal Service were as follows: $2,511 in 1886, $1,946 in 1887, $1,898 in 1888, $5,000 in 1889, $5,088 in 1890, $6,392 in 1891, $3,384 in 1892, $18,790 in 1893, and $16,914 in 1894 (sourced from Greely's annual reports). During this period, the US Army Signal Corps was in transition. On October 1, 1890, Congress mandated the transfer of the US Weather Service from the Signal Corps to the Department of Agriculture. However, this shift did not take place until July 1, 1891, and it would be a few more years before the Signal Service saw a substantial increase in funding.
45. Volkmar, *Report of General Practice of the Heliograph System*, 4, 18 (hereafter cited as *Report of the May 1890 Practice*). All the participating stations had at least one of the new service heliographs.
46. Grierson, *Annual Report of Brigadier General B. H. Grierson*, 2.
47. Greely, *Annual Report 1890*, 4; Volkmar, *Report of the May 1890 Practice*, 3–4.
48. Altshuler, *Cavalry Yellow & Infantry Blue*, 344.
49. Here and throughout this manuscript, *exercise* means a full or partial deployment of a heliograph system for training and evaluation.
50. Volkmar, "General Order No. 2," 2. The May 1–15 exercise was divided into two phases: a May 1–5 phase to practice "local [communication] between

contiguous stations, in perfecting the main line" and a May 6–15 general exercise phase "including the transmission of through messages between extreme stations of the main line and its branches." The heliograph divisions included Whipple Barracks, Fort Verde, Fort McDowell, San Carlos, Fort Grant, Fort Lowell, Fort Huachuca, Fort Bowie, Fort Bayard, and Fort Stanton.

51. Volkmar, "Heliograph System, Department of Arizona," 3–8.
52. Volkmar, "General Order No. 2," 2.
53. Volkmar, "Report of Concerted Practice Department Heliograph System," 34 (hereafter cited as "Report of the November 1889 Practice").
54. Volkmar, 35–36.
55. Volkmar, 1; Wharfield, "Apache Kid and the Record," 45. This was the infamous escape of the "Apache Kid."
56. The locations of Stein's Peak and Stein's Pass are both used by the lieutenants to describe the heliograph station on the bluffs above Stein's Pass. Stein's Peak, located farther north, was not used as a heliograph station. The name Stein's Pass will be used here to describe the heliograph station.
57. Volkmar, "Report of the November 1889 Practice," 12–28.
58. Volkmar, "Reports of Preliminary Reconnaissances and Concerted Practice between Arizona Divisions, 1 (hereafter cited as "Report of the April 1890 Practice").
59. Volkmar, *Report of the May 1890 Practice*, 4.
60. Greely, *Annual Report 1890*, 4.
61. Greely, "Report of the Chief Signal Officer, October 9, 1893," 652.
62. Greely, 652.
63. Greely, 665.
64. Greely, 666.
65. Greely, "Report of the Chief Signal Officer, September 7, 1887," 5.
66. Greely, 5.
67. Greely, 5.
68. Hazen, *Annual Report of the Chief Signal Officer, United States Army, to the Secretary of War for the Fiscal Year Ending June 30, 1882*, 4, 94.
69. Volkmar, *Report of the May 1890 Practice*, 5.
70. Greely, *Annual Report 1890*, 4; *Iron* (London) "Long Distance Heliograph Signalling," June 6, 1890; Greely, "Evolution of the Signal Corps," 18. Here

Greely remarks, "Foreign armies have repeatedly followed Signal Corps methods." He follows with an example of Spanish forces using the heliograph in Cuba.

71. Greely, *Annual Report 1891*, 41.
72. Greely, *Annual Report 1894*, 496.
73. Greely, 498, 646.
74. US Army, *Visual Signaling, Manual No. 6* (1905), 41; US Army, *Visual Signaling, Manual No. 6* (1910), 13.
75. Greely, "The Evolution of the Signal Corps," 14.
76. Finley, "The Situation in Mexico," 58–65; Scriven, *Report of the Chief Signal Officer*, 38. This was likely Field Communications Company D, which maintained communications with "land and naval stations by radio and visual signaling."
77. Fletcher, "Tar Heels Flirt with Precipices," 1–2. Their use here may have been only for training and practice.
78. Henry P. McCain, "Cable Number 1563, To General Headquarters, A. E. F. France, 21 June 1918" (adjutant general, US Army, 1918), NAID: 209265595, RG 120, Series: Cablegrams from General John J. Pershing to the Adjutant General, National Archives.
79. Fuller, "September 30, 1886, Report," 2.
80. Fuller, 8. Some of the 1886 stations may have had a more permanent wooden post on which the soldiers placed the heliograph.
81. Volkmar, *1890 Heliograph Lines and Stations*.
82. Volkmar, "Letter of April, 19 1890; Mills, *The Military Heliograph and Its Use in Arizona and New Mexico*, Appendix 2, Tab R, 1–7. Mills has transposed the azimuths and distances contained on the Volkmar map to create a much easier-to-read document.
83. Volkmar, "Letter of April, 19 1890," 1.
84. Volkmar, 1890 *Heliograph Lines and Stations*. While this map has a generally consistent scale, in parts the scale changes rapidly, especially with respect to the 1886 stations. For example, the measured (GIS) straight line distance between Fort Bowie and Old Camp Rucker is 27.2 miles, and the distance based on the Volkmar map is 40.2 miles.
85. Heitman, *Historical Register*, 910.

86. Spencer, *Spencer Map of 1886*; Spencer, *Report of the Map of the Field Operations*. In August 1886, Spencer interviewed (in Mexico) Lieutenant James Parker and Captains Henry W. Lawton, Charles A. P. Hatfield, Thomas C. Lebo, Alexander S. B. Keyes, and James G. MacAdams in its making.
87. Volkmar, "Report of the April 1890 Practice," 28 (report of Lieutenant Reichmann).
88. Glassford, "1887 Reconnaissance," 1–2.
89. US Geological Survey, "What Is a Digital Elevation Model (DEM)?"
90. Breed and Hosmer, *Higher Surveying*, 28. The instrument height is based on personal observations of heliographs.
91. The 5-meter models do not extend over the entire area of the heliograph system. A combination of the 1-meter and 10-meter models was used for added fidelity in the visibility diagrams. For the great majority of the heliograph sites, the 10-meter model's lack of micro-terrain fidelity has little impact. However, in some cases, where the stations relied on small elements of terrain, like rock outcrops, this detail becomes important. In these cases, the combined, higher-resolution model was applied. The 5-meter resolution model was made by the author using 1-meter and 10-meter digital elevation models.
92. Mitchell, *Esri Guide to GIS Analysis*, 3:69–71.
93. Knowles, *Past Time, Past Place*, 186. Knowles defines a GIS as "a collection of computer hardware, software, and geographic data for capturing, storing, updating, manipulating, analyzing, and displaying all forms of geographically referenced information."
94. Volkmar, "Letter of April, 19 1890"; American Society of Civil Engineers, et al., *Glossary of the Mapping Sciences*, 141. Volkmar's use of *departure* is apt. Departure is defined as "the orthogonal projection of a line, on the ground, onto an east west axis of reference." For example, using the law of sines, the subtended departure length of 1° at 40 miles is 3,686 feet.
95. In this case, there were five connected stations and an azimuth-distance line to the nearby railroad station, as referenced by Lieutenant Hart, who set up this particular station.
96. Mitchell, *ESRI Guide to GIS Analysis*, 2:39–43. The standard distance deviation statistic (called standard distance) indicates how much the distances between the mean center and the features (the azimuth-distance terminal

points) differ from the average. A larger standard distance value means the distances vary more from the average, showing that the features are more spread out around the center. The standard distance is depicted as a circle with a radius equal to the standard distance. Care was taken when using the mean center; if the standard distance is large and encompasses all the likely peaks (for a heliograph station), this broad circle reduces the relative importance of the mean center location, as it doesn't strongly favor one specific peak over the others. However, in some cases, the standard distance is small enough (or an alternative peak is far enough away) that the mean center is helpful in excluding that alternate peak as the heliograph location. In every case, though, the mean center is useful in indicating at least the general area of the peak being referenced.

97. Mitchell, *ESRI Guide to GIS Analysis*, 2:33.
98. There were three; for the example here I used two. All three are explained in more detail in the Fort Huachuca chapter.
99. The first of these is located at the Arizona Historical Society collections in Tucson. The second is in the special collections of the Phoenix Public Library.
100. Raines, *Getting the Message Through*.
101. Rolak, "General Miles' Mirrors."
102. Kelly, *Talking Mirrors at Fort Bowie*.
103. Mills, *The Military Heliograph and Its Use in Arizona and New Mexico*.
104. Miles, *Recollections*.
105. Grierson, *Annual Report of Brigadier General B. H. Grierson*.
106. Meribeth Price, *Mastering ArcGIS*, 302–3. I have used the North American Datum of 1983 (NAD83) Universal Transverse Mercator (UTM) system for the maps: NAD83 UTM Zone 12N for Arizona and the western edge of New Mexico, and NAD83 UTM Zone 13N for the rest of New Mexico. While no projection perfectly preserves both direction and distance, distortion within a UTM zone is minimal for both. The US Geological Survey also uses the UTM system for its 1:100,000 and 1:24,000 scale maps.

Chapter 2

1. Hazen, *Annual Report of 1882*, 94; Orlando B. Willcox, "Report of Bvt. Maj. Gen. O. B. Willcox," 152.
2. Altshuler, *Cavalry Yellow & Infantry Blue*, 225.
3. "Post Return for Fort Grant," October 1882, NAID: 561324, RG94, Microfilm Serial: M617, Microfilm Roll: 415, National Archives.
4. Hazen, *Annual Report of 1882*, 94.
5. Hazen, 94.
6. J. M. Hilton, *Fort Grant, Arizona*.
7. Hazen, *Annual Report of 1882*, 94.
8. Altshuler, *Cavalry Yellow & Infantry Blue*, 40; "Post Return for Fort Grant," July 1882, NAID: 561324, RG94, Microfilm Serial: M617, Microfilm Roll: 415, National Archives.
9. Willcox, "Report of Bvt. Maj. Gen. O. B. Willcox," 152. Willcox informs us that the connection to those Gila camps was indeed established.
10. Hazen, *Annual Report of 1882*, 94.
11. Willcox, "Report of Bvt. Maj. Gen. O. B. Willcox," 152.

Chapter 3

1. Utley, *Frontiersmen in Blue*, 161–63. In October 1860, Coyotero raiders kidnapped John Ward's stepson from his ranch in the Sonoita Valley. Mistakenly identifying the raiders as members of Cochise's band of Chiricahua Apaches, Ward reported the kidnapping to Colonel Morrison at Fort Buchanan. Once Morrison had gathered enough troops, he dispatched Lieutenant George N. Bascom to recover the boy. Bascom arrived at Bowie Pass, where a Butterfield stage station was located, on February 4. When Cochise and a small party, including some of his family members, visited Bascom's camp, Bascom demanded the boy's return. Cochise denied having the boy and a confrontation ensued, during which Cochise escaped, though his family members were captured. Bascom held the captured Chiricahua as hostages. In response, Cochise took a Butterfield employee, a Mr. Wallace, and two stage travelers hostage. Bascom refused to

exchange hostages until Ward's boy was produced. Over the following week, Cochise wreaked havoc near the pass, culminating in the killing of Wallace and the two stage travelers. When reinforcements arrived from Fort Breckenridge, they hanged some of the original Chiricahua prisoners, along with three Coyoteros. According to Utley, the "Bascom affair plunged Arizona into twenty-five years of hostility with the Chiricahua Apache."

2. National Park Service, "Arizona: Fort Bowie National Historic Site."
3. Utley, *Frontier Regulars*, 170.
4. Fuller, "September 30, 1886, Report," 1.
5. General Land Office, *GLO Records (Arizona Township 19 South, Range 27 East)* (Washington, DC: General Land Office, 1885). Emma Monk House is located a short distance south and west of the Swisshelm Mountains.
6. Fuller, "Consolidated Report of Heliograph Line."
7. Volkmar, "Report of the November 1889 Practice," 18 (report of Lieutenant Neall).
8. Fuller, "September 30, 1886, Report," 5.
9. Dravo, "Descriptive Notes," 1.
10. *US Army and Navy Journal, and Gazette of the Regular and Volunteer Forces* 18 (1881): 401; *Bennington Evening Banner*, "Death of Major Hovey," November 16, 1908, 1.
11. Volkmar, "Report of the November 1889 Practice," 14 (report of Lieutenant Hovey).
12. Kelly, *Talking Mirrors at Fort Bowie*, 9.
13. Fuller, "September 30, 1886, Report," 1.
14. Utley, *Historical Report on Fort Bowie*, Photo 1.
15. Greene, *Historic Structure Report*, map 2.
16. Fuller, "Consolidated Report of Heliograph Line."
17. General Land Office, *GLO Records (Arizona Township 13 South, Range 28 East)* (Washington, DC: General Land Office, 1882).
18. Clemensen, *Cattle, Copper, and Cactus*, 50.
19. Gale, "Report of Reconnaissance for a Central Station," 1.
20. Spencer, *Spencer Map of 1886*.
21. Gale, "Report of Reconnaissance for a Central Station," 2; National Park Service, *Historic American Landscapes Survey*, 6.

22. Gale, "Report of Reconnaissance for a Central Station," 1–4.
23. Gale, 4.
24. Gale, 1.
25. Fuller, "September 30, 1886, Report," 2.
26. Fuller, "Consolidated Report of Heliograph Line." Fuller's tabular report also describes the location as "one mile from camp on bluffs above road."
27. Fuller, "September 30, 1886, Report," 2.
28. Fuller, 2.
29. Fuller, "Consolidated Report of Heliograph Line."
30. Fuller, "September 30, 1886, Report," 2; Fuller, "Consolidated Report of Heliograph Line."
31. Dravo, "September 20, 1886, Report," 1.
32. Dravo, "Descriptive Notes," 1.
33. Dravo, 1.
34. Dravo, 1.
35. Altshuler, *Cavalry Yellow & Infantry Blue*, 164; Utley, *Geronimo*, 90. Austin Henely, originally from Ireland, commanded a company of Native American scouts in southern Arizona. (The army enlisted friendly Native Americans, called scouts, to search for hostile Native Americans.) Henely first enlisted in the US Army and graduated from West Point in 1872. He and Lieutenant John A. Rucker, who were brothers-in-law, drowned in a flash flood on July 11, 1878. Both men had camps named in their honor.
36. Although the Camp Henely heliograph station was established by Sherwood in 1886, it was not listed as a connection to Bowie Peak that year; however, it was listed as a connected station in 1890. Similarly, Colorado Peak was not part of the 1886 system, but Bowie Peak connected to it during the 1890 and 1893 exercises.
37. Spencer, *Spencer Map of 1886*.
38. A copy of this photo is available at the University of Arizona Library (as well as online at the National Archives). The back of this copy gives the following identifications: (1) "Lt A. M. Fuller, A.S.O., 2nd Cavalry"; (2) "Corpl Charles Swope, Co H, 8th Infantry"; (3) "Pvt Thomas Harvey, Co H, 8th Infantry"; (4) "Pvt W. H. Bowker, Co. B, 8th Infantry"; (5) "Sgt Wm S Wade, Co 'I' 1st Infantry"; (6) "Pvt Charles Laing, Co H, 8th Infantry"; (7) "Pvt F. W.

Allen, Co H, 8th Infantry"; (8) "Pvt Theo. Persee, Co H, 8th Infantry"; (9) "Adolph Miller, Co D, 10th Infantry."
39. Fuller, "September 30, 1886, Report," 2.
40. Bingham, *South Eastr'n Arizona*.
41. General Land Office, *GLO Records (Arizona Township 18 South, Range 28 East)* (Washington, DC: General Land Office, 1882).
42. General Land Office, *GLO Records (Arizona Township 23 South, Range 25 East)* (Washington, DC: General Land Office, May 11, 1885); Fuller, "September 30, 1886, Report," 5.
43. Fuller, "Consolidated Report of Heliograph Line."
44. Douglas, *Bisbee, Arizona*. Interestingly, this area was again used militarily in 1916. The 1916 US Geological Survey topographic map shows a US Army infantry regiment camped just east of Forrest's ranch.
45. Wood and Lane, *Chasing Geronimo*, 108; Utley, *Geronimo*, 216.
46. Glassford, "1887 Reconnaissance," 7.
47. Trail miles were computed using current-day roads and trails. These routes follow logical valleys and passes and are virtually the same as those shown on the 1900 GLO maps.
48. Utley, *Geronimo*, 310n2.
49. Grierson, *Annual Report of Brigadier General B. H. Grierson*, 15, 16. See the previous chapter about Maus's 1882 system. Both Fuller and Dravo were still stationed in the Department of Arizona during the 1890 exercise, though it appears they were not involved.

Chapter 4

1. US Army, "US Army Fort Huachuca." *Huachuca* is derived from a local Native American dialect and is commonly translated to mean "place of thunder." Fort Huachuca, situated at the base of the Huachuca Mountains in southeastern Arizona, was established in 1877 to protect settlers in the San Pedro and Santa Cruz Valleys.
2. Fuller, "September 30, 1886, Report," 3.
3. Altshuler, *Cavalry Yellow & Infantry Blue*, 86. Colonel Thomas Crittenden, a native of Kentucky, commanded the district of Tucson from May 1867 to 1869.

4. Fuller, "September 30, 1886, Report," 3.
5. Fuller, 3–4.
6. Spencer, *Spencer Map of 1886*.
7. Fuller, "September 30, 1886, Report," 4. Fuller never explicitly states that he connected with Huachuca from the Gardner's Ranch station. However, he was able to connect to Huachuca from nearby stations on the previous days and he did number this Garner's Ranch station (No. 8). This station would be the pivot from a future Patagonia station to Huachuca, so it is extremely unlikely that he did not connect with Huachuca from it prior to heading south to establish a Patagonia station.
8. "Post Return for Fort Huachuca," May 1886, NAID: 561324, RG94, Microfilm Serial: M617, Microfilm Roll: 490, National Archives; Endicott, *Secretary of War's 1886 Report to Congress*, 177. The post returns from Fort Huachuca report that Captain Lawton was absent "comdig battalion troops in the field" beginning May 5, 1886. Lawton's report states that his command was at Calabasas.
9. Endicott, *Secretary of War's 1886 Report to Congress*, 177. Report of Captain Lawton.
10. Wood and Lane, *Chasing Geronimo*, 45.
11. Fuller, "September 30, 1886, Report," 4.
12. Fuller, 4; Wood and Lane, *Chasing Geronimo*, 45. While Fuller is clear that he spoke to Miles on June 9, Lieutenant Wood reported that Miles left for Tucson in the morning of June 9.
13. Fuller, "September 30, 1886," Report, 4.
14. Fuller, "Consolidated Report of Heliograph Line."
15. US Geological Survey, "Mount Wrightson."
16. National Oceanic and Atmospheric Administration, "Station Datasheet—Baldy."
17. Spencer, *Spencer Map of 1886*.
18. *Times Leader*, "Indian Fighter Death Victim: William W. Neifert, 86, Helped Take Geronimo," March 27, 1951, 13.
19. Neifert, "Trailing Geronimo by Heliograph," 560.
20. Neifert, 560.
21. Eckhoff and Riecker, *Official Map of the Territory of Arizona*; Douglas, *Arizona (Santa Cruz County) Nogales Quadrangle*.
22. Fuller, "September 30, 1886, Report," 5–6.

23. General Land Office, *GLO Records (Arizona Township 17 South, Range 22 East)* (Washington, DC: General Land Office, 1888); Spencer, *Spencer Map of 1886*; Hodges, *St. David Quadrangle, Revision 1922*.
24. My initial inclination was to map the heliograph on the north side of the road leading into the canyon, where the rising slope from the canyon floor would provide the necessary elevation for an observer to see Fort Huachuca. However, Spencer's map clearly shows the station situated to the east of the Dragoon–Tombstone road (parts of which still exist) and south of the road leading into the canyon.
25. Fuller, "Consolidated Report of Heliograph Line."
26. Rolak, "General Miles' Mirrors," 150.
27. Fuller, "September 30, 1886, Report," 3.
28. Stephen Gregory grew up around Fort Huachuca in a military family. He has a bachelor's degree in American history from Northern Arizona University and a master's degree in library and information science from the University of Arizona.
29. Cullum, *Biographical Register*, 6-A:484–85.
30. Volkmar, "Report of the November 1889 Practice," 7. Volkmar states that he was able to connect with the Huachuca station, which was under the direction of Lieutenant Albright, on November 20, 1889. Albright likely set up the station at the start of the exercise, on November 17.
31. Gale, "Report of Reconnaissance for a Central Station," 3. Gale noted that the station was "2,000 feet higher than their flash, and reached in 2 hours." I assume he meant higher than the post.
32. Gale, 5.
33. Gale, 3.
34. Volkmar, *Report of the May 1890 Practice*, 54 (report of Lieutenant W. H. Hart).
35. Volkmar, "Report of the November 1889 Practice," 24 (report of Lieutenant Hart); Murry et al., *Oxford English Dictionary*, 16:373. In this context a spur is "a range, ridge, mountain, hill, or part of this, projecting for some distance from the main system or mass; an offshoot or offset."
36. Volkmar, "Report of the November 1889 Practice," 26 (report of Lieutenant Hart).

37. Volkmar, 26.
38. Volkmar, *1890 Heliograph Lines and Stations*.
39. Fuller, "September 30, 1886, Report," 3. Fuller reported that the Huachuca station was set up on May 23, after the station at Antelope Springs. So whom was he signaling to before then? We know he was coordinating with Fort Huachuca, as soldiers from the fort joined him at White's Ranch on May 19 or 20. This suggests that Fuller must have established a temporary heliograph station at Fort Huachuca to communicate with Antelope Springs before the permanent station was set up.
40. The name Signal Hill does not automatically indicate the presence of a heliograph station. There are over three dozen Signal Hills, past and present, distributed throughout the United States, one dating as far back as the late 1700s.
41. Fuller, "September 30, 1886, Report," 3.
42. Fuller, 3.
43. Greely, *Annual Report 1886*, 32. This measurement is likely a typographical error. Fuller consistently measured elevations with precision, and his recorded elevations closely correlate with present-day US Geological Survey data. (Without going into the details of the analysis, it is worth noting that the correlation between Fuller's measurements and modern US Geological Survey elevations produces an impressive R^2 value of 0.92.) In 1886 Greely sent Miles thirty-four heliographs, ten telescopes, thirty marine glasses, and one aneroid barometer. Fuller likely had that barometer.
44. Fuller, "Consolidated Report of Heliograph Line."
45. In addition to Fuller's data, Volkmar's 1890 map provides estimated magnetic azimuths from both Emma Monk and Fort Huachuca to Antelope Springs, but these diverge and do not intersect. As a result, evaluating those azimuths does not provide any useful information.
46. There are two peaks on present-day Signal Hill, one to the north and one to the south. The hill to the south, a bit lower than its northern sister, is the only one offering a view of Antelope Springs proper.

Chapter 5

1. I used the 1890 spelling for Cook's Spring and Cook's Peak. The current spelling is "Cooke's" or "Cookes."
2. Staski, *Research on the American West*, 1.
3. Dravo, "Descriptive Notes," 2.
4. Staski, *Research on the American West*, 9; Volkmar, *Report of the May 1890 Practice*, 7. Staski states that the heliograph station was atop Cook's Peak. I find this unlikely for three reasons: (1) Cook's Peak, located 7 miles to the northwest and 3,600 feet higher than the fort, does not match Dravo's description of a hill almost due south and easily accessible from the fort. (2) Given the distance and elevation to Cook's Peak, the soldiers manning the heliograph station would not have camped at the fort, as Dravo suggests, but would have likely established a camp closer to the peak. (3) The geometry of both the Volkmar and Spencer maps does not support the station being located so far north and west. Since Staski's report lacks references or analysis supporting the assertion that the heliograph station was on Cook's Peak, I am inclined to discount this claim and rely on Dravo's description for the station's location. Near the end of the May 1890 exercise, as an experiment, Lieutenant Henry W. Hovey left the regular Fort Cummings station and traveled to Cook's Peak in an attempt to send a nearly 140-mile flash to Mount Graham. The effort failed for various reasons.
5. Dravo, "Report of Heliograph Lines"; Dravo, "Monthly Report of Heliograph Lines." Both the Volkmar and Spencer maps show a direct connection between Fort Cummings and Hillsboro to the north. However, Dravo repeatedly lists only four connecting stations: Henely, Lockhart's Well, Lake Valley, and Deming. A direct connection to Hillsboro would be unlikely due to intervening terrain.
6. Cullum, *Biographical Register*, 7:293–94.
7. Volkmar, *Report of the May 1890 Practice*, 63 (report of Lieutenant Joseph D. Leitch).
8. Dravo, "September 20, 1886, Report," 2.
9. Henry W. Hovey and Richard B. Paddock, "Reports of Concerted Reconnaissances for Short Line Connecting Bayard and Stanton Heliograph Divisions, N.M.," 2 (hereafter cited as "Connecting Bayard and Stanton").

10. "The City Column," *Las Vegas Gazette*, December 4, 1885. Hovey mentions the sale in his report as well.
11. Hovey and Paddock, "Connecting Bayard and Stanton," 2.
12. General Land Office, *GLO Records (New Mexico Township 27 South, Range 8 West)* (Washington, DC: General Land Office, 1887).
13. Dravo, "September 20, 1886, Report," 2. Dravo reported 35 miles.
14. Bureau of Land Management, "Lake Valley Historic Townsite."
15. Sanborn Map Company, *Sanborn Fire Insurance Map from Lake Valley*.
16. Adler, *History of the Seventy Seventh Division*, 166–67; *Fort Collins Coloradoan*, "Gen E. M. Johnson, US Army Officer Is Dead in Paris," October 15, 1923.
17. Dravo, "September 20, 1886, Report," 2.
18. Dravo, "Descriptive Notes," 2.
19. The GIS-measured azimuth to the hilltop is 92°.
20. Hillsboro, New Mexico, "Welcome to Hillsboro, New Mexico."
21. Dravo, "September 20, 1886, Report," 2.
22. Dravo, "Descriptive Notes," 2.
23. General Land Office, *GLO Records (New Mexico Township 16 South, Range 7 West)* (Washington, DC: General Land Office, 1882).
24. The measured azimuth using GIS is 84.8°. Here also I identify an unnamed hill by its elevation shown on topographic maps.
25. Another remote possibility is Cook's Peak, 7 miles to the northeast. I find this unlikely for the reasons given in note 4.
26. Spencer, *Spencer Map of 1886*; Spencer, "Report of the Map of the Field Operations against the Hostile Bands of Chiricahua Indians under Natchez and Geronimo"; Volkmar, *1890 Heliograph Lines and Stations*. While Lieutenant Spencer traveled extensively in preparation for his 1886 map, he did not report visiting the Fort Cummings, Lake Valley, or Hillsboro areas. Additionally, Lake Valley and Hillsboro were not used in subsequent exercises, and Volkmar indicated that the azimuth between Fort Cummings and Hillsboro was estimated rather than accurately measured.
27. Dravo, "Descriptive Notes," 2.
28. Sanborn Map Company, *Sanborn Fire Insurance Map from Deming*.
29. Dravo, "Descriptive Notes," 2.
30. Julyan, *The Place Names of New Mexico*, 331; US Army, *Annual Report of the Quartermaster-General*, 193; Brown, *History of the Ninth US Infantry*,

149–50. The quartermaster general reports on contracts given to distribute water and other goods. According to Brown, men and equipment would disembark and embark at Separ.
31. Dravo, "September 20, 1886, Report," 1; Dravo, "Report of Heliograph Lines"; Utley, *Geronimo*, 116.
32. Dravo, "September 20, 1886, Report," 3.

Chapter 6

1. Fort Bayard Historic Preservation Society, "Fort Bayard History"; Heitman, *Historical Register*, 200.
2. Dravo, "Descriptive Notes," 2.
3. US War Department, *Fort Bayard, New Mexico*.
4. Volkmar, *Report of the May 1890 Practice*, 62 (report of Lieutenant William Black).
5. Dravo, "Descriptive Notes," 2.
6. Dravo, "September 20, 1886, Report," 3.
7. Dravo, 3–4.
8. Dravo, 3–4.
9. Dravo, 4.
10. National Park Service, *Prospector, Cowhand, and Sodbuster*, 222.
11. National Park Service, L.C. Ranch Headquarters National Register of Historic Places nomination form, 2–3.
12. Dravo, "September 20, 1886, Report," 3.
13. Klip, *Southwestern New Mexico*.
14. Dravo, "September 20, 1886, Report," 4. Isaac Siggins's name is spelled various ways. Dravo spells it "Siggin" and "Siggins" (on the same page). Volkmar spells it "Siggen." I use "Siggins."
15. Brasher, "The Francisco Vázquez de Coronado Expedition in Tierra Doblada," 182; General Land Office, *GLO Records (New Mexico Township 13 South, Range 19 West)* (Washington, DC: General Land Office, 1898). The 1898 GLO plat map places the John D. Lee house and stage station in the northwest quadrant of Section 5, Township 13, South Range 19 west. "Siggins Corral"

is marked just over an eighth of a mile to the northwest of this stage station. In his map, Brasher depicts the Siggins homestead house about a half mile west and a bit north of the Lee homestead house.

16. Brasher, "The Francisco Vázquez de Coronado Expedition in Tierra Doblada," 183. Brasher notes that this is an attractive campsite with evidence of chipped obsidian, indicating a prehistoric Native American camp; Klip, *Southwestern New Mexico*.
17. Dravo, "September 20, 1886, Report," 4.
18. General Land Office, *GLO Records (New Mexico Township 27 South, Range 16 West)* (Washington, DC: General Land Office, 1913).
19. Dravo, "September 20, 1886, Report," 2.

Chapter 7

1. Bureau of Reclamation, "News Archive: Fort McRae Placed on National Register."
2. Vandiver, *Black Jack*, 59.
3. Dravo, "Descriptive Notes," 2.
4. Dravo, "Descriptive Notes," 2.
5. Dravo, 2. The description reads: "Located ... on a butte 1 1/2 miles 25° E. of N. from the fort and 2 1/2 miles 11° West of N. from the A. T. and S. F. pump house. Butte just visible from pump house looking northward through arroya [*sic*]." (A. T. stands for Atchison and Topeka, and S. F. stands for Santa Fe.) Dravo also describes the direction to the heliograph station from a point along the railroad (5 miles east of the fort) as almost due north, "11° West of N. from the A. T. and S. F. pump house." Eleven degrees west of north is 349° magnetic, or 1.3° true in 1886. This is perplexing, as one might expect the direction to be westward from a point along the railroad. With that in mind, it makes more sense if the azimuth were 11° north of west. Even this remains unclear, as no reference to a pump house from 1886 can be found.
6. I am aware that the location of Dripping Spring has not yet been identified. However, I maintain the sequence in which the stations were set up. The description of the Dripping Spring station is sufficient to determine its location without knowing the position of Fort McRae's heliograph station—that

is, visibility from the McRae station was not a factor in determining the Dripping Spring location.
7. Dravo, "Descriptive Notes," 2.
8. General Land Office, *GLO Records (New Mexico Township 11 South, Range 5 East)* (Washington, DC: General Land Office, 1883).
9. Dravo, "Descriptive Notes," 3.
10. Measuring with GIS, the eastern hill is 2.5 miles at 134.5° from the spring, while the other, more western hill is 2.4 miles at 153.4°. The azimuth is what discriminates the two; the eastern hilltop is just over 7° different from Dravo's 127.2°, while the western hill is just over 26°.
11. Dravo, "Descriptive Notes," 3.
12. King, *National Register of Historic Places Inventory*.
13. Dravo, "Descriptive Notes," 3.
14. Sheridan and Gillespie, *Outline Descriptions of the Posts in the Military Division of the Missouri*, 153; Diddy, *Fort Stanton Historic Site Cultural Landscape Report*, 178, 188. Sheridan and Gillespie suggest that during this time, the building at the east corner of the parade field was the hospital, as recorded on the accompanying map in 1876. However, the more recent *Fort Stanton Historic Site Cultural Landscape Report* identifies the same building—though much larger—as a barracks and lists the adjacent building, constructed in 1938, as the hospital. Both Sheridan and Dravo believed there was a hospital at the post in 1876 and 1886, located at the east corner of the parade field.
15. On this hill is a pile of stones that may have been a cairn at some point.
16. Hovey and Paddock, "Connecting Bayard and Stanton," 1, 7.
17. Hovey and Paddock, 1.
18. Hovey and Paddock, 8.
19. Volkmar, *Report of the May 1890 Practice*, 18.
20. Diddy, *Fort Stanton Historic Site Cultural Landscape Report*, 33, 250.
21. Hovey and Paddock, "Connecting Bayard and Stanton," 7.
22. Paddock also gives an azimuth to Sierra Blanca Peak proper (spelling it "Blanco") of west 39° south, or 231° magnetic from Fort Stanton, which becomes 242.9° true after applying the magnetic declination of 11.9° east. This is very close to the GIS direction of 243.4° to Sierra Blanca Peak.
23. Volkmar, *Report of the May 1890 Practice*, 66 (report of Lieutenant Brewster).

24. Patterson, "Andre Walker Brewster."
25. US Geological Survey, "Little San Nicolas Canyon."
26. Hovey and Paddock, "Connecting Bayard and Stanton," 7.
27. Grierson, *Annual Report of Brigadier General B. H. Grierson*, 9; "Post Return for Fort Selden," December 1889, NAID: 561324, RG 94, Microfilm Serial: M617, Microfilm Roll: 1147, National Archives. The post return indicates one officer and twenty-three enlisted men at the post during December 1889.
28. Hovey and Paddock, "Connecting Bayard and Stanton," 2.
29. Hovey and Paddock, 7–8.
30. Hovey and Paddock, 2.
31. Hovey and Paddock, 3.
32. Hovey and Paddock, 3–4.
33. Hovey and Paddock, 8–9; General Land Office, *GLO Records (New Mexico Township 19 South, Range 2 West)* (Washington, DC: General Land Office, 1882); Price, "The Railroad, Rincon, and the River," 453; General Land Office, *GLO Records (New Mexico Township 19 South, Range 2 West) Rincon Post-1884* (Washington, DC: General Land Office, 1919). The 1882 GLO plat has Rincon located about a mile to the west. According to Price, the town was moved to its present location after it flooded in 1884. The buildings shown on the included map are from the 1919 GLO survey.
34. Hovey and Paddock, "Connecting Bayard and Stanton," 5.
35. Hovey and Paddock, 4.
36. Hovey and Paddock, 4; McMiller, "The Formation of the Tenth Cavalry Regiment," 75.
37. Hovey and Paddock, "Connecting Bayard and Stanton," 5.
38. Hovey and Paddock, 8; Volkmar, *Report of the May 1890 Practice*, 65 (report of Lieutenant Paddock). Paddock provided complex directions to the peak, based on trails and places that do not appear on contemporary maps, GLO maps, or US Geological Survey archived topographic maps. In 1890 he also supplied additional period names, which are equally obscure. I tried to follow these names on maps as best I could, but I had to make assumptions to trace the route all the way to the peak. Fortunately, the visibility diagrams and azimuths offer a logical basis for determining the location.

39. Volkmar, *Report of the May 1890 Practice*, 65 (report of Lieutenant Paddock). This is clearly a misreported distance. Paddock stated that the Tularosa station was on a hill about 7 miles west of the town of Tularosa. However, from any reasonable point on the San Andres ridge, Tularosa is approximately 35 miles to the east. A distance of 37 miles from the ridge would place the station 2 miles east of Tularosa, not 7 miles west.
40. If we assume that the Tularosa heliograph station's reported distance is too long by 10 miles (believing 37 miles to be a typo), Tularosa's azimuth-distance terminal point would shift northeast of the visible peak north of the canyon. When the standard distance is recalculated using this corrected point, the resulting radius is small enough to exclude any peak to the south of Dead Man's Canyon.
41. Hovey and Paddock, "Connecting Bayard and Stanton," 9.
42. Volkmar, *Report of the May 1890 Practice*, 65 (report of Lieutenant Paddock).
43. I realize the Tularosa station is out of sequence, as it was used in the analysis of the San Andreas station. I structured the book this way because the chapter describes the flow of events as Hovey and Paddock attempted to build the line during the winter of 1889–1890. Including the Tularosa, established much later, in the midst of this would have disrupted that flow. Since the Tularosa station can be located independently of other station data, I felt comfortable placing it later in the narrative.

Chapter 8

1. General Land Office, *GLO Records (Arizona Township 6 North, Range 10 East)* (Washington, DC: General Land Office, 1881).
2. Glassford, "1887 Reconnaissance," 4.
3. Volkmar, "Report of the April 1890 Practice," 11 (report of Lieutenant Overton); Overton, "Report of Heliograph Stations in Tonto Basin," 3–4. Overton climbed Reno in mid-September 1889. He set up a heliograph but was unable to reach back to Fort McDowell as those there were unaware of his attempt.
4. Volkmar, "Report of the April 1890 Practice," 16 (report of Lieutenant Wittenmyer); "Post Return for Fort McDowell," May 1890, NAID: 561324, RG94, Microfilm Serial: M617, Microfilm Roll: 669, National Archives.
5. Volkmar, "Report of the April 1890 Practice," 9 (report of Lieutenant Ramsey).

6. Volkmar, *1890 Heliograph Lines and Stations*.
7. Wood, "1888 Reconnaissance," 1.
8. Wood, 2.
9. Wood, 1-2.
10. Overton, "Report of Heliograph Stations in Tonto Basin," 2. The name here is confusing. In the late 1800s (and today as well) there were two mountains in this area with the name Baker: Baker Mountain, located near Aztec Peak, and Baker's Butte, located about 37 miles to the north. Wood's "Lookout Peak" is near Baker Mountain (a name likely unknown to him). Overton's "Baker's Butte" is Baker Mountain as well.
11. Overton, 2.
12. Mazatzal Peak was not used in the common view calculations, as determining its location requires analysis based on Lookout Peak.
13. US Geological Survey, *Aztec Peak, AZ*. The US Geological Survey includes the label "Murphy" on only a few editions of its 1:24,000 series maps. This peak is unnamed on current maps.
14. Overton, "Report of Heliograph Stations in Tonto Basin," 4.
15. Volkmar, *Report of the May 1890 Practice*, 33 (report of Lieutenant Reichmann).
16. Pine Mountain is not visible from the Pinal Mountains.
17. Volkmar, "Report of the April 1890 Practice," 12 (report of Lieutenant Overton).
18. Volkmar, 12 (report of Lieutenant Overton).
19. Volkmar, 12 (report of Lieutenant Overton).
20. Volkmar, 13 (report of Lieutenant Overton).
21. Volkmar, 13 (report of Lieutenant Overton).
22. Volkmar, 17 (report of Lieutenant Wittenmyer).
23. Volkmar, 13 (report of Lieutenant Overton).
24. Volkmar, 13 (report of Lieutenant Overton).
25. Volkmar, *Report of the May 1890 Practice*, 28 (report of Lieutenant Overton).
26. Volkmar, 32 (report of Lieutenant Campbell).
27. General Land Office, *GLO Records (Arizona Township 3 North, Range 7 East)* (Washington, DC: General Land Office, 1902).
28. US Geological Survey, *Fort McDowell*.
29. Volkmar, "Report of the April 1890 Practice," 13 (report of Lieutenant Overton).
30. Volkmar, *Report of the May 1890 Practice*, 32 (report of Lieutenant Campbell).
31. Volkmar, "Report of the April 1890 Practice," 11 (report of Lieutenant Overton).

Chapter 9

1. Slavens, *San Carlos Arizona in the Eighties*, 3.
2. Inter Tribal Council of Arizona, "San Carlos Apache Tribe"; Utley, *Frontier Regulars*, 194.
3. General Land Office, *GLO Records (Arizona Township 3 South, Range 19 East)* (Washington, DC: General Land Office, 1916).
4. Eggleston, "Reconnaissance for Stations in San Carlos Heliograph Division," 1.
5. Slavens, *San Carlos Arizona in the Eighties*, 3.
6. Using digital elevation models, computers can generate sketch views of the horizon in any direction and from any location. By using such a system from peakfinder.org, I placed the viewpoint atop the bluff north of the agency. The horizon sketch created by the system closely matches the horizon in the photograph.
7. Eggleston, "Reconnaissance for Stations in San Carlos Heliograph Division," 1.
8. Utley, *Frontier Regulars*, 358; "Post Return for Camp Thomas," August 1876, NAID: 561324, RG94, Microfilm Serial: M617, Microfilm Roll: 1265, National Archives.
9. E. D. Thomas, *The Military Reserve at Camp Thomas, Arizona*.
10. Eggleston, "Reconnaissance for Stations in San Carlos Heliograph Division," 2.
11. Glassford, "1887 Reconnaissance," 5.
12. Glassford, 6.
13. Glassford, 6.
14. Eggleston, "Reconnaissance for Stations in San Carlos Heliograph Division," 3.
15. Eggleston, 2.
16. Volkmar, *Report of the May 1890 Practice*, 18.
17. Cullum, *Biographical Register*, 3:371; Cullum and Holden, *Biographical Register*, 375.
18. Volkmar, "Report of the April 1890 Practice," 24 (report of Lieutenant Reichmann).
19. Volkmar, 23–24 (report of Lieutenant Reichmann).
20. Volkmar, 29 (report of Lieutenant Reichmann).
21. Volkmar, 21–22 (report of Lieutenant Watson). Watson was sent to evaluate Saddle Peak "near mouth of San Pedro" for a heliograph station. He accurately reported the direction and distance to Saddle Mountain from Saddle

Peak. Watson's commanding officer, Captain Lewis Johnson, recommend that Saddle Peak (at the mouth of the San Pedro) not be used as a station because of "its limited height and range of vision."

22. Cullum, *Biographical Register*, 3:413; Cullum and Holden, *Biographical Register*, 454; Cullum, *Biographical Register*, 5:411.
23. Dade, "Report of a Reconnaissance of Table Mountain," 2–3.
24. Volkmar, "Report of the April 1890 Practice," 20 (report of Lieutenant Smith).
25. Volkmar, 28 (report of Lieutenant Reichmann).
26. Volkmar, *Report of the May 1890 Practice*, 43 (report of Lieutenant Littlebrant).
27. Unfortunately, none of the reported directions between Saddle Mountain and Table Mountain are consistent with one another. From Table Mountain, Lieutenant Smith gave a direction of only 9° (when converted to true) to Saddle Mountain, meaning a back direction of 189°. Lieutenant Reichman's direction to the station at Table Mountain (the table *on the left*, he pointed out) is 213°, which almost exactly points to the top of the table *on the right*. The Volkmar map, in its tabular data, gives a direction of 201° from Saddle Mountain to Table Mountain, which is the closest to the measured direction of 206°.
28. Volkmar, "Report of the April 1890 Practice," 30 (report of Lieutenant Murray).
29. As we will see, there were four stations at Mount Graham (or the Pinaleños). During May, this station would be the one established by Captain Murray.
30. I am aware that the sixth station is located on Mount Graham, for which I have not yet established an exact location, so it was excluded from this part of the analysis. The Pinal Peak location will later be used to help identify the Mount Graham station. If the Mount Graham station were included now, the mean center would shift only about 400 yards to the southwest, with a negligible increase in the distance from Pinal Peak (for this analysis) to just over 1,500 yards.
31. Cullum, *Biographical Register*, 6-A:232–33.
32. Volkmar, "Report of the April 1890 Practice," 30 (report of Lieutenant Murray).
33. Volkmar, 31 (report of Lieutenant Murray).
34. This description is based on an 1880s Arizona State Library photograph of Bremen's sawmill, Arizona Historical Society.
35. Douglas, *Ariz. Globe*.

36. Author's telephone discussion with Gila County Historical Museum staff, April 27, 2023.
37. Volkmar, *Report of the May 1890 Practice*, 35 (report of Lieutenant Read).
38. Volkmar, 36 (report of Lieutenant Read).
39. Volkmar, 36–37 (report of Lieutenant Read). The geometry suggests that this location is Ferndell Spring. After climbing on foot from the sawmill to the top of the ridge, Read turned left and hiked approximately 1.25 miles to Keene's camp. Depending on where he reached the ridge from the sawmill, the measured distance is about 0.9 miles. The next day, Read reported that Keene had set up a heliograph station on a hilltop 800 yards east of the camp, on the adjacent eastern ridgeline, which is about 1,100 yards east of the spring. Based on Read's account of the camp being situated between the ascent from the sawmill and the eastern ridgeline (and farther from the ascent), this aligns with the location of today's Ferndell Spring, though the camp's exact location remains uncertain.
40. Volkmar, *Report of the May 1890 Practice*, 36 (report of Lieutenant Read). Although both Read and Murray place the station southwest of the camp, the uncertainty surrounding the camp's exact location complicates interpretation.
41. Volkmar, 36 (report of Lieutenant Read).
42. Volkmar, 37 (report of Lieutenant Read).
43. Volkmar, 37 (report of Lieutenant Read).
44. Volkmar, 37 (report of Lieutenant Read).
45. US Geological Survey, elevation application, September 20, 2024, https://apps.nationalmap.gov/viewer/.
46. Volkmar, *Report of the May 1890 Practice*, 37–38 (report of Lieutenant Read).

Chapter 10

1. Dillinger, "Fort Whipple."
2. Fieberger, *Military Reservation at Whipple Barracks Arizona*.
3. Collins, *Map of the Military Reservation at Whipple Barracks*.
4. Office of the Quartermaster General, *Whipple Barracks*.
5. Altshuler, *Cavalry Yellow & Infantry Blue*, 230. General Miles moved the Department of Arizona to Los Angeles in January 1887.

6. Volkmar, *Report of the May 1890 Practice*, 22 (report of Lieutenant Tyson).
7. Volkmar, 23 (report of Lieutenant Tyson).
8. Fenton, "Reconnaissance for Intermediate Stations Connecting Whipple and Verde Heliograph Divisions," 2.
9. Glassford, "1887 Reconnaissance," 3.
10. Fenton, "Reconnaissance for Intermediate Stations Connecting Whipple and Verde Heliograph Divisions," 1–2.
11. Glassford, "1887 Reconnaissance," 3; Wood, "1888 Reconnaissance," 3.
12. Volkmar, "Report of the April 1890 Practice," 4 (report of Lieutenant Ramsey).
13. Vegors, *National Register of Historic Places Inventory*, 3.
14. Volkmar, "Report of the April 1890 Practice," 10 (report of Lieutenant Duncan).
15. Volkmar, *Report of the May 1890 Practice*, 18, 25–26. Pages 25–26 are the report of Lieutenant Ramsey.
16. Glassford, "1887 Reconnaissance," 3; Wood, "1888 Reconnaissance," 3.
17. "Post Return for Fort Verde," April 1890, NAID: 561324, RG94, Microfilm Serial: M617, Microfilm Roll: 1326, National Archives; Volkmar, "Report of the April 1890 Practice," 9 (report of Lieutenant Duncan). The post return identifies Lieutenant Richards as the commanding officer.
18. Volkmar, "Report of the April 1890 Practice," 9 (report of Lieutenant Duncan).
19. Volkmar, 9–10 (report of Lieutenant Duncan).
20. Volkmar, *Report of the May 1890 Practice*, 26 (report of Sergeant Robinson). The Volkmar map and accompanying tabular data both show a connection between Little Squaw Peak and Mazatzal Peak to the south. Mazatzal Peak would be occupied as a heliograph station in May, but to reach the station, Robinson, commander of the Squaw Peak heliograph stations in May, had to climb up to Squaw Peak. Despite this, he and his crew spent the majority of their time at Little Squaw Peak.
21. US Geological Survey, *Arnold Mesa*; US Geological Survey, *Middle Verde*.
22. Volkmar, "Report of the April 1890 Practice," 9 (report of Lieutenant Duncan).
23. Volkmar, 8–9 (report of Lieutenant Ramsey). Ramsey gave a list of magnetic azimuths to Squaw Peak, Little Squaw Peak, Bald Mountain, the San Francisco Mountains, Lookout Peak, Granite Mountain, and Mount Reno—all of which are back azimuths. The declination from Baker's Butte in 1890 was 13.6°.
24. Greely, *Annual Report 1890*, 41; Volkmar, *Report of the May 1890 Practice*, 18.

25. Glassford, "1887 Reconnaissance," 3.
26. Fenton, "Reconnaissance for Intermediate Stations Connecting Whipple and Verde Heliograph Divisions," 2; "Post Return for Fort Verde," March 1890, NAID: 561324, RG94, Microfilm Serial: M617, Microfilm Roll: 1326, National Archives. Lieutenant Ramsey was Fort Verde's ASO.
27. Volkmar, *Report of the May 1890 Practice*, 26 (report of Sergeant Robinson). Robinson stated the he connected with North Peak and Mount Reno on these dates (April 30 and May 15, 1890). Mazatzal was sometimes referred to as North Peak.

Chapter 11

1. Volkmar, "Report of the November 1889 Practice," 2.
2. Mills, *My Story*, 192; Hilton, *Fort Grant, Arizona*; Arizona State Land Commission, *Report on Fort Grant Military Reservation*, 9–10. There was a large concrete-lined reservoir on the north end of the parade deck, fed via a 5-inch steel pipe from another reservoir 1:5 miles upstream (the Hilton map records the distance at 8,564 feet) and 400 feet in elevation above in today's Grant Creek. The pressure delivered by 400 feet of water head is about 175 psi.
3. Murry et al., *Oxford English Dictionary*, 14:463. Here, *sanitary* refers to something "intended or tending to promote health."
4. Coronado National Forest, "Hospital Flat Campground"; Mills, *My Story*, 195. Mills refers to a letter his wife, Nannie, sent to her mother on July 18, 1888. In it, she described the ride up the mountain to the camp and mentions the log cabin being built.
5. Neall, "Report Upon Establishment of Station Upon Mt. Graham," 2.
6. Willcox, "Report of Bvt. Maj. Gen. O. B. Willcox," 152.
7. Glassford, "1887 Reconnaissance," 6–7.
8. Volkmar, "Report of the November 1889 Practice," 2; Grierson, *Annual Report of Brigadier General B. H. Grierson*, 2. General Grierson had two aides, Lieutenant Perry and his son, Lieutenant Charles H. Grierson.
9. Volkmar, "Report of the November 1889 Practice," 3. In the telegraph messages between Keene and Volkmar, Keene signs his messages "A.S.O." However,

Fort Grant records list Dade as the ASO. In March 1890 Keene and his company were transferred to San Carlos, where he participated in the April heliograph exercise on Lookout Peak and the May heliograph exercise in the Pinal Mountains. He also took part in the 1893 heliograph exercise.

10. Volkmar, 3, 5.
11. Volkmar, 22 (report of Lieutenant John Perry). Perry sheds more light on Pearson's faulty reasoning for the lack of trained personnel: "the general inability of the colored troops to learn."
12. "Post Return for Fort Grant," November 1889, NAID 561324, RG 94, Microfilm Serial: M617, Microfilm Roll: 415, National Archives. Keene was in the field November 6–16; the exercise started on November 17.
13. Eggleston, "Report of Reconnaissance for Connecting Mt. Graham with the Bowie and San Carlos Heliograph Divisions," 6.
14. Eggleston, 7–9.
15. Winter weather is severe enough that Arizona 366, the road leading into and atop the Pinaleño Range, is closed between mid-November and mid-April each year.
16. Eggleston, "Report of Reconnaissance for Connecting Mt. Graham with the Bowie and San Carlos Heliograph Divisions," 7.
17. Neall, "Report Upon Establishment of Station Upon Mt. Graham," 2.
18. Neall, 2.
19. Eggleston, "Report of Reconnaissance for Connecting Mt. Graham with the Bowie and San Carlos Heliograph Divisions," 9.
20. Eggleston, 8–9; Neall, "Report Upon Establishment of Station Upon Mt. Graham," 2. Neall records 8,650 feet for the same location.
21. Neall, "Report Upon Establishment of Station Upon Mt. Graham," 2. When describing this station, Neall comments that it is close enough to Fort Grant to use flags for communication.
22. Eggleston, "Report of Reconnaissance for Connecting Mt. Graham with the Bowie and San Carlos Heliograph Divisions," 9.
23. Neall, "Report Upon Establishment of Station Upon Mt. Graham," 2.
24. Dade, "Report of a Reconnaissance of Table Mountain, Galiuro Range," 1. Neither Eggleston nor Neall provided a specific date when Keene used this peak to connect with Table Mountain. Since Fort Grant and Keene did not

fully participate in the November exercise, the connection likely occurred afterward, when someone was present on Table Mountain for communication. Lieutenant Dade was positioned at Table Mountain on December 9, so it is possible the connection was made at that time, though this remains uncertain.

25. Eggleston, "Report of Reconnaissance for Connecting Mt. Graham with the Bowie and San Carlos Heliograph Divisions," 8.
26. Eggleston, 8.
27. Eggleston, 8. He is incorrect in his assertion that there are not any places west of the canyon where Fort Bowie can be seen. As we will see, there are several areas where this is possible.
28. Eggleston, 9.
29. Eggleston, 9.
30. Eggleston, 9.
31. Hilton, *Reservation, Fort Grant, Arizona*.
32. US Geological Survey, *Mt. Graham Quadrangle*.
33. Eggleston, "Report of Reconnaissance for Connecting Mt. Graham with the Bowie and San Carlos Heliograph Divisions," 9.
34. Weech, *A History of Mount Graham*, 77, 107, 200 (map). According to Weech, "A road was constructed up Goudy Creek and along Jesus Ridge in order to haul logs to the post."
35. Hilton, *Reservation, Fort Grant, Arizona*.
36. Is it possible that Eggleston took a road farther west than Goudy Canyon? The next canyon to the west is Jesus Canyon. Following the shape of the road from the 1911 map up Jesus Canyon would place the top of the road near present-day Riggs Flat, an ideal location for a wood camp. In either case—whether Goudy or Jesus Canyon—Merrill Peak remains the dominant feature at the top. If Eggleston had traveled farther west than Jesus Canyon, he would have ascended one of the Babcock canyons, with the top of the road leading to Clark Peak. However, this would place him much closer to Taylor Creek—just over 1 mile, rather than the 2 or 3 miles Eggleston described—making this route far less likely. Additionally, some of the peaks Eggleston described as visible from the station he established are not visible from Clark Peak.

37. Eggleston, "Report of Reconnaissance for Connecting Mt. Graham with the Bowie and San Carlos Heliograph Divisions," 9–10.
38. They are Mount Thomas (today's Mount Baldy in the White Mountains), Gila Peak, the Triplets, Mount Turnbull, Turnbull Station, Lookout Peak, Four Mountains (Four Peaks), Table Mountain, Saddle Mountain, the high point in the Pinal Mountains, the distinct high mountain near Tubac (today's Mount Wrightson), the highest point in the Huachuca Mountains (Miller Peak), Cochise Stronghold (as Eggleston probably had not seen Lieutenant Hart's report, this is probably the high point on the Dragoon Range), the highest point of the Catalina Range (Mount Lemmon), Tucson Mountain (there is no line of sight between Tucson Mountain and the Pinaleño Mountains because the Catalina Mountains are in the way; this may be the Sierrita Mountains in the Altar Valley, visible through present-day Reddington Pass), the Whetstone Mountains, Rincon Mountain, and a high point near the San Pedro River. Eggleston put a question mark next to some of the names, including Mount Thomas, the High Mountains near Tubac, and Lookout Peak; these were not used in the analysis. Tucson Mountain and the high point near San Pedro were also excluded, as the descriptions are not specific enough. Also, Eggleston does not mention Fort Grant or Alpina, which were included in the analysis.
39. Eggleston, "Report of Reconnaissance for Connecting Mt. Graham with the Bowie and San Carlos Heliograph Divisions," 11.
40. Volkmar, "Report of the April 1890 Practice," 34 (report of Captain Murray). Murray later learned that Bowie Peak was blocked from view by an intervening mountain, Apache Peak, which is likely today's Government Peak near Apache Pass.
41. Eggleston, "Report of Reconnaissance for Connecting Mt. Graham with the Bowie and San Carlos Heliograph Divisions," 13. This is a January 6, 1890, follow-up report describing the difficulty of seeing Lookout Peak from Eggleston's RM station.
42. Volkmar, "Report of the April 1890 Practice," 32–33 (report of Captain Murray). This is in reference to the report of Lieutenant Dade (pp. 35–37), who was largely unsuccessful communicating from Eggleston's RM station to most of the required stations during the April exercise.
43. Volkmar, 33 (report of Captain Murray).

44. Volkmar, 38.
45. Volkmar, 34 (report of Captain Murray).
46. Volkmar, 34 (report of Captain Murray).
47. Volkmar, 34 (report of Captain Murray).
48. Volkmar, 34 (report of Captain Murray). While a heliograph station called Bowie Station was located in the town of Bowie in 1886, it was not there in 1889 or 1890. Murray is referring to the Bowie Peak station.
49. Volkmar, 34 (report of Captain Murray).
50. US Geological Survey, "Grand View Peak"; US Geological Survey, *Mt. Graham Quadrangle*.
51. Volkmar, "Heliograph System, Department of Arizona"; Volkmar, *Report of the May 1890 Practice*, 8, 44–46 (report of Lieutenant Dade); Volkmar, "Report of the April 1890 Practice," 35 (report of Captain Murray). The eleven stations were Mount Reno, Lookout Peak, Pinal Mountains, Saddle Mountain, Table Mountain, Fort Grant, Rincon Mountain (Rock C), Colorado Peak, Fourr's Ranch North, Fort Huachuca 1890, and Bowie Peak. Dade reports communicating with all the required stations during the May exercise except Saddle Mountain and Fort Grant. However, Murray, in his April 24 report, confirms visibility of Saddle Mountain, and Volkmar, in his 1890 report, confirms communication with Fort Grant.
52. Murray also stated that he could see Mount Turnbull, the Triplets, and Four Peaks from his new station. Including those in the visibility analysis has very little impact on the visible areas.
53. By this point Murray had witnessed two failed attempts to establish a heliographic bridge across the Pinaleño Range during the November and April exercises. Given the significance of this connection, he must have believed that the effort to thoroughly explore potential stations was worth the time, leading him to choose carefully. It's important to acknowledge, however, that the lieutenants who had previously established these stations, Keene and Eggleston, were operating in the dead of winter at elevations hovering around 10,000 feet, making their efforts considerably more challenging.
54. Volkmar, "Report of the April 1890 Practice," 34 (report of Captain Murray). Mount Graham is the highest peak in the Pinaleño Range, and the range as a whole is often called Mount Graham.

55. Volkmar, *Report of the May 1890 Practice*, 46 (report of Lieutenant Dade), 47 (report of Lieutenant Peterson), 48 (report of Lieutenant Perry). Peterson describes the peak a mile northeast of the camp as Alpina. He is mistaken about the name.
56. Hilton, *Reservation, Fort Grant, Arizona*. Several maps show the road leading to the area near present-day Merrill Peak. Hilton's map shows the "wood camp" at the end of the road.
57. Volkmar, *Report of the May 1890 Practice*, 46 (report of Lieutenant Dade).
58. There is a gap in this 127° of about 30° starting to the north of Webb Peak. From Bowie Peak the view is blocked for about 75° by Webb Peak and other nearby hills. Then there is the gap of about 30° and then the view is blocked for another 34° to the north and northeast by the northern portion of Grand View Peak itself. However, if one stands on the northern rock outcrop of the southern peak, the area blocked from view appears to be 75°.
59. Volkmar, *Report of the May 1890 Practice*, 8.
60. Volkmar, *Report of the May 1890 Practice*, 48 (report of Lieutenant Perry); Mary and Charley Miller, email communication with author, April 16, 2023. The Millers, owners of Elkhorn Ranch, a generations-old dude ranch in southern Arizona, estimate horse travel in mountain terrain to be between 2.5 and 3 miles per hour.
61. The lieutenants also reported the size and attributes of the peak as well as a better camping location, closer and to the southeast of the peak. These factors are generally the same for both Webb and the southern Grand View Peak and therefore did not impact the analysis.
62. Today these peaks are covered with trees. Nevertheless, there are places where visibility is clear (atop large boulders). The soldiers of 1890 had the advantage of axing trees with impunity, unhindered by present-day regulations. The views would be much better if some trees were removed.
63. Weech, *A History of Mount Graham*; US Geological Survey, "Heliograph Peak." The Geographic Names Information System decision card, authored by the US Forest Service, states that Heliograph Peak was used by the US Army for heliographic communication starting in the 1860s. This is incorrect.

Chapter 12

1. Clemensen, *Cattle, Copper, and Cactus*, 37.
2. Thiel et al., *Cultural Resources Assessment for the Fort Lowell-Adkins Steel Property*; Thiel, *Archaeological Investigations at Fort Lowell-Atkins Steel Property*.
3. Camp Lowell map, February 23, 1876, NAID: 241570846, RG 77, National Archives; Thiel et al., *Cultural Resources Assessment for the Fort Lowell-Adkins Steel Property*, 14.
4. Volkmar, "Report of the November 1889 Practice," 27 (report of Lieutenant Gale).
5. National Park Service, National Register of Historic Places registration form, 8-3, 8-4.
6. National Park Service, *Historic American Landscapes Survey*, 6.
7. Douglas, *Tucson*.
8. Roskruge, *Official Map of Pima County*
9. Gale, "Report of Reconnaissance for Stations in Lowell Heliograph Division," 2.
10. Volkmar, "Report of the November 1889 Practice," 27 (report of Lieutenant Gale).
11. Volkmar, 27 (report of Lieutenant Gale).
12. Gale, "Report of Reconnaissance for Stations in Lowell Heliograph Division," 2. Gale also provides directions to the visible peaks and stations. While many of these are quite distant and generally aligned, which complicates their convergence points, two locations—Fort Lowell and Wrightson Peak—are relatively close and nearly form a right angle. The convergence of the azimuth lines from these two points (converted to true azimuths) is only 0.3 miles from the summit of Pistol Hill.
13. Volkmar, *Report of the May 1890 Practice*, 56 (report of Corporal Gouldman).
14. Volkmar, 58 (report of Lieutenant Gale).
15. Volkmar, 56, 58 (reports of Corporal Gouldman and Lieutenant Gale).
16. Volkmar, 57 (report of Corporal Gouldman).
17. Volkmar, 8, 46. Lieutenant Dade from Murray's RM station reports connecting with Rincon on May 15, the day Gouldman was at Rock C.
18. The Fort Huachuca parade ground and original buildings are located at the mouth of Huachuca Canyon. The May 1890 heliograph station was about 1.5 miles south of the fort, along a long ridge, at 6,800 feet, on the east side of Huachuca Canyon. A similar position, used earlier, is located to the west

of the canyon (see chapter 4). Spud Rock is visible from the western position but not from the eastern one.

19. Volkmar, *Report of the May 1890 Practice*, 58 (report of Lieutenant Gale).
20. Volkmar, 58 (report of Lieutenant Gale).
21. Volkmar, 52 (report of Lieutenant Hart).
22. Volkmar, 56 (report of Corporal Gouldman).
23. Volkmar, 58 (report of Lieutenant Gale).

Chapter 13

1. Kelly, *Talking Mirrors at Fort Bowie*, Appendix C.
2. Greely, *Annual Report 1890*, 45.
3. US War Department, *Annual Report of the Secretary of War*, 602, 626–27.
4. This information was provided in the various chief signal officer reports from 1886 to 1891.
5. Greely, *Annual Report 1891*, 3, 34. While the shift of the US Weather Service was approved by Congress on October 1, 1890, the move itself occurred on July 1, 1891. In his report ending June 30, 1891, Greely says it was necessary to discuss the shift before waiting another year.
6. Greely, "Report of the Chief Signal Officer, October 9, 1893," 671.
7. Greely, 650.
8. Mills, *The Military Heliograph and Its Use in Arizona and New Mexico*, 8; Library of Congress, "US Soldiers Heliographing on the Top of the Tartar City Wall"; Henry P. McCain, "Cable Number 1563, To General Headquarters, A. E. F. France, 21 June 1918" (adjutant general, US Army, 1918), NAID: 209265595, RG 120, Series: Cablegrams from General John J. Pershing to the Adjutant General, National Archives.
9. MacArthur, *Annual Report of Major General Arthur MacArthur*, Appendix K, 11.
10. Fletcher, "Tar Heels Flirt with Precipices," 1–2.
11. *Graham Guardian*, "Heliograph Test," October 31, 1913, 1.
12. Henry P. McCain, "Cable Number 1563, To General Headquarters, A. E. F. France, 21 June 1918" (adjutant general, US Army, 1918), NAID: 209265595, RG 120, Series: Cablegrams from General John J. Pershing to the Adjutant General, National Archives.

13. Greely, *Annual Report 1898*, 28-29.
14. Greely, "Report of The Chief Signal Officer [1901]," 937; *Brooklyn Daily Eagle*, "Wireless Telegraphy," September 30, 1899, 13; Raines, *Getting the Message Through*, 105–6.
15. Greely, *Annual Report 1905*, 3; Allen, *Report of the Chief Signal Officer*, 5; Raines, *Getting the Message Through*, 126. These portable systems weighed more than 400 pounds.
16. Allen, *Report of the Chief Signal Officer*, 7–8.
17. Raines, *Getting the Message Through*, 126.
18. Raines, 126.
19. Allen, *Report of the Chief Signal Officer*, 32.
20. Greely, "The Evolution of the Signal Corps," 20.
21. Raines, *Getting the Message Through*, 126.
22. Greely, *Annual Report 1899*, 28.

Appendix 1

1. Military careers are complex. In my experience, it is not uncommon for service members to move twice (or more) in the same year—for instance, to a new posting, then to a deployment, and then back. Many of these soldiers had long careers, involving numerous moves, deployments, wars, campaigns, promotions, demotions, and even transfers between branches of the army. Accordingly, I have tried to focus only on the highlights of their careers.
2. Cullum, *Biographical Register*, 6-A:484–85.
3. Cullum, 6-A:485.
4. Cullum, 6-A:485.
5. *Los Angeles Times*, "General Albright Rites Will Be Friday," July 24, 1940, 10.
6. Heitman, *Historical Register*, 221.
7. "Post Return for Fort Bayard," May 1890, NAID: 561324, RG94, Microfilm Serial: M617, Microfilm Roll: 88, National Archives.
8. *Daily Gate City*, "Maj. William Black Dead; Military Funeral Here," November 27, 1911.
9. Patterson, "Andre Walker Brewster."

10. "Post Return for Fort Union," December 1885, NAID: 561324, RG94, Microfilm Serial: M617, Microfilm Roll: 1308, National Archives.
11. Patterson, "Andre Walker Brewster."
12. Heitman, *Historical Register*, 279.
13. *Omaha Daily Bee*, "News for the Army," June 25, 1893, 3.
14. "Post Return for Frankford Arsenal," October 1893, NAID: 561324, RG94, Microfilm Serial: M617, Microfilm Roll: 374, National Archives.
15. "Post Return for Fort Crook," November 1898, NAID: 561324, RG94, Microfilm Serial: M617, Microfilm Roll: 272, National Archives.
16. "Post Return for Arayat Pampanga Luzon, P. I.," July 1900, NAID: 561324, RG94, Microfilm Serial: M617, Microfilm Roll: 1493, National Archives.
17. *Detroit Free Press*, "United Spanish Veterans to Honor Their Fallen Heroes," May 27, 1906.
18. Utley, *Frontier Regulars*, 376–77.
19. US Army Center of Military History, "US Army Chiefs of Staff."
20. Cullum, *Biographical Register*, 3:413.
21. Cullum and Holden, *Biographical Register*, 454.
22. Cullum, *Biographical Register*, 7:262.
23. Cullum, *Biographical Register*, 6-A:474.
24. Cullum, *Biographical Register*, 7:262.
25. Altshuler, *Cavalry Yellow & Infantry Blue*, 108.
26. Altshuler, 108.
27. Cullum, *Biographical Register*, 3:403.
28. Cullum, *Biographical Register*, 5:396–97
29. *Lexington Herald*, "Gen. George B. Duncan," March 17, 1950.
30. Cullum, *Biographical Register*, 3:290.
31. Cullum, *Biographical Register*, 6-A:243; Oregon State Legislature, "Chronological List."
32. "Post Returns for Various Posts," Various, NAID: 561324, RG94, National Archives; *New-York Tribune*, "Col. Charles W. Fenton Dies of Meningitis," January 16, 1918.
33. Heitman, *Historical Register*, 439.
34. *Catholic Advance*, "Maj. Alvarado M. Fuller Promoted to Be Colonel," September 28, 1918, 4.

NOTES TO PAGES 201–203

35. *Evening Star*, "Indian Fighter Dead," January 9, 1924, 11.
36. Cullum, *Biographical Register*, 3:311–12.
37. Altshuler, *Cavalry Yellow & Infantry Blue*, 136.
38. Thrapp, *The Conquest of Apacheria*, 182.
39. Endicott, *Secretary of War's 1886 Report to Congress*, 179 (report of Captain Lawton).
40. Endicott, 179 (report of Captain Lawton).
41. Altshuler, *Cavalry Yellow & Infantry Blue*, 138–39.
42. Cullum, *Biographical Register*, 3:222.
43. Cullum and Holden, *Biographical Register*, 240.
44. "Post Return for Field Camp near Hillsboro, New Mexico (Camp Boyd)," January 1886, NAID: 561324, RG94, Microfilm Serial: M617, Microfilm Roll: 1497, National Archives.
45. Cullum and Holden, *Biographical Register*, 240.
46. Cullum, *Biographical Register*, 6-A:188.
47. Heitman, *Historical Register*, 459.
48. Glassford, "1887 Reconnaissance."
49. Raines, *Getting the Message Through*, 85.
50. *News-Pilot*, "Famous Indian Fighter Dies," August 7, 1931, 5.
51. Greely, *Annual Report 1890*, 41.
52. *Evening Star*, "City and District," April 2, 1898, 12.
53. Greely, "Report of the Chief Signal Officer, October 9, 1893," 652.
54. Heitman, *Historical Register*, 475.
55. *Philadelphia Inquirer*, "Caught on the Fly," October 8, 1898, 8; *Atlanta Journal*, "Army Orders," October 20, 1912, 11.
56. *San Francisco Examiner*, "Lieut. Col. Greene Goes on Staff of General Murray," December 23, 1917, 7.
57. *Los Angeles Times*, "Col. Frank Green, Veteran of Army, Dies," August 10, 1924.
58. Heitman, *Historical Register*, 482.
59. "Post Return for Fort Whipple, Virginia," September 1873, NAID: 561324, RG94, Microfilm Serial: M617, Microfilm Roll: 822, National Archives.
60. Heitman, *Historical Register*, 482.
61. United States Patent Office, "Frank C. Grugan."
62. Greely, *Annual Report 1888*, 4–5.

63. *New York Times*, "Major F. C. Grugan Dead," November 7, 1917, 13.
64. US Army Quartermaster Corps, "26th Quartermaster Commandant."
65. US Army Quartermaster Corps, "26th Quartermaster Commandant."
66. *Los Angeles Times*, "Gen. Hart Dies in Washington," January 3, 1926, 3.
67. *Journal of the Armed Forces* 18 (December 18, 1880):401.
68. Heitman, *Historical Register*, 545.
69. "Post Returns for Various Posts," Various, NAID: 561324, RG94, National Archives.
70. *Bennington Evening Banner*, "Death of Major Hovey," November 16, 1908, 1.
71. Adler, *History of the Seventy Seventh Division*, 166–67; *Fort Collins Coloradoan*, "Gen E. M. Johnson, US Army Officer Is Dead in Paris," October 15, 1923, 7.
72. Cullum, *Biographical Register*, 3:402.
73. Cullum and Holden, *Biographical Register* 431; Cullum, *Biographical Register*, 5:394.
74. Cullum, *Biographical Register*, 5:394; *Boston Globe*, "Capt Keene Enters US Guard as Major," July 7, 1918, 2.
75. *Boston Globe*, "Double Funeral Services for Maj. and Mrs. Keene," March 25, 1940, 15.
76. Cullum and Holden, *Biographical Register*, 488.
77. Cullum, *Biographical Register*, 5:441.
78. Cullum, *Biographical Register*, 6-A:534; Cullum, *Biographical Register*, 7:293.
79. *San Francisco Examiner*, "General Leitch Dies Here," October 27, 1938, 13.
80. Cullum, *Biographical Register*, 7:293–94.
81. *San Francisco Examiner*, "General Leitch Dies Here," October 27, 1938, 13.
82. Cullum, *Biographical Register*, 3:423.
83. Cullum and Holden, *Biographical Register*, 472.
84. Cullum, Biographical Register, 5:427; 6-A:506
85. Cullum, *Biographical Register*, 6-A:506–7.
86. Altshuler, *Cavalry Yellow & Infantry Blue*, 225.
87. Utley, *Frontier Regulars*, 384–85.
88. Congressional Medal of Honor Society, "Marion Perry Maus, Indian Campaigns."
89. Altshuler, *Cavalry Yellow & Infantry Blue*, 225.
90. Altshuler, 229.

91. Altshuler, 229.
92. Altshuler, 230.
93. Utley, *Frontier Regulars*, 389.
94. Altshuler, *Cavalry Yellow & Infantry Blue*, 230.
95. Altshuler, 230.
96. Utley, *Encyclopedia of the American West*, 285–86.
97. Altshuler, *Cavalry Yellow & Infantry Blue*, 230.
98. Cullum, *Biographical Register*, 3:278; Cullum, *Biographical Register*, 8:53.
99. Cullum, *Biographical Register*, 6-A:233.
100. Cullum, 6-A:232–33.
101. Cullum, *Biographical Register*, 8:53.
102. Cullum, *Biographical Register*, 3:371.
103. Cullum and Holden, *Biographical Register*, 375.
104. Cullum and Holden, 375.
105. Cullum and Holden, 375.
106. US War Department, *General Orders and Circulars*, General Order 126:1–8.
107. *Times Leader*, "Indian Fighter Death Victim: William W. Neifert, 86, Helped Take Geronimo," March 27, 1951, 13.
108. Neifert, "Trailing Geronimo by Heliograph," 558.
109. *Times Leader*, "Indian Fighter Death Victim: William W. Neifert, 86, Helped Take Geronimo," March 27, 1951, 13.
110. Cullum, *Biographical Register*, 3:420.
111. Cullum and Holden, *Biographical Register*, 465.
112. "Post Return for Fort Washakie, Wyoming," October 1898, NAID: 561324, RG94, Microfilm Serial: M617, Microfilm Roll: 1365, National Archives.
113. "Post Returns for Various Posts," Various, NAID: 561324, RG94, National Archives; Cullum, *Biographical Register*, 5:421.
114. *Boston Globe*, "In Bolo Rush US Officer and Private Were Killed, Capt Clough Overton and Harry Noyes Cut Down," May 16, 1903; Cullum, *Biographical Register*, 5:421. The *Boston Globe* and Cullum give different dates for Overton's death. Cullum's date is used here.
115. Hitt, "Amphibious Infantry."
116. *Evening Star*, "Assigned to Duty," October 10, 1883; Heitman, *Historical Register*, 764.

117. Grierson, *Annual Report of Brigadier General B. H. Grierson*, 10.
118. *Lincoln Star*, "The Week in Chicago," April 30, 1904, 7; Heitman, *Historical Register*, 764.
119. Cullum, *Biographical Register*, 3:397.
120. Cullum and Holden, *Biographical Register*, 422.
121. Cullum, *Biographical Register*, 6-A:423.
122. Cullum, *Biographical Register*, 6-A:423–24.
123. Cullum, *Biographical Register*, 9:49.
124. Jacoby, "General John J. Pershing Facts and Biography."
125. Cullum, *Biographical Register*, 3:388.
126. Cullum and Holden, *Biographical Register*, 405.
127. Cullum, *Biographical Register*, 5:372.
128. *Los Angeles Times*, "Colonel Read Dies," December 16, 1919, 17.
129. Cullum, *Biographical Register*, 3:288.
130. Cullum, *Biographical Register*, 5:266.
131. Cullum, *Biographical Register*, 6-A:241.
132. Cullum, 6-A:241.
133. *Los Angeles Times*, "Colonel Read Dies," December 16, 1919, 17.
134. Phillips, "Your Country Is My Country," 262–63,
135. Phillips, 262–63.
136. Heitman, *Historical Register*, 822.
137. "Post Returns for Various Posts," Various, NAID: 561324, RG94, National Archives.
138. Phillips, "Your Country Is My Country," 259–60.
139. *Star Tribune*, "Colonel Will Retire After Long Service," April 3, 1925.
140. *Minneapolis Star*, "Col. Carl Reichmann," October 28, 1936.
141. Brown, *History of the Ninth US Infantry*, 156.
142. Cullum, *Biographical Register* 3, 409; Cullum and Holden, *Biographical Register*, 446.
143. Cullum, *Biographical Register*, 5:405–6.
144. Cullum, *Biographical Register*, 6-A:463.
145. Cullum, 6-A:463.
146. Cullum, *Biographical Register*, 7:257.
147. US Department of Veterans Affairs, "Nationwide Gravesite Locator."

148. *South Haven New Era*, "Soldiers of High Merit," November 17, 1894.
149. *San Francisco Call and Post*, "Army Orders," March 2, 1913, 40.
150. Cullum, *Biographical Register*, 3:371.
151. Cullum and Holden, *Biographical Register*, 375.
152. Cullum and Holden, 375.
153. Cullum, *Biographical Register*, 3:422; Cullum and Holden, *Biographical Register*, 471.
154. Cullum and Holden, *Biographical Register*, 471; Cullum, *Biographical Register*, 5:425.
155. Cullum, *Biographical Register*, 5:425.
156. Cullum, *Biographical Register*, 6-A:504–5.
157. Cullum, 6-A:505.
158. *Spokane Chronicle*, "First Professor at W.S.C. Taken," March 28, 1933.
159. Altshuler, *Cavalry Yellow & Infantry Blue*, 338.
160. Altshuler, 344.
161. Cullum, *Biographical Register*, 3:118.
162. Cullum and Holden, *Biographical Register*, 182.
163. Cullum, *Biographical Register*, 5:153.
164. Altshuler, *Cavalry Yellow & Infantry Blue*, 344.
165. McMiller, "Buffalo Soldiers," 65.
166. Spanish American War Centennial Website. "10th US Cavalry Roster, 1898."
167. Cullum, *Biographical Register*, 3:415.
168. Cullum and Holden, *Biographical Register*, 457.
169. Cullum and Holden, 457; Cullum, *Biographical Register*, 5:413.
170. Cullum, *Biographical Register*, 5:413.
171. Cullum, *Biographical Register*, 6-A:481.
172. Cullum, 6-A:481.
173. Cullum, *Biographical Register*, 7:265; *Columbus Ledger*, "Retired General Dies in Capital," July 6, 1937.
174. Heitman, *Historical Register*, 1055.
175. Army Historical Foundation, "Major General Leonard Wood."
176. Altshuler, *Cavalry Yellow & Infantry Blue*, 377.
177. Army Historical Foundation, "Major General Leonard Wood."

Appendix 2

1. Burt Jenness, "Fort Selden Indian Outpost," *New Mexico Magazine*, January 1963, 24.
2. Forsyth, "Radio Replaces Cavalry Signaling Device on Rio Grande Project," 10.
3. National Park Service, Fort Selden National Register of Historic Places nomination form, 3.
4. Peterson and Bilbo, *Investigation of a Possible US Army Heliograph Station*, 3.
5. Holmes, *Fort Selden*, 77.
6. White House, "Presidential Proclamation."
7. Hazen, *Annual Report of the Chief Signal Officer 1882*, 94.
8. Miles, *Recollections*, 484.
9. Dravo, "September 20, 1886, Report."
10. Hovey and Paddock, "Connecting Bayard and Stanton."
11. Volkmar, "Report of the April 1890 Practice"; Volkmar, *Report of the May 1890 Practice*; Greely, "Report of the Chief Signal Officer, October 9, 1893," 652.
12. Spencer, *Spencer Map of 1886*; Volkmar, *1890 Heliograph Lines and Stations*.
13. Spencer, *Spencer Map of 1886*; Kelly, *Talking Mirrors at Fort Bowie*. Included are some excerpts of Colonel E. E. Dravo's 1926 correspondence explaining the use heliographs by all the outposts labeled on the Spencer map.
14. Hovey and Paddock, "Connecting Bayard and Stanton," 8–9.
15. Holmes, *Fort Selden*, 77.
16. "Post Returns for Fort Selden, New Mexico," Various, NAID: 561324, RG94, National Archives.
17. Volkmar, *Report of the May 1890 Practice*, 60 (report of Lieutenant Hovey).
18. Dravo, "Descriptive Notes," 2.
19. Cottam et al., "The 1882 Transit of Venus and the Popularization of Astronomy in the USA," 188, 190.
20. Cottam et al., "The 1874 Transit of Venus and the Popularisation of Astronomy in the USA," 229–31.
21. Dick, "The American Transit of Venus Expeditions of 1874 and 1882," 102.

References

Adler, Julius Ochs, ed. *History of the Seventy Seventh Division, August 25th, 1917, November 11th, 1918*. New York: W. H. Crawford, 1919.

Allen, James. *Report of the Chief Signal Officer US Army to the Secretary of War 1908*. Washington, DC: GPO, 1908.

Altshuler, Constance W. *Cavalry Yellow & Infantry Blue: Army Officers in Arizona between 1851 and 1886*. Tucson: Arizona Historical Society, 1991.

American Society of Civil Engineers, American Congress on Surveying and Mapping, American Society for Photogrammetry and Remote Sensing. *Glossary of the Mapping Sciences*. New York: American Society of Civil Engineers, American Congress on Surveying and Mapping, American Society for Photogrammetry and Remote Sensing, 1994.

Army Historical Foundation. "Major General Leonard Wood." https://armyhistory.org/major-general-leonard-wood/. Accessed May 5, 2023.

Arizona State Land Commission. *Report on Fort Grant Military Reservation*. Phoenix: Arizona State Land Commission, 1912.

Bigelow, John. *On the Bloody Trail of Geronimo*. Los Angeles: Westernlore Press, 1968.

Bigelow, John. "Tenth Regiment of Cavalry." In *The Army of the US Historical Sketches of Staff and Line with Portraits of Generals-in-Chief*, edited by Theophilus F. Rodenbough and William L. Haskin. New York: Maynard, Merrill, 1896.

Bingham, Theodore, A. *South Eastr'n Arizona*, August 1, 1884. General reference map. 6 miles to 1 inch. Arizona Historical Society.

Brasher, Nugent. "The Francisco Vázquez de Coronado Expedition in Tierra Doblada: The 2013 Report on Artifacts and Isotopes of the Minnie Bell Site at Big Dry Creek, Catron County, New Mexico." *New Mexico Historical Review* 88, no. 2 (2013).

Breed, Charles B., and George L Hosmer. *Higher Surveying*. Vol. 2, *Principles and Practice of Surveying*. 8th ed. New York: John Wiley & Sons, 1962.

Brown, Fred Radford. *History of the Ninth US Infantry, 1799–1909*. Chicago: Donnelley & Sons, 1909.

REFERENCES

Bureau of Land Management. "Lake Valley Historic Townsite." https://www.blm.gov/visit/lake-valley-historic-townsite. Accessed September 20, 2023.

Bureau of Reclamation. "News Archive: Fort McRae Placed on National Register." https://www.usbr.gov/newsroomold/newsrelease/detail.cfm?RecordID=5221. Accessed March 4, 2023.

Clemensen, A. Berle. *Cattle, Copper, and Cactus: The History of Saguaro National Monument, Arizona.* Denver: National Park Service, 1987.

Collins, Charles L. *Map of the Military Reservation at Whipple Barracks.* Plat map. Whipple Barracks, AZ: 24th Infantry, 1890.

Congressional Medal of Honor Society. "Marion Perry Maus, Indian Campaigns." https://www.cmohs.org/recipients/marion-p-maus. Accessed May 6, 2023.

Coronado National Forest. "Hospital Flat Campground." https://www.fs.usda.gov/recarea/coronado/recarea/?recid=25580. Accessed May 22, 2023.

Cottam, Stella, Wayne Orchiston, and Richard Stephenson. "The 1874 Transit of Venus and the Popularisation of Astronomy in the USA as Reflected in the New York Times." In *Highlighting the History of Astronomy in the Asia-Pacific Region,* edited by Wayne Orchiston, Richard Strom, and Tsuko Nakamura, pp. 225–41. New York: Springer, 2011.

Cottam, Stella, Wayne Orchiston, and Richard F. Stephenson. "The 1882 Transit of Venus and the Popularization of Astronomy in the USA as Reflected in the New York Times." *Journal of Astronomical History and Heritage* 15, no. 3 (2012): 183–99.

Cullum, George W. *Biographical Register of the Officers and Graduates of the US Military Academy at West Point, New York since Its Establishment in 1802.* Vol. 3. 3rd ed. Boston: Houghton Mifflin, 1891.

Cullum, George W. *Biographical Register of the Officers and Graduates of the US Military Academy at West Point, New York since Its Establishment in 1802.* Vol. 5. Saginaw, MI: Seemann & Peters, 1910.

Cullum, George W. *Biographical Register of the Officers and Graduates of the US Military Academy at West Point, New York since Its Establishment in 1802,* Vol. 6-A. Edited by Writ Robinson. Saginaw, MI: Seemann & Peters, 1920.

Cullum, George W. *Biographical Register of the Officers and Graduates of the US Military Academy at West Point, New York since Its Establishment in 1802.* Vol. 7. Chicago: R. R. Donnelley & Sons, 1930.

REFERENCES

Cullum, George W. *Biographical Register of the Officers and Graduates of the US Military Academy at West Point, New York since Its Establishment in 1802*. Vol. 8. Chicago: R. R. Donnelley & Sons, 1940.

Cullum, George W. *Biographical Register of the Officers and Graduates of the US Military Academy at West Point, New York since Its Establishment in 1802*. Vol. 9. Edited by Colonel Charles Branham. West Point, NY: Association of Graduates, US Military Academy, 1950.

Cullum, George W., and Edward S. Holden. *Biographical Register of the Officers and Graduates of the US Military Academy at West Point, New York since Its Establishment in 1802*. Vol. 4. Cambridge, MA: Riverside Press, 1901.

Dade, Alexander L. "Report of a Reconnaissance of Table Mountain, Galiuro Range." In *Instructions for Signal Officers of Heliograph Divisions & Stations*, by William J. Volkmar. Los Angeles: Department of Arizona, 1890.

Dempsey, General Martin E. *Doctrine for the Armed Forces of the United States*. Arlington, VA: US Department of Defense, 2017.

Dick, Steven J. "The American Transit of Venus Expeditions of 1874 and 1882." *Proceedings of the International Astronomical Union* 2004, no. 196: 100–10. https://doi.org/10.1017/S1743921305001304.

Diddy, Miriam. *Fort Stanton Historic Site Cultural Landscape Report*. Santa Fe: New Mexico Historic Sites and New Mexico Historic Preservation Division, 2023.

Dillinger, Mary. "Fort Whipple—Historic VA Medical Center Started as Army Post." US Department of Veterans Affairs. 2021. https://department.va.gov/history/featured-stories/fort-whipple/.

Douglas, E. M. *Ariz. Globe (Surveyed in 1900–1901)*. Topographic map. 1:62,500. Washington, DC: US Geological Survey, 1902.

Douglas, E. M. *Arizona (Santa Cruz County) Nogales Quadrangle*. Topographic map. 1:125,000. Washington, DC: US Geological Survey, 1905.

Douglas, E. M. *Bisbee, Arizona*. Topographic map. 1:62,500. Washington, DC: US Geological Survey, 1916.

Douglas, E. M. *Tucson*. Topographic map. 1:125,000. Washington, DC: US Geological Survey, 1904.

Dravo, Edward E. "Heliograph Stations in New Mexico. Descriptive Notes by 1st Lieut. E. E. Dravo, 6th Cavalry, A. S. O. January 26, 1887." In *Instructions for Signal Officers of Heliograph Divisions & Stations*, by William J. Volkmar. Los Angeles: Department of Arizona, 1890.

Dravo, Edward E. "Monthly Report of Heliograph Lines. District New Mexico July 31, to August 31, 1886." In *Instructions for Signal Officers of Heliograph Divisions & Stations*, by William J. Volkmar. Los Angeles: Department of Arizona, 1890.

Dravo, Edward E. "Report of Heliograph Lines. District of New Mexico, to July 31, 1886." In *Instructions for Signal Officers of Heliograph Divisions & Stations*, by William J. Volkmar. Los Angeles: Department of Arizona, 1890.

Dravo, Edward E. "Report of the Heliograph System, District of New Mexico. E. E. Dravo, 1st Lieut. 6th Cavalry, A. S. O., Officer in Charge. Fort Bayard, New Mexico, September 20, 1886." In *Instructions for Signal Officers of Heliograph Divisions & Stations*, by William J. Volkmar. Los Angeles: Department of Arizona, 1890.

Eckhoff, Emil A., and Paul Riecker. *Official Map of the Territory of Arizona.* Topographic map. 1:823,680. New York: Graphic Co. Photo-Lith., 1880. https://davidrumsey.oldmapsonline.org/maps/140bbccb-21a0-54cc-8f26-01537cda9661/view

Eggleston, Millard F. "Reconnaissance for Stations in San Carlos Heliograph Division, by 1st Lieut. M. F. Eggleston, 10th Cav., A. S. O. San Carlos, A. T., November 30, 1889." In *Instructions for Signal Officers of Heliograph Divisions & Stations*, by William J. Volkmar. Los Angeles: Department of Arizona, 1890.

Eggleston, Millard F. "Report of Reconnaissance for Connecting Mt. Graham with the Bowie and San Carlos Heliograph Divisions." In *Instructions for Signal Officers of Heliograph Divisions & Stations*, by William J. Volkmar. Los Angeles: Department of Arizona, 1890.

Endicott, William C. *Report of the Secretary of War; Being Part of the Message and Documents Communicated to the Two Houses of Congress at the Beginning of the Second Session of the Forty-Ninth Congress,* Vol. 1. Washington, DC: GPO, 1886.

Fenton, Charles W. "Reconnaissance for Intermediate Stations Connecting Whipple and Verde Heliograph Divisions, by 2d Lieut. C. W. Fenton, 9th Infantry, A. S. O. Office of Acting Signal Officer, Post of Whipple Barracks, A. T.,

March 15th, 1890." In *Instructions for Signal Officers of Heliograph Divisions & Stations*, by William J. Volkmar. Los Angeles: US Army, 1890.

Fieberger, Gustav J. *Military Reservation at Whipple Barracks Arizona*. Plat map. Large scale. Arizona Territory: Department of Arizona, February 1883. https://exhibits.stanford.edu/ruderman/catalog/zc850vg7563.

Finley, James P. "The Situation in Mexico." *Huachuca Illustrated* 1 (1993): 58–65.

Fletcher, A. L. "Tar Heels Flirt with Precipices." *News and Observer*. November 21, 1916.

Forsyth, Gordon J., ed. "Radio Replaces Cavalry Signaling Device on Rio Grande Project." *Reclamation Era* 55, no. 2 (February 1969): 10–11.

Fort Bayard Historic Preservation Society. "Fort Bayard History." 2023. https://www.historicfortbayard.org/fort-bayard-history/.

Fuller, Alvarado M. "Consolidated Report of Heliograph Line. Division of Arizona from May 1st to Sept 30th, 1886, during the Continuance of the Line. 2d Lieutenant Alvarado M. Fuller, 2d Cavalry, Acting Signal Officer in Charge of Heliograph Service." In *Instructions for Signal Officers of Heliograph Divisions & Stations*, by William J. Volkmar. Los Angeles: Department of Arizona, 1890.

Fuller, Alvarado M. "Heliograph Report of A. M. Fuller, 2d Lieut. 2d Cavalry, A. S. O., in Charge of Line. Fort Huachuca, A. T., September 30, 1886." In *Instructions for Signal Officers of Heliograph Divisions & Stations*, by William J. Volkmar. Los Angeles: Department of Arizona, 1890.

Gale, George H. G. "Report of Reconnaissance for a Central Station, Connecting Lowell, Huachuca, Bowie and Grant Heliograph Divisions." In *Instructions for Signal Officers of Heliograph Divisions & Stations*, by William J. Volkmar. Los Angeles: Department of Arizona, 1890.

Gale, George H. G. "Report of Reconnaissance for Stations in Lowell Heliograph Division." In *Instructions for Signal Officers of Heliograph Divisions & Stations*, by William J. Volkmar. Los Angeles: Department of Arizona, 1890.

Glass, Edward L. N. *The History of the 10th Cavalry*. Tucson: Acme Printing Company, 1921.

Glassford, William A. "Heliograph and Signal Work. Extracts from Report of Lieut. Glassford, Signal Corps. July 17, 1887." In *Instructions for Signal Officers of Heliograph Divisions & Stations*, by William J. Volkmar. Los Angeles: Department of Arizona, 1890.

Greely, Adolphus W. *Annual Report of the Chief Signal Officer of the Army to the Secretary of War for the Year 1886.* Washington, DC: GPO, 1886.

Greely, Adolphus W. *Annual Report of the Chief Signal Officer of the Army to the Secretary of War for the Year 1888.* Washington, DC: GPO, 1889.

Greely, Adolphus W. *Annual Report of the Chief Signal Officer of the Army to the Secretary of War for the Year 1889.* Washington, DC: GPO, 1890.

Greely, Adolphus W. *Annual Report of the Chief Signal Officer of the Army to the Secretary of War for the Year 1890.* Washington, DC: GPO, 1890.

Greely, Adolphus W. *Annual Report of the Chief Signal Officer of the Army to the Secretary of War for the Year 1891.* Washington, DC: GPO, 1892.

Greely, Adolphus W. *Annual Report of the Chief Signal Officer of the Army to the Secretary of War for the Year 1894.* Washington, DC: GPO, 1894.

Greely, Adolphus W. *Annual Report of the Chief Signal Officer to the Secretary of War for the Fiscal Year Ending June 30, 1898.* Washington, DC: GPO, 1898.

Greely, Adolphus W. *Annual Report of the Chief to the Secretary of War for the Fiscal Year Ending June 30, 1899.* Washington, DC: GPO, 1899.

Greely, Adolphus W. *Annual Report of the Chief Signal Officer, US Army for the Fiscal Year Ending June 30, 1905.* Washington, DC: GPO, 1905.

Greely, Adolphus W. "The Evolution of the Signal Corps." *Ainslee's Magazine* 4, no. 1 (August 1899): 12–21.

Greely, Adolphus W. "Report of The Chief Signal Officer." In *Annual Reports of the War Department for the Fiscal Year Ended June 30, 1901*, by Elihu Root. Washington, DC: GPO, 1901.

Greely, Adolphus W. "Report of the Chief Signal Officer, October 9, 1893." In *Annual Report of the Secretary of War for the Year 1893, Vol. 1.*, by Daniel D Lamont. Washington, DC: GPO, 1893.

Greely, Adolphus W. "Report of the Chief Signal Officer, September 7, 1887." In *Annual Report of the Secretary of War for the Year 1887, Vol. 4.*, by William C. Endicott. Washington, DC: GPO, 1887.

Greene, Jerome A. *Historic Structure Report, Historical Data Section. Fort Bowie Its Physical Evolution, 1862–1894. Fort Bowie National Historic Site, Arizona.* Denver: National Park Service, 1980. https://npshistory.com/publications/fobo/greene.pdf.

Grierson, Benjamin H. *Annual Report of Brigadier General B. H. Grierson, Brevet Major General, US Army, Commanding Department of Arizona 1890*. Los Angeles: US Army, 1890.

Hazen, William B. *Annual Report of the Chief Signal-Officer to the Secretary of War for the Fiscal Year Ending June 30, 1881*. Washington, DC: GPO, 1881.

Hazen, William B. *Annual Report of the Chief Signal Officer, United States Army, to the Secretary of War for the Fiscal Year Ending June 30, 1882*. Washington, DC: GPO, 1883.

Heitman, Francis B. *Historical Register and Dictionary of the United States Army, from Its Organization, September 29, 1789*. Vol. 1. Washington, DC: GPO, 1903.

Hillsboro, New Mexico. "Welcome to Hillsboro, New Mexico." https://www.hillsboronm.com/. Accessed March 16, 2023.

Hilton, J. M. *Fort Grant, Arizona*. Plat map. War Department, Office of the Chief of Engineers, April 22, 1910. NAID: 241570384. National Archives.

Hilton, J. M. *Reservation, Fort Grant, Arizona*. War Department, Office of the Chief of Engineers, October 11, 1911. National Archives. https://catalog.archives.gov/id/241570386.

Hitt, Parker. "Amphibious Infantry—A Fleet on Lake Lanao." *US Naval Institute Proceedings* 64, no. 2/420 (February 1938). https://www.usni.org/magazines/proceedings/1938/february/amphibious-infantry-fleet-lake-lanao.

Hodges, J. N. *St. David Quadrangle, Revision 1922*. Tactical map. 1:62,500. Fort Sam Houston, TX: Engineer Eighth Corps, 1916.

Holmes, Allan, J. *Fort Selden, 1865–1891: The Birth, Life, and Death of a Frontier Fort in New Mexico*. La Vergne, TN: Sunstone Press, 2009.

Horn, Tom. *Life of Tom Horn: Government Scout and Interpreter*. Denver: Louthan Book Company, 1904.

Hovey, Henry W., and Richard B. Paddock. "Reports of Concerted Reconnaissances for Short Line Connecting Bayard and Stanton Heliograph Divisions, N.M." In *Instructions for Signal Officers of Heliograph Divisions & Stations*, by William J. Volkmar. Los Angeles: Department of Arizona, 1890.

Inter Tribal Council of Arizona. "San Carlos Apache Tribe." https://itcaonline.com/member-tribes/san-carlos-apache-tribe/. Accessed February 17, 2024.

Jacoby, Erika. "General John J. Pershing Facts and Biography." Historic Tours of America. May 22, 2017. https://www.arlingtontours.com/john-j-pershing.

REFERENCES

Jenness, Burt. "Fort Selden Indian Outpost." *New Mexico Magazine*, January 1963.

Julyan, Robert. *The Place Names of New Mexico*. Albuquerque: University of New Mexico Press, 1996.

Kelly, Roger E. *Talking Mirrors at Fort Bowie: Military Heliograph Communication in the Southwest*. Report for the National Park Service, Chiricahua National Monument. Flagstaff: Department of Anthropology, Northern Arizona University, 1967.

King, David. *National Register of Historic Places Inventory: Fort Stanton*. Nomination form. Washington, DC: National Park Service, 1973.

Klip, W. *Southwestern New Mexico*. Topographic map. Office of the Chief of Engineers, 1883. https://www.davidrumsey.com/luna/servlet/s/17j2u5. David Rumsey.

Knowles, Anne Kelly, ed. *Past Time, Past Place: GIS for History*. Redlands, CA: ESRI Press, 2002.

Library of Congress. "US Soldiers Heliographing on the Top of the Tartar City Wall, Peking. China," https://www.loc.gov/resource/stereo.1s47941. Accessed January 22, 2025.

MacArthur, General Arthur. *Annual Report of Major General Arthur MacArthur, US Volunteers, Commanding, Division of the Philippines*. Manila: US Army, 1900.

McMiller, Anita W. "Buffalo Soldiers: The Formation of the Tenth Cavalry Regiment." Master's thesis, US Army Command and General Staff College, 1990.

Miles, Nelson A. *Personal Recollections and Observations of General Nelson A. Miles, Embracing a Brief View of the Civil War or from New England to the Golden Gate, and the Story of His Indian Campaigns with Comments on the Exploration, Development, and Progress of Our Great Western Empire*. Chicago: Werner, 1896.

Miller, Lester L. *The Heliograph, a Bibliography*. Fort Sill, OK: US Army Field Artillery School, 1984. https://apps.dtic.mil/sti/tr/pdf/ADA136731.pdf.

Mills, Anson. *My Story*. Edited by C. H. Claudy. Washington, DC: Published by the author, 1918.

Mills, John. *The Military Heliograph and Its Use in Arizona and New Mexico*. Washington, DC: US Army Signal Corps Intelligence Agency, 1954.

Mitchell, Andy. *The ESRI Guide to GIS Analysis*. Vol. 2, *Spatial Measurements and Statistics*. 2nd ed. Redlands, CA: ESRI Press, 2009.

REFERENCES

Mitchell, Andy. *The Esri Guide to GIS Analysis.* Vol. 3, *Modeling Suitability, Movement, and Interaction.* Redlands, CA: ESRI Press, 2012.

Murry, James, A. H., Henry Bradley, W. A. Craigie, and C. T. Onions, eds. *The Oxford English Dictionary.* 2nd ed. 24 vols. Oxford: Clarendon Press, 1991.

Myer, Albert J. *A Manual of Signals: For the Use of Signal Officers in the Field, and for Military and Naval Students, Military Schools, Etc.* New York: D. Van Nostrand, 1872.

Myer, Albert J. *A Manual of Signals: For the Use of Signal Officers in the Field, and for Military and Naval Students, Military Schools, Etc.* Washington, DC: GPO, 1879.

Myer, Albert J. *Annual Report of the Chief Signal Officer to the Secretary of War for the Year 1877.* Washington, DC: GPO, 1877.

National Oceanic and Atmospheric Administration. "Station Datasheet—Baldy." https://www.ngs.noaa.gov/cgi-bin/ds_mark.prl?PidBox=CG1050. Accessed February 16, 2023.

National Park Service. "Arizona: Fort Bowie National Historic Site." https://www.nps.gov/articles/ftbowie.htm. Accessed January 26, 2024.

National Park Service. Fort Selden National Register of Historic Places nomination form. July 8, 1970. https://npgallery.nps.gov/GetAsset/e8dd7662-15c6-478a-90eb-8017d4c05d63.

National Park Service. L.C. Ranch Headquarters National Register of Historic Places nomination form. October 19, 1978. https://npgallery.nps.gov/NRHP/GetAsset/NRHP/78001816_text.

National Park Service. *Historic American Landscapes Survey, Colossal Cave Mountain Park (Colossal Cave County Park).* http://www.briannalehman.com/uploads/8/8/0/3/8803722/hals_colossal_cave_mountain_park_final_073114.pdf. Accessed January 3, 2025.

National Park Service. National Register of Historic Places registration form. July 16, 1992. https://npgallery.nps.gov/GetAsset/2a78ffdd-e3ad-4586-91a6-04ea0f390433.

National Park Service. *Prospector, Cowhand, and Sodbuster: Historic Places Associated with the Mining, Ranching, and Farming Frontiers in the Trans-Mississippi West.* Washington, DC: GPO, 1967.

REFERENCES

Neall, John M. "Report Upon Establishment of Station Upon Mt. Graham, for Operating with Bowie Peak, by 2d Lieut. J. M. Neall, 4th Cavalry, A. S. O., Fort Bowie, A. T., December 21, 1889." In *Instructions for Signal Officers of Heliograph Divisions & Stations*, by William J. Volkmar. Los Angeles: US Army, 1890.

Neifert, William W. "Trailing Geronimo by Heliograph." In *Eyewitnesses to the Indian Wars 1865–1890*. Vol. 1, *The Struggle for Apacheria*, edited by Peter Cozzens. Mechanicsburg, PA: Stackpole Books, 2001.

Office of the Quartermaster General. *Whipple Barracks, Arizona*. Plat map. Washington, DC: Office of the Quartermaster Generals, 1902.

Oregon State Legislature. "Chronological List of Oregon Legislatures from 1841 to Present." 2008. https://www.oregonlegislature.gov/legislators-chronological.

Overton, Clough. "Report of Heliograph Stations in Tonto Basin, Visited or Observed by 2d Lieut. Clough Overton, 4th Cav., A. S. O." In *Instructions for Signal Officers. Heliograph Divisions & Stations. Department of Arizona*. Los Angeles: US Army, 1890.

Patterson, Michael Robert. "Andre Walker Brewster—Major General, United States Army." Arlington National Cemetery, March 3, 2024. https://www.arlingtoncemetery.net/awbrewst.htm.

Peterson, John E., and Michael J. Bilbo. *Investigation of a Possible US Army Heliograph Station ca. 1880s (Redacted)*. El Paso: El Paso Centennial Museum, 1975.

Phillips, P. Michael. "Your Country Is My Country: Civil–Military Relations as Social Reproduction, 1880–1920." PhD thesis, Kings College London, 2021. https://kclpure.kcl.ac.uk/portal/files/156514270/2021_Phillips_Paul_1350005_ethesis.pd.

Price, Maribeth. *Mastering ArcGIS*. 6th ed. New York: McGraw-Hill, 2014.

Price, Paxton P. "The Railroad, Rincon, and the River." *New Mexico Historical Review* 65, no. 4 (October 1, 1990): 437–54.

Raines, Rebecca Robbins. *Getting the Message Through: A Branch History of the US Army Signal Corps*. Washington, DC: Center of Military History, 1996.

Reade, Philip. "About Heliographs." *United Service: A Monthly Review of Military and Naval Affairs* 2 (January 1880): 91–108.

Rolak, Bruno J. "General Miles' Mirrors: The Heliograph in the Geronimo Campaign of 1886." *Journal of Arizona History* 16, no. 2 (1975): 145–60.

REFERENCES

Roskruge, George J. *Official Map of Pima County, Arizona, Authorized by Board of Supervisors*. Topographic map. 1:250,000. Tucson: Pima County Board of Supervisors, 1893.

Sanborn Map Company. *Sanborn Fire Insurance Map from Deming, Luna County, New Mexico*. New York: Sanborn Map Company, 1886. https://www.loc.gov/item/sanborn05681_001/.

Sanborn Map Company. *Sanborn Fire Insurance Map from Lake Valley, Sierra County, New Mexico*. New York: Sanborn Map Company, 1893. https://www.loc.gov/resource/g4324lm.g4324lm_g056961893?r=-0.27,0.531,1.578,0.824,0.

Scriven, George P. *Report of the Chief Signal Officer, United States Army to the Secretary of War 1915*. Washington, DC: GPO, 1915.

Sheridan, Philip H., and G. L. Gillespie. *Outline Descriptions of the Posts in the Military Division of the Missouri*. Chicago: Headquarters Military Division of the Missouri, 1876.

Slavens, T. H. *San Carlos Arizona in the Eighties: Land of the Apache*. Washington, DC: Order of Indian Wars, 1944.

Smith, Cornelius C. *Fort Huachuca: The Story of a Frontier Post*. Fort Huachuca, AZ: Fort Huachuca, 1978.

Spanish American War Centennial Website. "10th US Cavalry Roster, 1898." https://www.spanamwar.com/10thcav.htm. Accessed October 4, 2023.

Spencer, Eugene J. *Outline Map of the Field of Operations Against Hostile Chiricahua Indians: Showing Operations from April 12, 1886 to the Date of Their Surrender September 4th, 1886*. Tactical map, 1886. NAID: 146555752, images 121–29, RG 94, National Archives.

Spencer, Eugene J. "Report of the Map of the Field Operations against the Hostile Bands of Chiricahua Indians under Natchez and Geronimo." Fort Whipple, Prescott, Arizona Territory, September 25, 1886. Arizona State Historical Society.

Staski, Edward. *Research on the American West: Archaeology at Forts Cummings and Filmore*. Cultural Resources Series. Santa Fe: Bureau of Land Management, 1995.

Thomas, E. D. *The Military Reserve at Camp Thomas, Arizona, 1877*. 1:5000. NAID: 232923934, RG 75. National Archives. https://catalog.archives.gov/id/232923934.

REFERENCES

Thiel, J. Homer. *Archaeological Investigations at Fort Lowell-Atkins Steel Property Locus of Fort Lowell, AZ BB:9:40 (ASM), Tucson, Pima County, Arizona.* Tucson: Desert Archaeology, 2013.

Thiel, J. Homer, M. L. Brack, and Tyler S. Theriot. *Cultural Resources Assessment for the Fort Lowell-Adkins Steel Property within Historic Fort Lowell, Tucson Pima County, Arizona.* Tucson: Desert Archaeology, 2008.

Thrapp, Dan L. *The Conquest of Apacheria.* 5th ed. Norman: University of Oklahoma Press, 1988.

United States Patent Office. "Frank C. Grugan, of the United States Army, Assignor to the Chief Signal Officer of the United States Army." https://patents.google.com/patent/US239095A/en. Accessed January 3, 2025.

US Air Force. *Air Force Doctrine Publication (AFDP) 3–30, Command and Control.* Washington, DC: US Air Force, 2020.

US Army. *Annual Report of the Quartermaster-General, USA, for the Fiscal Year Ending June 30, 1887.* Washington, DC: GPO, 1887.

US Army. *FM 6–0, Commander and Staff Organization and Operations.* Washington, DC: Department of the Army, 2022.

US Army. "US Army Fort Huachuca." https://home.army.mil/huachuca/index.php. Accessed January 27, 2024.

US Army. *Visual Signaling, Manual No. 6.* Washington, DC: GPO, 1905.

US Army. *Visual Signaling, Manual No. 6.* Washington, DC: GPO, 1910.

US Army Center of Military History. "Fourth Regiment of Cavalry—The Army of the US Historical Sketches of Staff and Line with Portraits of Generals-in-Chief." https://history.army.mil/books/r&h/R&H-4CV.htm. Accessed September 14, 2023.

US Army Center of Military History. "US Army Chiefs of Staff." https://history.army.mil/faq/FAQ-CSA.htm. Accessed February 8, 2023.

US Army Quartermaster Corps. "26th Quartermaster Commandant, Major General William H. Hart." https://quartermaster.army.mil/bios/previous-qm-generals/quartermaster_general_bio-hart.html. Accessed September 14, 2023.

US Department of Veterans Affairs. "Nationwide Gravesite Locator." https://www.cem.va.gov/nationwide-gravesite-locator/. Accessed April 23, 2023.

US Geological Survey. *Arnold Mesa.* Topographic map. 1:24,000. Washington, DC: US Geological Survey, 1967.

REFERENCES

US Geological Survey. *Aztec Peak, AZ.* Topographic map. 1:24,000. Washington, DC: US Geological Survey, 1986.

US Geological Survey. *Fort McDowell, AZ.* 1:24,000. Washington, DC: US Geological Survey, 1965.

US Geological Survey. "Grand View Peak." https://edits.nationalmap.gov/apps/gaz-domestic/public/gaz-record/5224. Accessed October 7, 2024.

US Geological Survey. "Heliograph Peak." https://edits.nationalmap.gov/apps/gaz-domestic/public/gaz-record/5224. Accessed January 3, 2025.

US Geological Survey. "Little San Nicolas Canyon." https://edits.nationalmap.gov/apps/gaz-domestic/public/search/names/920633. Accessed September 18, 2024.

US Geological Survey. "Mount Wrightson." https://edits.nationalmap.gov/apps/gaz-domestic/public/search/names/e29395d5-d975–57bb-82c9–27ce1f8bee32/summary. Accessed February 16, 2023.

US Geological Survey. *Middle Verde.* Topographic map. Washington, DC: US Geological Survey, 1969.

US Geological Survey. *Mt. Graham Quadrangle 15-Minute Series.* Topographic map. 1:62,500. Washington, DC: US Geological Survey, 1942.

US Geological Survey. "What Is a Digital Elevation Model (DEM)?" https://www.usgs.gov/faqs/what-digital-elevation-model-dem. Accessed July 4, 2023.

US War Department. *Annual Report of the Secretary of War*, Vol. 1. Washington, DC: GPO, 1892.

US War Department. *Fort Bayard, New Mexico.* Post plat map. 1 inch to 48 feet. US Army Chief of Engineers, 1884. NAID 205135495, RG 77. National Archives.

US War Department. *General Orders and Circulars, Adjutant General's Office, 1899.* Washington, DC: GPO, 1900.

Utley, Robert M., ed. *Encyclopedia of the American West.* New York: Wings Books, 1997.

Utley, Robert M. *Frontier Regulars.* Lincoln: University of Nebraska Press, 1984.

Utley, Robert M. *Frontiersmen in Blue.* Lincoln: University of Nebraska Press, 1981.

Utley, Robert M. *Geronimo.* New Haven, CT: Yale University Press, 2012.

Utley, Robert M. *Historical Report on Fort Bowie, Arizona.* Washington, DC: National Park Service, 1958.

Vandiver, Frank E. *Black Jack: The Life and Times of John J. Pershing,* Vol. 1. College Station: Texas A&M University Press, 1977.

Vegors, Wallace. *National Register of Historic Places Inventory: Fort Verde.* Washington, DC: National Park Service, 1971.

Volkmar, William J. *1890 Heliograph Lines and Stations in Arizona & New Mexico, Prepared from Reports and Reconnaissances of Acting Signal Officers.* General reference map. Los Angeles: US Army, Department of Arizona, 1890. https://www.flickr.com/photos/signalmirror/24201276232/.

Volkmar, William J. "General Order No. 2." In *Instructions for Signal Officers of Heliograph Divisions & Stations,* by William J. Volkmar. Los Angeles: Department of Arizona, 1890.

Volkmar, William J. "Heliograph System, Department of Arizona." In *Instructions for Signal Officers of Heliograph Divisions & Stations,* by William J. Volkmar. Los Angeles: Department of Arizona, 1890.

Volkmar, William J. *Instructions for Signal Officers of Heliograph Divisions & Stations.* Los Angeles: Department of Arizona, 1890.

Volkmar, William J. "Letter of April, 19 1890." In *Instructions for Signal Officers of Heliograph Divisions & Stations,* by William J. Volkmar. Los Angeles: Department of Arizona, 1890.

Volkmar, William J. "Report of Concerted Practice Department Heliograph System, in November 1889." In *Instructions for Signal Officers of Heliograph Divisions & Stations,* by William J. Volkmar. Los Angeles: Department of Arizona, 1890.

Volkmar, William J. *Report of General Practice of the Heliograph System, Dept. of Arizona, in May, 1890: Under Direction of Wm. J. Volkmar, Assistant Adjutant General, Chief Signal Officer.* Washington, DC: Signal Office, 1890.

Volkmar, William J. "Reports of Preliminary Reconnaissances and Concerted Practice between Arizona Divisions, of the Department Heliograph System, during April, 1890." In *Instructions for Signal Officers of Heliograph Divisions & Stations,* by William J. Volkmar. Los Angles: Department of Arizona, 1890.

Weech, Allen Bertell. *A History of Mount Graham.* Safford, AZ: Privately published, 2003.

Wharfield, H. B. "Apache Kid and the Record." *Journal of Arizona History* 6, no. 1 (1965): 37–46.

REFERENCES

White House. "Presidential Proclamation—Organ Mountains-Desert Peaks National Monument." May 21, 2014. https://obamawhitehouse.archives.gov/the-press-office/2014/05/21/presidential-proclamation-organ-mountains-desert-peaks-national-monument.

Willcox, Orlando B. "Report of Bvt. Maj. Gen. O. B. Willcox." In *Annual Report of the Secretary of War. 1882/83*, Vol. 1, by Robert T. Lincoln. Washington, DC: GPO, 1882.

Wood, Leonard. "Reconnaissance for Heliograph Stations in Arizona and New Mexico. Extracts from Report of 1st Lieut. Leonard Wood, Med. Dept., Fort McDowell, A. T., October 20, 1888." In *Instructions for Signal Officers of Heliograph Divisions & Stations*, by William J. Volkmar. Los Angeles: Department of Arizona, 1890.

Wood, Leonard, and Jack C. Lane. *Chasing Geronimo: The Journal of Leonard Wood, May–September 1886*. Lincoln: University of Nebraska Press, 2009.

Index

acting signal officers (ASOs), 14
A.D. 2000 (Fuller), 6
adjutant's office, 42
Albright, Frank H., 69, 197
Alma (Camp Maddox), 91, 95, 96
Alpina, 166, 170, 171, 177, 258n4
Antelope Springs, 233n33, 245n46; Apaches spotted from, 12; Fuller at, 72; Fuller sketches of, 68; heliograph station at, 75; visibility analysis near, 74–76
Apache Peak, 261n40
Apaches: Antelope Springs looking for, 12; Fort Thomas as protection against, 140; heliograph networks and hostilities by, 12; Mexico return of, 10; Miles's information on hostile, 10; U.S. Armies conflict with, xiv; Ward reporting kidnapping by, 239n1; Whetstone Mountains exited by, 233n33
Arizona (1882-1893): Bisbee, 191; Bowie, 42; Fuller establishing heliograph stations in, 38; heliograph networks messages in, 8; heliographs display in, 4; heliograph stations connections in, 2–3, 6; Overton exploring, 12–13
Arizona 366 road, 259n15
Army, U.S.: Apaches conflict with, xiv; heliograph communications by, xiv; heliographs training by, 5; heliograph use by, 196; *Instructions for Signal Officers Heliograph Divisions & Stations, Department of Arizona 1890*, 26–27; *Instructions for Using the Heliograph of the United States Signal Service* by, 13; *A Manual for Signals*, 5; maps, 95; *Report of General Practice of the Heliograph System, Department of Arizona, in May, 1890*, 26–27
Army and Navy Journal of 1918, 200

Army Signal Service, 234n44
Ash Butte station, 32, 34, 36–37
Ash Peak, 32
ASOs. *See* acting signal officers
azimuth and distance, 237n96; Baker's Butte readings of, 123–24; from Fort Bayard, 91; from Fort Grant, 179; magnetic, 104, 238n106, 245n45, 257n23; to Murray's RM station, 178–79; Ramsey's readings of, 123; of terminal points, 71; of Tularosa station, 252nn39–40; in visibility diagrams, 24–25; on Volkmar map, 21, 106, 126, 236n82; of White House ranch, 250n10
Aztec Peak, 126–28

Baker Mountain, 124, 126, 253n10
Baker's Butte, 253n10; azimuth readings of, 123–24; excellent views from, 156–60; heliograph stations at, 159, 160; Ramsey stationed on, 156–57, 162
Bald Mountain (Glassford Hill), 156, 158
Bascom, George N., 239n1
Battle of Verdun, 200
Beaumont, Eugene B., 233n32
Big Squaw Peak, 164
Bingham map (1884), 54
Biographical Register of the Officers and Graduates of the United States Military Academy (Cullum), 27
Bisbee, Arizona, 191
Black, William, 88, 197–98
BLM. *See* Bureau of Land Management
blueprint map, 17
Board on Geographic Names, 63
Bowie, Arizona (1882 map), 42
Bowie, George Washington, 38
Bowie Mountain, 24, 38

INDEX

Bowie Peak, 170, 241n36; connection with, 177; heliograph stations at, 58–59; Merrill Peak and visibility of, 173–74; Murray not reaching, 174; Pinaleño Range visibility of, 40; range ring analysis from, 53; station, 177, 262n48; views blocked of, 261n40; visibility analysis from, 40, 50, 173; Webb Peak obscuring view of, 179

Boxer Rebellion, 112, 198, 208

Brackett, Albert G., 18, 32

Bremen's sawmill, 149–50

Brett, James E., 217

Brewster, Andre W., 112, 198

Bureau of Land Management (BLM), 28, 42

Butterfield stage route, 47, 77, 79

buzzer (in reports), *194*, 195

C2. *See* command and control

Calabasas, 10, 12, 60, 243n8

Campbell, Angus, 91

Campbell, William A., 132, 198

Camp Henely: clear description of, 40; heliograph station at, 48, 50, 100, 241n36; visibility from, 24

Camp Overton, 208

Camp Price, 37

Camp Reno, 123

Chaffee, Adna R., 7, 81, 198–99

Chiricahua Mountains, *51*, 54

Cloverdale, New Mexico, 191

Cochise, 239n1

Collins, Charles L., 154

Collins map (1890), 154–56

Colorado Peak (Rincon Peak), 43; Gale stationed on, 69; heliograph station on, 185; mirroring capabilities of, 187; sightlines from, 45; west view of, 44

command and control (C2), 10, 16–18, 232n22

communications: command and control using, 16–18; Dravo and Fuller building network for, 10, 18; 1885 system of, 229n9; Fort McDowell, 134; heliograph, xiv; between heliograph stations, 234n50; methods of, 192–96; stations costs for, 9

compass readings, 25

Cook's Peak, 217, 246n4

coverage, of heliograph networks, *11*

Coyotero raiders, 239n1

Crawford, Emmet, 205–6

Crittenden Station, 62

Crook, George, 205–6, 229n9

Cullum, George W., 27

Cummings, Joseph, 77

Dade, Alexander L., 15, 146, 178, 199

Davis, Robert E. C., xiv

Dead Man's Canyon, 120, 252n40

Deming, New Mexico, 86, 87

Department of Arizona, 242n49

departure, term, 237n94

distribution analysis, 71

Doane, Gustavus C., 7

Dos Cabezas Range, 30–32, *35*, 174

Dragoon Mountains, 70, 244n35, 261n38

Dragoon-Tombstone road, 64, 244n24

Dravo, Edward E., 40, 191, 216; background of, 199; Camp Henely station by, 100; communications network built by, 10, 18; Deming station report from, 86; at Department of Arizona, 242n49; fever bout of, 88; heliograph station by, 80–84; heliograph stations described by, 84–86; magnetic azimuth reading by, 104; messages sent provided by, 8; Miles appointing, 6

Dripping Spring (Silver Top Mountain), 104–6, *107*, 249n6

D-type kit, 195

Duckbill, 190

Duck Creek, 91

Duncan, George Brand, 160–61; background of, 199–200; Little Squaw Peak visibility for, 162

Eckhoff map, 62

Eggleston, Mallard F., 15; background of, 200; elevation differences recorded by, 170; Fort Grant arrival of, 169; Gila valley connections by, 229n8; at Hill B,

290

142; Merrill Peak station by, 166–68; Murray moving RM station of, 168; near Merrill Peak, 173; old road traveled by, 173, 260n36; RM station of, 168, *175*; at San Carlos, 136

Eggleston Station. *See* Mount Turnbull

1882 heliograph networks, *31*

1882 map, of Bowie, Arizona, *42*

1890 heliograph line, 154

electrical systems, 192–95

elevation differences, 170

elevation measurements, 245n43

elevation model, 22, 24, 50, 237n89, 237n91, 254n6

Elgin, 59

Elkhorn Ranch, 263n60

Emma Monk: heliograph station at, *47*; range ring analysis from, 53, *54*; Swisshelm Mountains distance from, 240n5; visibility from, 58

Encyclopedia of the American West (Utley), 206

engine house, 110–12

Eureka. *See* Hachita Mining Camp

exercises, 14, 234n49

expenditures, of Signal Corps, 191–92, *193*, 234n44

Fairview, New Mexico, 191

Fenton, Charles W., 15, 164, 200

Ferndell Spring, 150, 256n39

Fieberger, Gustav J., 154

Fieberger map (1883), 154

Field Communications Company D, 19, 236n76

Field of Operations against Hostile Chiricahua Indians (Spencer), 21

First North Carolina Infantry, *193*

flag-based systems, 9, 230n13, 231n18

flash-to-flash connections, 20

Fort Apache, 191

Fort Bayard, 191; heliograph station at, 88, *92*, 249n5; Pinos Altos Mountain connection with, 90–91; Volkmar map with azimuth from, *91*

Fort Bliss, 215, 217

Fort Bowie, 166; adjutant's office near, *42*; Fort Grant messages exchanged with, 30–32; heliograph station at, 32, 38–58; Lawton sending message to, 58; stations near, *39*; Webb Peak views to, 179

Fort Cummings (1886), 7, 246n5; connections with, 81; heliograph station at, 77–80, *79*; Hovey leaving heliograph at, 116

Fort Cummings (1890), 122, 217

Fort Grant: azimuth and distance from, 179; Eggleston arriving at, 169; Fort Bowie messages exchanged with, 30–32; heliograph station at, *33*; Neall on, 169

Fort Huachuca: Fuller on station at, 245n39; Fuller seeking connection with, 59, 72; Gale making connection with, 69, 244n31; Hart at, 190; heliograph station at, 64–69, *67*, 264n18; not visible, 187–88; as place of thunder, 242n1; Rock A and, 188; visibility analysis of, 69, 187–88; on Volkmar map, 26, *70*; Volkmar's connection with, 244n30

Fort Lowell, 182, *184*

Fort Marcy, 191

Fort McDowell, 132–34, *135*, 252n3

Fort McRae, 102, 104, *105*, 249n6

Fort Mojave, 191

Fort Selden: Fort Bliss messages with, 215; heliograph message sent from, 217; heliograph networks not mentioning, 215–16; heliograph station at, *219*; Maus not mentioning, 215–16; as National Historic Landmark, 215; San Andreas connection with, 216

Fort Stanton (1886), 250n22; heliograph stations at, *103, 109, 111*; Pershing, J., setting up station at, 106

Fort Stanton (1889), 112, 217

Fort Stanton Historic Site Cultural Landscape Report, 250n14

Fort Thomas: Apache attack protection by, 140; heliograph stations at, *141*, 142; Mount Turnbull view from, *143*

Fort Verde, 156, 160, *161*

Fort Whipple, 18
Fourr's Ranch, 70; North station, 73; sight-line azimuths of terminal points at, 71; South, 64, 66, 68; visibility analysis of, 71–72
France, US Army heliographs used in, 19
Frost's Ranch, Bisbee Cañon. *See* Henry Forrest's Ranch
Fuller, Alvarado M., 200–201, 242n49; A.D. 2000, 6; at Antelope Springs, 72; Antelope Springs sketches by, 68; Arizona heliograph stations established by, 38; Bowie Peak report from, 40; communications network built by, 10, 18; elevation measurements by, 245n43; Fort Huachuca connection sought by, 59, 72; on Fort Huachuca station, 245n39; Gardner's Ranch and, 59–60, 233n33, 243n7; heliograph messages recorded by, 8; on Henry Forrest's Ranch, 54–56; Old Camp Rucker station from, 49; Patagonia Mountains reached by, 60; Stein's Peak station by, 64

Gale, George H. G., 43, 49, 190; on Colorado Peak, 69; on Fort Huachuca connection, 244n31; Fort Huachuca connection by, 69; peaks near Mountain Spring, 185, 264n12; Rincon Mountains ascended by, 15, 187; in World War I, 201
Galiuro Mountains, 146
Gardner's Ranch, 59–60, 233n33, 243n7
Gatewood, Charles, 42, 43, 201
Gatewood, Charles (son), 6, 191
General Field Order No. 7, 12
General Land Office (GLO) maps, 98–100, 248n15; Camp Reno on, 123; Dripping Springs on, 104; station identifying process using, 26–28; White's Ranch mapped using, 54
"General Miles' Mirrors" (Rolak), 27
General Order No. 2, 27, 234n50

Geographic Information System (GIS) software, 24, 237n93
georeferencing, 25–26, 154
Geronimo: Gatewood's peace discussions with, 201; Lawton camp visited by, 201; Maus expedition to capture, 205–6; Miles accepting surrender by, 206; Miles involved in campaign against, 5; Wood's involvement in campaign against, 56, 213
Geronimo campaign (1886), 5
Getting the Message Through (Rains), 27
Gila Valley, 166, 229n8
Gillespie, G. L., 250n14
Gillmore, Quincy O'Maher, 84, 201–2
GIS. *See* Geographic Information System software
Glassford, William A., 12; background of, 202; Fort Thomas distances from, 140–42; Granite Mountain suggested by, 156; heliograph stations surveyed by, 13; Mount Reno described by, 123; Old Camp Rucker report by, 56; Squaw Peak potential from, 160, 164; successful station attributes from, 22; Taylor Canyon links proposed by, 229n8
Glassford Hill. *See* Bald Mountain
GLO. *See* General Land Office maps
Goudy Creek, 172–73, 260n34, 260n36
Gouldman, L. P., 202; Colorado Peak station led by, 185; Rincon Mountains ascended by, 187; on Rock A and B, 190; on Rock C, 187–88
Grand View Peak, 179, 263n58, 263n62
Granite Mountain, 156
Grant Creek, 172–73, 258n2
Grant Hill, 170, 259n24
Greely, Adolphus W, 245n43; standard heliograph design with, 13; on Weather Service shift, 192, 265n5; on wireless telegraphy, 195
Greene, Frank, 13, 16, 191, 202–3, 210
Gregory, Stephen, 68, 244n28
Grierson, Benjamin H., 14, 27, 168
Grugan, Frank C., 5, 13, 203

INDEX

Hachita Mining Camp (Eureka), 98, 100, *101*
Hart, William H., 70–71, 190, 203
Hawk Peak, 173–74, 179
Hay Mountain, 74–76
Heatley's Well, 77, 81, *82*
Helen's Dome, 32, 38–40
Helen's Dome (near Fort Lowell), 187–88, 190
heliograph networks: Apache hostilities and, 12; Arizona messages by, 8; blueprint map of, *17*; chronology of, 29; in 1882, *31*; established per month, 8; Fort Selden not mentioned in, 215–16; heliograph stations within, 19–20; Maus creating, 18, 30; Miles creating, 6; Miles's transmitting, 7; military operations using, 230n15; in New Mexico, 50; number of stations involved in, 18; officers designing, 18; operational coverage of, *11*; from Sherwood, 88; signal flags bridging links in, 9, 231n18; southwestern, 18–19; Volkmar outlining, 14–15; water-hole campaign using, 229n9
Heliograph Peak, 169, *171*, 180, 263n63
heliographs: Army's use of, 196; Beaumont messaging using, 233n32; benefits of, 18–19; display of, *4*; Fort Selden sending message by, 217; France's use of, 19; Fuller recording messages from, 8; Greely and design of standard, 13; Hovey dropping off, 116; later uses of, 192; Lookout Peak (near Fort Selden) with possible, 217–18; Maus on board to create standard, 13; median range of, 230n13; mobility of, 22; on Mount Franklin, 192; Murray on failure of bridge connection, 262n53; Overton and perfect map of, 191; service, 13; Signal Corps signaling using, 16; in signal officer reports, *194*; Spencer map with camps using, 191; sunlight and line of sight required for, 196; technology making obsolete, 192–96; U.S. Army training on, 5

heliograph stations: at Alma, 91, *96*; at Alpina and Heliograph Peak, *171*; at Antelope Springs, 75; Arizona connections of, 2–3, 6; at Ash Butte, 34; at Baker's Butte, *159*, 160; at Bald Mountain, 156, *158*; barriers influencing, 5–6; at Bowie Peak, 58–59; at Camp Henely, 48, 50, 100, 241n36; at Colorado Peak, 45, 185; communications between, 234n50; compass readings locating, 25; connections of, 2–3, 28, 36, 61, 78, 89, 120, 125, 137, 155, 167, 183; at Crittenden Station, 62; at Deming, New Mexico, 86, 87; at Dos Cabezas, 35; at Dragoon Mountain, 70; by Dravo, 80–84; Dravo describing, 84–86; at Dripping Spring, 104–6, 107, 249n6; 1886 line of, 102; at Emma Monk, 47; exploration for potential, 12; flash-to-flash connections between, 20; at Fort Bayard, 88, 92, 249n5; at Fort Bowie, 32, 38–58; at Fort Cummings, 77–80, 79; at Fort Grant, 33; at Fort Huachuca, 64–69, 67, 264n18; at Fort Lowell, *184*; at Fort McDowell, 132–34, *135*; at Fort McRae, 105; at Fort Selden, 219; at Fort Stanton, 103, *109*, *111*; at Fort Thomas, 141, *142*; at Fort Verde, *161*; at Fourr's Ranch North, 73; at Fourr's Ranch South, 66; Fuller establishing, 38; Glassford on successful attributes for, 22; Glassford surveying, 13; at Hachita Mining Camp, *101*; at Heatley's Well, *82*; within heliograph networks, 19–20; at Henry Forrest's Ranch, 55; at Hillsboro, New Mexico, 85; Hovey mapping out, 15, 80, 216; Keene setting up, 256n39; at Lake Valley, 83; at Limestone Mountain, 57, 58; at Little Florida, 115; locations of, 16; at Lookout Peak, 128, 129; at Lyda Spring, 99; at Mazatzal Peak, *133*; Miles on personnel requirements for, 231n19; Morse codes used by, 1–5; at Mountain Spring, *186*;

INDEX

heliograph stations (*continued*)
Mount Graham station connections of, 177, 262nn51–52; at Mount Reno, 126–28, *127*; at Mount Turnbull, *144*; at Mule Spring, 98; near Fort Bowie, *39*; New Mexico connections of, 6; Nogal, *108*; at Old Camp Rucker, 46, 49; Paddock mapping out, 15; at Pinaleño Mountains, 166; at Pinal Mountain, *153*; at Pinal Peak, 148–49; at Pinos Altos, *93*; range ring analysis from, 53; at Rincon, *118*; at Rincon Peak, *189*; at Saddle Mountain, *145*; at San Andreas, *119*, 122; at San Carlos, *139*; at San Nicholas Peak, *114*, 116–17; Sherwood setting up, 50, 210; at Sierra Blanca, *111*, 112, *113*; at Siggins's Ranch, 95, 97, 98; signal mirrors used in, 1; on Spencer map, 80–81; at Spur, *133*; at Squaw Peak, *163*; Stein's Pass, 40–43, *41*; at Stein's Pass, *41*; surveying of, 13; at Table Mountain, 146, *147*; temporary, 231n17; Triplets, *138*, 140; at Tularosa, *121*, 122; U.S. Armies communications using, xiv; visibility analysis from, 30, 70; visibility information from, 22–24; Volkmar map of, 20–21; at Whipple Barracks, *157*; at White House, 91, *94*; at White's Ranch, 20, 50–54, *51*–*52*, 241n38; whole system of, xiii
heliostat, 218
Henely, Austin, 241n35
Henry Forrest's Ranch, 54–56, *55*
Hill 5485, 162–64
Hill A, Mount Turnbull, 140
Hill B, Mount Turnbull, 140, 142
Hillsboro (Camp Boyd), 84, *85*, 246n5
Hilton, J. M., 173
Hilton map (1911), 173
Historic Structure Report (National Park Services), 43
History of the Ninth US Infantry, 210
Holmes, Allan J., 215
Horse Spring, New Mexico, 191
Hospital Flat, 166, 170
Hovey, Henry W., 110; background of, 203–4; on Cook's Peak, 246n4; heliograph left at Fort Cummings by, 116; heliograph stations mapped out by, 15, 80, 216; Rincon station connection by, 120
Huachuca Canyon, 69–70

Icehouse Canyon, 149–50
impedance buzzer, 195
Instructions for Signal Officers Heliograph Divisions & Stations, Department of Arizona 1890 (U.S. Army), 26–27
Instructions for Using the Heliograph of the United States Signal Service (U.S. Army), 13

Jenness, Burt, 215
Jesus Canyon, 165, 172, 260n36
John D. Lee house, 248n15
Johnson, Evan Malbone, Jr., 81, 204
Johnson, Lewis, 254n21
Josephine Peak, 62, 68. *See also* Little Baldy Peak

Keene, Henry C., Jr., 149–50, 168–69; background of, 204; Grant Hill and, 259n24; heliograph stations set up by, 256n39; Volkmar getting message from, 258n9
Kelly, Roger, 27

Lake Constance, 165–66, 258n2
Lake Valley, 81–84, *83*
Lava Station, 104
Lawton, Henry W.: battalion of, 68; as Calabasas battalion leader, 10; Calabasas report by, 12, 243n8; Fort Bowie message from, 58; Geronimo visiting camp of, 201; military service of, 232n29
Leitch, Joseph D., 80, 204–5
Limestone Mountain station, 57, 58
Little Baldy Peak, 60–64, *65*
Littlebrant, William T., 146, 205
Little Florida Mountains, 110, *115*, 116–17, 217

INDEX

Little Squaw Peak, 160–62, 164
Lookout Peak: difficulties connecting to, 176–77; heliograph stations at, 128, *129*; Mazatzal Peak location determined by, 253n12; RM station not located by, 174; RM station troubleshooting connection with, 180; Wood at, 124–26, 253n10
Lookout Peak (near Fort Selden): possible heliograph on, 217–18; Rincon station and, 216; Signal Peak and, 215; steel and iron pipes on, 218; visibility from, 217
Lowell, Charles Russell, 182
Lyda Spring (Mule Spring), 90, 95–98, *99*
Lyons, Tom, 91
Lyons & Campbell (LC) Ranch, 91

MacArthur, Arthur, 192, 210
magnetic azimuth, 104, 238n106, 245n45, 257n23
"Maj. Alvarado M. Fuller Promoted to Be Colonel" (article), 200
Mance, Henry C., 5
A Manual for Signals (U.S. Army), 5
Maus, Marian P.: background of, 205–6; 1882 heliograph network created by, 18; Fort Selden not mentioned by, 215–16; Geronimo capture expedition of, 205–6; Gila Valley connection achieved by, 166; heliograph networks created by, 30; Medal of Honor for, 5, 206; standard heliograph design and, 13; work overlooked of, 6
Mazatzal Peak, 257n20; heliograph station at, *133*; Lookout Peak determining location of, 253n12; Mount Reno view of, *130*; Overton occupying, 132
McCook, Alexander M., 16
McRae, Alexander, 102
mean centers, 71–72; analysis, 152; formula determining, 25; to Webb Peak, 178–79; weighted, 120
Medal of Honor, 5, 198, 206, 213
Merrill Peak station: Bowie Peak not visible from, 173–74; Eggleston establishing, 166–68; Mount Turnbull's alignment with, 174; RM station on, *175*
Mexican Punitive Expedition, 192, *193*, 208, 210
Mexico, Apaches returning to, 10
Mica Mountain, 188
Michigan Signal Corps, *193*
Miles, Nelson A., 1, 216; at Calabasas, 60; Dravo appointed by, 6; Geronimo campaign involvement of, 5; Geronimo surrender accepted by, 206; heliograph networks created by, 6; heliograph networks transmitting and, 7; heliograph stations personal requirements from, 231n19; hostile Apaches information, 10; *Personal Recollections and Observations*, 5, 27; Wood reporting on, 243n12
military careers, 266n1
military operations, 230n15
military service, of Lawton, 232n29
military signals, 191–92, *193*
Miller, Charley, 263n60
Miller, Mary, 263n60
Mills, Anson, 165–66
Mills, John V., 27
Mills, Nannie, 258n4
mobility, of heliographs, 22
Mogollon Mountains, 88
Morrison, Colonel, 239n1
Morse codes, 1–5
Mountain Spring (Pistol Hill): Canyon, 49; Gale on peaks near, 185, 264n12; heliograph stations at, *186*; as stage station, 182; visibility from peaks of, 185
Mount Franklin heliograph, 192
Mount Glenn, 71
Mount Graham, 255nn29–30; on 1911 map, 173; as Pinaleño Range's highest peak, 262n54; station, 177, 262nn51–52; visibility diagram including, 177
Mount Reno (Mount Ord), 124; Glassford describing, 123; heliograph stations at, 126–28, *127*; Mazatzal Peak viewed from, *130*; Overton reaching, 128–31, 252n3
Mount Robledo, 215

Mount Thomas, 13
Mount Turnbull (Eggleston Station): flash observed from, 176; Fort Thomas view to, *143*; heliograph stations at, *144*; Hill A and B on, 140, 144; Merrill Peak alignment with, 174; San Carlos view to, *143*; visibility analysis from, 140, 142
Mount Wrightson, *63*, *65*
Mt. Baldy, 60
Mule Springs, 95–98, *99*
Murphy Peak, 126, 253n13
Murray, Cunliffe H., 148–50; azimuth and distance to RM station of, 178–79; background of, 206–7; Bowie Peak not reached by, 174; Bowie Peak views blocked and, 261n40; common visibility areas, *151*; Eggleston's RM station moved by, 168; Gila valley connections by, 229n8; heliograph bridge connection failure of, 262n53; RM station location, 177–79, *181*, 190; visibility requirements of, 152

Naiche, 10, 60
National Archives, 26
National Historic Landmark, 215
National Oceanic and Atmospheric Administration (NOAA), 28
National Park Services, 43
Neall, John M., 15, 40; background of, 207; court-martial and dismissal of, 207; on Fort Grant, 169
Neifert, William W., 62, 207
New Mexico (1882-1893), 2–3, 6, 106, 191; Alma, 91, 95, *96*; Deming, 86; heliograph networks in, 50, 210; Hillsboro, *85*; Tularosa, *121*, 122; White House, 91, *94*
NOAA. *See* National Oceanic and Atmospheric Administration
Nogal (Vera Cruz Mountain), 106, *108*
Nogales Ranch. *See* White House Ranch
Nogales station, 62
North Peak, 128, 131

officer training, 5
Official Map of Pima County (1893), 185
Old Camp Rucker: Fuller and station at, 49; Glassford's report on, 56; heliograph station at, 46, 49; range ring analysis from, *53*, *54*; visibility analysis of, 56–58
Outlaw Mountain, 74
Overton, Clough, 123–24; Arizona explored by, 12–13; background of, 207–8; on Baker's Butte (Baker Mountain), 126; Fort McDowell communications with, 134; Mazatzal Peak description by, 132; Mazatzal Peak misidentified by, *130*; Mazatzal Peak occupied by, 132; Mount Reno reached by, 128–31, 252n3; near North Peak, 131; perfect heliograph map and, 191

Paddock, Richard, xiii, 110–11; background of, 208; Fort Stanton distances recorded by, 112; heliograph stations mapped by, 15; Rincon station described by, 117; San Andreas Mountains described by, 251n38; San Andreas Mountains destination of, 117–20; San Nicholas Peak connection by, 117; on Sierra Blanca Peak, 112, 250n22; southern line established by, 122, 216; on Tularosa station, 252nn39–40; Williams communicating with, 120
Patagonia Mountains, 60
peakfinder.org, 254n6
Pearson, Edward P., 168
Perry, John A., 168, 263n60
Pershing, Grace, xiii
Pershing, John J., 12, 102, 106, 208–9
Personal Recollections and Observations (Miles), 5, 27
Pinaleño Mountains, 69, 165; Arizona 366 road into, 259n15; Bowie Peak visibility from, 40; Eggleston's RM station on, *175*; heliograph stations at, 166; Mount Graham highest peak in, 262n54; visibility from, 169–70
Pinal Mountain station, 148, *153*
Pinal Peak, 148–49, 152, 255n30

INDEX

Pinos Altos Mountain, 90–91, *93*
Pistol Hill. *See* Mountain Spring
place of thunder, Fort Huachuca as, 242n1
Pope, General, 212
Proctor, Redfield, 16

railroad station, *41*, 81
Rains, Rebecca, 27
Ramsey, Frank DeWitt: azimuth readings by, 123; background of, 209; on Baker's Butte, 156–57, 162; Fenton connecting with, 164; magnetic azimuths from, 257n23
range ring analysis, 53, 54
Read, Robert D., 149–52, 209, 256n39
Recollections (Miles). *See Personal Recollections and Observations*
Reichmann, Carl, 126–28, 148; background of, 209–10; remained on Saddle Mountain, 142; Table Mountain direction from, 146, 255n27; Table Mountain report from, 22
Report of General Practice of the Heliograph System, Department of Arizona, in May, 1890 (U.S. Army), 26–27
Reservoir Cañon, 172
Richards, J. R., Jr., 160
Rincon Mountains, 15, 43, 49, 187
Rincon Peak, *189*
Rincon station, 251n33; Fort Cummings connecting with, 122; Hovey connecting with, 120; Lookout Peak (near Fort Selden) referred to as, 216; map of, *118*; Paddock's description of, 117
RM station: excellent views from, 178–79; location of, 177–79, *181*, 190; Lookout Peak connection analyzed at, 180; Lookout Peak not locating, 174; on Merrill Peak station, *175*; of Murray, 177, 190; Murray moving, 168
Robinson, A. J., 164, 210
Rock A (Helen's Dome), 187–88, 190
Rock B (Duckbill), 190
Rock C (Spud Rock), 187–88
Rolak, Bruno, 27
Rucker, John A., 241n35

Saddle Mountain, 255n27; heliograph stations at, *145*; Reichmann remaining on, 142; Stanley Butte on, 146; Watson on distance to, 254n21
Salamis Canyon, 124–26
Salome Creek, 124–26
San Andreas Mountains: Fort Selden connection with, 216; heliograph stations at, *119*, 122; Paddock describing, 251n38; Paddock's destination of, 117–20
San Bernardino Ranch, 56
San Carlos: agency, 136; Eggleston stationed at, 136; heliograph stations at, *139*; Mount Turnbull view from, *143*
San Carlos Arizona in the Eighties (Slavens), 136
San Nicholas Peak, 110, *114*, 116–17
Santa Rita Mountains, 59
Separ station, 86
service heliographs, 13
Shaw's Ranch, 182
Sheridan, Philip H., 9, 212, 250n14
Sherwood, Robert: heliograph network from, 88; Lyda Spring troubles encountered by, 95–98; New Mexico heliograph station and, 50, 210; telegraph lines and, 86
Sierra Blanca: Peak, 112, 250n22; station, *111*, 112, *113*; visibility analysis from, 112
Siggins, Isaac, 95
Siggins's Ranch, 95, *97*, 98, 248n15
sightlines, 21, *45*, 71, 196
Signal Corps: expenditures of, 191–92, *193*, 234n44; funding increase for, 13; heliograph signaling by, 16; Michigan, *193*; officer training by, 5; pillars of, 19; Weather Service responsibilities shed by, 19; wireless telegraphy and, 195
signal flag stations, 9, 231n18
Signal Hill, 74–76, 245n40, 245n46
signal mirrors, 1
signal officer reports, *194*
Signal Peak, 215
silver, 81
Silver Top Mountain. *See* Dripping Spring

297

INDEX

Six Shooter Canyon, 148
Slavens, Thomas H., 15, 136, 210–11, 254n6
small-scale maps, xiii, 1
Smart, Alexander, 211
Smith, William H., 142, 146, 211, 230n15
southern line, 122, 216
southwestern heliograph networks, 18–19
Spencer, Eugene J., 21, 216
Spencer map, 54, 84, 98; camps using heliographs on, 191; Dragoon-Tombstone road on, 244n24; Fort Cummings connections on, 246n5; heliograph stations on, 80–81; Mt. Baldy on, 60; Pinos Altos Mountain on, 90–91; Stronghold Canyon on, 64; Volkmar map differences with, 21, 236n84
Spud Rock, 187–88
spur, on Dragoon Mountain, 244n35
Spur station, *133*
Squaw Peak, 160, *163*, 164, 257n20
Squire, George, 195
stage station, 77, 182
standard distance deviation, 237n96
Stanley Butte, 146
Stanton, Henry W., 106
Staski's report, 246n4
steel and iron pipes, 218
Stein's Pass station, 40–43, 235n56; establishing, 15–16; Helen's Dome connecting, 38; heliograph station at, *41*
Stein's Peak station, 64, 235n56
Stockle, George E., 142, 211–12
Stronghold Canyon, 64
Sundial Mountain, 97, 98
Swisshelm Mountains, 49–50, 240n5

Table Mountain, 142, 187; Reichmann on direction to, 146, 255n27; Reichmann's report on, 22; station, 146, *147*; on Volkmar map, 255n27
tabular data, 71–74
"Talking Mirrors at Fort Bowie" (Kelly), 27
Taylor Canyon links, 229n8
Taylor Creek, 260n36
technology, 192–96

telegraph lines, 86
telegraph office, in Deming, *87*
telegraphy, wired, 192–96
telegraphy, wireless, 192–96, *194*
temporary stations, 231n17
terminal points, 71
Thomas, E. D., 140
Thompson, Lieutenant, 18
Tonto Basin stations, 154
trail miles, 56, 242n47
Triplets, heliograph stations, *138*, 140
Tubac, 59–60
Tularosa, New Mexico, *121*, 122
Tularosa station, 120, 252n43, 252nn39–40
Tyson, Lawrence Davis, 212

United States, wireless telegraphy in, 195
Universal Transverse Mercator (UTM) system, 238n106
US Geological Survey map, 80, 172, 182, *184*, 245n43
Utley, Robert, 206
UTM. *See* Universal Transverse Mercator system

Venus, transit of, 218
Vera Cruz Mountain, 106
visibility analysis, 104; of Ash Butte, 36–37; from Bowie Peak, 40, 50, 173; of Duncan's Little Squaw Peak, 162; from Emma Monk, 58; of Fort Huachuca, 69, 187–88; of Fourr's Ranch, 71–72; from heliograph stations, 30, 70; to Little Baldy Peak, 62–64; from Lookout Peak (near Fort Selden), 217; from Mount Graham, 177; of Mount Turnbull, 140, 142; Murray's common areas in, *151*; Murray's requirements for, 152; near Antelope Springs, 74–76; near Pinal Peak, 149; of North Peak, 128; of Old Camp Rucker, 56–58; from peaks near Mountain Spring, 185; from Pinaleño Mountains, 169–70; from Pinal Peak, 152; from Pinos Altos, 91; from Sierra Blanca, 112

298

INDEX

visibility diagrams: azimuths and distances in, 24–25; defining, 22; elevation models and, 24; including Mount Graham, 177; map of, 23; mean centers formula for, 25

visibility information, 22–24

Volkmar, William J., 18, 191; ASOs directives from, 14; background of, 212–13; departure used by, 237n94; Fort Huachuca connection by, 244n30; General Order No. 2 by, 234n50; heliograph network outlined by, 14–15; Keene sending message to, 258n9; as mapmaker, 216; map preparations by, 247n26

Volkmar map, 84, 95; azimuth and distance on, 106, 126, 236n82; Fort Bayard azimuth on, 91; Fort Cummings connections on, 246n5; Fort Huachuca on, 26, 70; of heliograph stations, 20–21; Little Squaw Peak on, 164; magnetic azimuths on, 245n45; Mazatzal Peak on, 257n20; measured distances on, 106; Saddle to Table Mountain on, 255n27; sightline azimuths on, 21; Spencer map differences with, 21, 236n84; tabular data from, 71–74; Volkmar's preparations for, 247n26

Walsh, Robert D., 10
Ward, John, 239n1
water-hole campaign, 229n9

Watson, Lieutenant, 254n21
Weather Service responsibilities, 19, 192, 234n44, 265n5
Webb Peak, 178–79, 263n58
Weech, Allen Bertell, 260n34
weighted mean centers, 120
Whetstone Mountains, 233n33
Whipple, Amiel Weeks, 154
Whipple Barracks, 154–56, *157*
White House, New Mexico, 91, *94*
White House ranch, 91, 250n10
White's Ranch station, 20, 50–54, *51–52*, 241n38
whole system, of heliograph stations, xiii
Willcox, Orlando B., 37
Williams, Daniel, 120, 213
Wilson, Woodrow, 208–9
wired telegraphy, 192–96
wireless telegraphy, 192–96, *194*
Wittenmyer, Edmond, 123, 128, 213
Wood, Leonard, 12, 229n9, 234n40; background of, 213–14; Baker's Butte potential recognized by, 156; Geronimo campaign with, 56, 213; at Lookout Peak, 124–26, 253n10; Medal of Honor for, 213; Miles report from, 243n12; Mount Thomas potential report by, 13; Squaw Peak potential recognized by, 160
World War I, xiv, 19, 192, 209

www.ingramcontent.com/pod-product-compliance
Lightning Source LLC
Chambersburg PA
CBHW020638230426
43665CB00008B/222